BENOÎT MICHEL

DIGITAL STEREOSCOPY

Scene to Screen 3D Production Workflow

DIGITAL STEREOSCOPY
Scene to Screen 3D Production Workflow
by Benoit Michel

Cover credits
Kommer Kleijn SBC, cinematographer/stereographer (right), and Tim Mendler, assistant (left), shooting "Le Huitième Continent" for Alterface / Futuroscope Poitiers. Picture: Jacques Campens.

Disclaimer
All product names and services identified throughout this book are trademarks or registered trademarks of their respective companies. They are used throughout this book in editorial fashion only and for the benefit of such companies. No suche uses, or the use of any trade name, is intended to convey endorsement or other affiliation with the book.
The information and material contained in this book are provided «AS IS», without warranty of any kind, express or implied, including without limitation any warranty concerning the accuracy, adequacy, or completeness of such information or material or th results to be obtained from using such information or material. Neither StereoscopyNews nor the author shall be responsible for any claims attributable to errors, omissions, or other inaccuracies in the information or material contained in this book, and in no event shall StereoscopyNews or the author be liable for direct, indirect, special, incidental, or consequential damages arising out the use of such information or material.

Translation of « La stéréoscopie numérique », © Editions EYROLLES, 2012.
ISBN 978-2-212-12988-5.
Translation Gabrielle Leyden
Layout Colette Michel

Published by StereoscopyNews
Rue de Sendrogne, 100, 4141 Sprimont, Belgium.
www.stereoscopynews.com

ISBN 978-1-48015709-5

Foreword

Will the current renaissance of stereoscopy have lasting effects? The question is raging. After the major box-office successes of films such as *Avatar*, some back-pedaling is now being seen. What is its cause? The public's fickleness? Disappointment due to poor movies and shooting? The added expense of viewing devices? the discomfort or weight of 3D glasses? The immaturity of the filming, postproduction, and/or broadcasting tools? The brain strain caused by uncorrected or poorly correctly datastreams? The rough edges that are linked to technicians and filmmakers' lack of experience in staging stereoscopic productions?

A long list of questions that attests to the diversity of experiences involved. These questions can be seen as the sum of errors made or an unavoidable learning phase. This book describes in great detail the huge contribution made by equipment manufacturers and research laboratories to mastering stereoscopic film technique. Methods and tools are multiplying and competing with each other. Improving the viewer's comfort is at the heart of all these efforts, and such improvement is necessary.

The main issue, besides that of mastering the stereoscopic technique, is that of artistic excellence, of inventing a new language and, at the end of the day, a new cinematographic art that is both different from yet close to the old one and that has no reasons to be replaced.

This book will join the shelf of the few indispensable volumes for preparing stereoscopic projects, for it will enable producers and filmmakers to think about the various opportunities for choosing the right shooting technology or workflow for a film, shoot, or any other audiovisual creation. It will contribute to building up the experience that is necessary for the artistic inventiveness that is the heart and goal of all this research.

This volume thus holds the greatest interest for these times of invention in which we are lucky enough to live. The technical mastery that it helps to establish lets us believe that, this time, stereoscopy has a real possibility of existing as a new art form. This is a rare opportunity that must not be missed.

Yves Pupulin

Yves Pupulin is a stereographer and one of the founders of Binocle

Introduction

We live in, move around in, and interact with a three-dimensional world. To enable us to act effectively, our brains are constantly producing internal representations of this three-dimensional world. To do this, the brain uses the five senses, but primarily vision. The images that our eyes see are constantly analyzed and fed into our mental database. Like all terrestrial predators, human beings have two eyes with overlapping fields of vision. In analyzing unconsciously the differences between these two views of the world, the brain deduces the distances of nearby objects with great accuracy. This is called "stereoscopy." It should be pointed out, however, that binocular vision is far from the only source of depth information about the surrounding world. We build our mental representations of the three-dimensional world from other visual clues as well, such as occlusion: An object that hides another one is necessarily located in front of it.

Even when certain visual clues are lacking, the human brain still almost always manages to generate a correct representation of the world. The greater the number of visual cues contributing to a representation of the world, the better the illusion. By adding stereoscopic vision to a moving picture, we bolster the illusion of realism shown on the screen considerably. Under ideal conditions, this leads to what is called "immersion in the action." After a few minutes of watching a 3D movie projected under optimal conditions, the viewer truly identifies with a part of the fictional world being presented. This is where "suspension of belief" kicks in. This transition from the role of an outside observer to that of a participant in the action creates a greater degree of satisfaction and induces much better recall of the action seen. That is what explains a 3D movie's success!

The underlying technology is obviously complex and having perfect images is an indispensable condition for a successful immersive experience. Until now, this perfection had never been achieved in the history of 3D imagery. Technical flaws dampened audiences' interest, even causing them to turn away. However, with the advent of digital cinema, constant quality has become a commercial reality that has completely changed the game.

Why Digital is Changing the Game

A slight technical flaw in an image may be bearable. What moviegoer has never seen the series of vertical lines sweeping across the screen from top to bottom with each reel change? The mechanical handling of film reels ends up scratching the celluloid, and with each showing, the least little scratch or bit of dust that is added to the picture remains. The buildup of such problems means that a film printed on celluloid cannot be shown more than a hundred or so times. Now, however, with the advent of digital film, each frame remains perfect, showing after showing, and remains exactly as good as during the screening in the studio that produced the movie.

We shall see in this book that a great many factors can make stereoscopic 3D images unfit for consumption. However, when all these pitfalls are avoided or corrected and quality control has approved a picture, the latter will reach viewers who are the farthest from the studio in pristine condition every time. In the heyday of 3D movies on celluloid, great care was taken in making the reels for theater projection. However, the projectionists' inexperience, mechanical complexity of the projectors, and inevitable chemical, optical, and technical vagaries to which the reels were subjected tipped the balance towards the minus side. As a result, quality suitable for guaranteeing immersion in the action could not be provided after the first few projections of a reel. The success of stereoscopic films prompted the movie industry at the time to produce at a frantic pace. This led each time to a drop in quality and ultimately to a dead-end, with audiences losing all interest in the process.

The digital revolution offers two major benefits: ease of image production and correction and perfect quality throughout the production chain, down to the thousandth showing of a film, and even beyond!

Whom is this Book For?

This book is for anyone who wants to get involved in stereoscopic imagery or those who already work in 3D but want to hone their skills. Broadcasting students are not forgotten: The first chapters give the basic principles needed to discover 3D imagery, while the subsequent chapters contain the practical information that they will need to put them into practice.

Whether your focus is photography, amateur videos, or professional moviemaking and whether you are a student, creator, or end user, this book will provide the necessary foundations for making your first stereoscopic images, and if you have already taken the first steps, you will find in this book the information you need to tackle more complex projects.

That being said, this book has not been written for the total beginner; readers are assumed to know the basics of taking pictures or video filming. Its orientation is also technical. So, the creative and artistic aspects of filming and photography will not be covered in depth. However, since introducing 3D into the composition of stills and movies calls for a new grammar of images, these more artistic aspects will of course be touched upon.

Complementary Web Page

Information and illustrations connected to this book are available online on the author's website at www.stereoscopynews.com/digital-stereoscopy.html. You can access the site directly by scanning the QR code below. Several pictures in this book offer also a QR code; scanning the code will redirect you to a high resolution color version of the same figure.

Table of Contents

History of Stereoscopy

1

Does stereoscopic 3D predate photography?

When was the first 3D picture produced?

Why has 3D cinema died several deaths?

When can we expect 3D movies without glasses?

16th and 17th Centuries

The idea of stereoscopy and binocular depth perception arose well before the invention of photography. The first traces of the study of 3D vision date back to 300 years Before the Common Era (B.C.E.). Euclid, in his treatise on optics, defined depth as the two eyes' views of two dissimilar images of the same subject. Historians then found binocular drawings by Giovanni Battista della Porta (circa 1535-1615) and Jacopo Chimenti, called Jacopo da Empoli (1551-1640), that showed clearly that certain 16th- and 17th-century scholars understood binocular vision and the notion of parallax.

In 1613 the Jesuit François d'Aguilon (1567-1617) published a treatise on optics in which he coined the word "stereoscopic." He also invented the word "horopter" to describe the place where a series of points are perceived as one. Today we would say that the horopter is the set of points with zero parallax.

So, the notions of stereoscopy were known but raised not the slightest interest in the public at large. This is because a crucial element was lacking, namely, a way to reproduce 3D vision from two images depicting the same scene from slightly different points of view. Another two centuries would go by before discoveries in the field of optics would lead to this invention.

19th Century: the First Wave

Discoveries

Go See the First Stereoscope
Wheatstone's original stereoscope still exists. It is displayed at the London Museum of Science. Charles Wheatstone was a prolific inventor. Besides the stereoscope, he invented the famous Wheatstone bridge, which is well known to electricians, and patented several models of telegraph.

In 1838 two Englishmen, Charles Wheatstone and David Brewster, found, independently but almost simultaneously, a way to reproduce three-dimensional vision from two images. However, Charles Wheatstone was first to communicate this finding publicly. He did so at the Royal Scottish Society of Arts' meeting of June 1838, at which he broached his theory of binocular vision. He explained how two different visions characterized by retinal disparities produced the sense of depth. Using drawings designed to be ambiguous with regard to all the other depth cues, he showed that the brain could reconstruct the depth of a scene based solely on these retinal disparities. To prove this, he placed his two drawings in a mirror device – which he had designed six years earlier, in 1832 – that showed each eye a different image and in so doing recreated a view of a three-dimensional solid. He called this device a "stereoscope" and filed a patent. No one prior to Wheatstone had as yet understood that these tiny differences in images seen from different points of view gave the brain depth information.

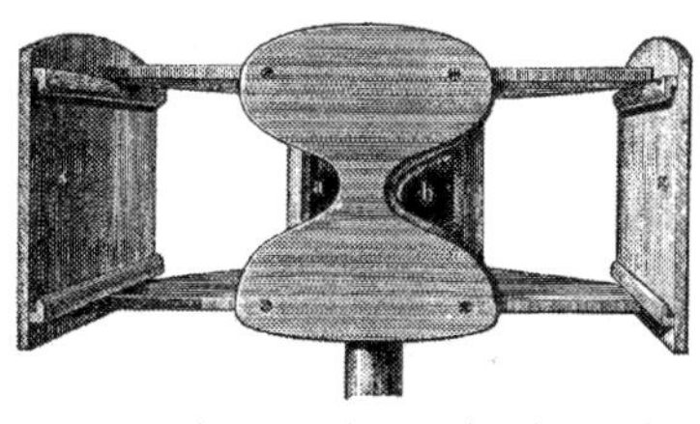

© Popular Science 1882

Wheatstone's stereoscope: mirrors set in a yard-wide wooden apparatus

Photography was invented a few months later in 1839 by Niépce and Daguerre in France and Talbot in England, once again independently. Wheatstone met with Talbot as early as 1840 and asked him to take stereoscopic views to be able to view them in his stereoscope.

Brewster invented a stereoscope fitted with lenses that was much smaller and easier to use than Wheatstone's in 1844. He presented it to the Royal Society of Edinburgh in 1849. With his apparatus, the two printed or transparent views of a stereoscopic pair could be mounted on a single card, which could be inserted into the apparatus very easily without further adjustment. One then looked at them through a pair of prisms that doubled as magnifying glasses. All of the stereoscopes invented thereafter, including the very widespread View-Master™, were no more than variations of this original design.

Brewster and Duboscq's stereoscope – a portable apparatus with lenses. The stereoscopic cards are inserted sideways through the slit in the rear. The upper hatch lets in light to light up the pictures.

Also in 1849 Brewster described the first stereoscopic camera. The next year the public could contemplate the first 3D photographs thanks to the lenticular stereoscope, which Brewster had made by the French optician Jules Duboscq. The pioneers of stereoscopy also include Antoine Claudet, who filed several patents for stereoscopes in 1853. One of his models could be folded up and was designed for observing a pair of daguerreotypes. Another had an adjustable width and thus could be adjusted perfectly to changes in viewers' interocular distances. Finally, a third one was practically a cabinet enclosing a rotating magazine that could hold 100 pairs of pictures.

The cameras of the day were built to order by cabinetmakers. The stereoscopic cameras, which were less common, were built according to the same principles but with two chambers and two lens mounts instead of one. The photographer had to prepare his emulsion on a light-sensitive plate himself and the camera exposed the two parts of the plate simultaneously when the shutter was opened. The plates were generally about 8 cm across, which was also the distance between the two views. The 3D effect was thus exaggerated by this slightly too great axial distance. The 8-cm cards and plates matched the dimensions of most of the stereoscopes made, although no standard had been chosen.

A stereoscopic card from 1891

Most of the stereoscopes developed in the second half of the 19th century were made of wood and had adjustable lenses. Some of them, like Claudet's, were cabinets containing up to 100 pairs of stereoscopic views mounted side by side on cards.

The stereoscope's popularity literally skyrocketed in 1851, notably because of the publicity that the press gave to Queen Victoria's fascination with 3D photographs during an exhibition in the Crystal Palace. Brewster gave the queen a stereoscope and the public followed suit: Half a million stereoscopes were scooped up in the United Kingdom in less than five years. The stereoscopy craze of the time can be likened to the craze of the first years of television. Movies had not yet been invented, but the public had the possibility for the first time in history to see the great places of history and current events with unequalled realism, as if they were there. The most popular stereoscopic cards depicted distant lands, such as Egypt, and the world's major cities: New York, Paris, and San Francisco, for example. A large proportion of the sales also covered current events, such as the international expos and major natural disasters of the time: the Johnstown floods, San Francisco earthquake, and so on.

The stereoscopic cards and stereoscopes business took on a new dimension in 1854. At the time, an estimated one million households had stereoscopes in the United Kingdom. George Swann Nottage created the London Stereoscopic and Photographic Company, with a catalog of 10,000 3D views, in 1854. Four years later, there were more than 100,000 views in its catalog. The company's avowed goal was to put a stereoscope in every home in the UK, and it effectively sold half a million of them in just two years. At the peak of this craze, in 1862, more than one million stereoscopic cards were sold in a single year, at the average price of half a crown for a pack of three. Having made his fortune thanks to the success of the stereoscope and cards, Nottage went on to become Lord Mayor of London in 1884. The London Stereoscopic and Photographic Company survived until 1920.

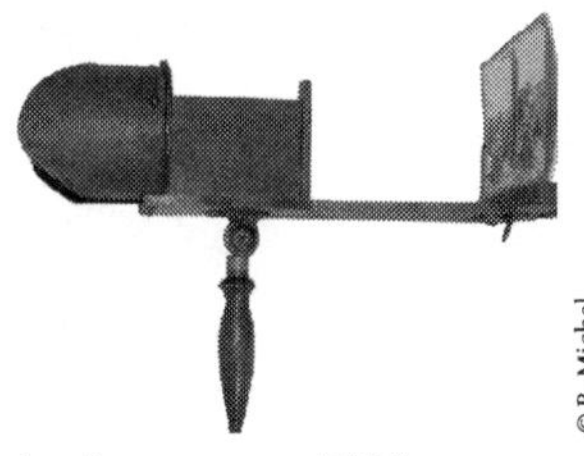

© B. Michel

Holmes's stereoscope (1860)

One of the most popular stereoscopes was the one invented by the American Oliver Wendell Holmes in 1860. It owed its success to its ease of use: You simply held it in front of your eyes with a handle, like a carnival mask, and a simple pair of slides let you change the views quite easily.

Gaumont 25-view basket stereoscopes, 1910

Stereoscopy took stores by storm and a great many "fan clubs" were created as of the late 19th century. The most famous one, the Stereoscopic Society, was created in London in 1893 and continues to bring together hundreds of thousands of 3D imagery fans (see *www.stereoscopicsociety.org.uk*). The first French club, Stéréo-Club français, was created in 1903. It, too, still exists (see *www.stereo-club.fr*).

More Discoveries

Anaglyph

Still in 1858, Charles d'Almeida suggested using red and blue colored filters on the lenses of two projectors, superimposing the images on a screen, and looking at them through glasses with the same filters. In 1891 the French photographer Louis Ducos du Hauron, who was one of the pioneers of color photography, invented a method for printing the two colored images on the same photographic paper. This made it possible to look at 3D images directly with colored glasses as the only accessory. The commercialization of this invention began around 1920. It was first used in movie theaters in 1921.

Polarization

Nicol's prisms were already known to polarize light in 1852, when William Bird Herepath, in the UK, discovered polarizing sheets. This material, called "herepatite," had a neutral color. The small elongated crystals could be deposited on sheets that could be cut to order. The quality and transparency of the resulting film limited the output's reproducibility and the sizes of the available sheets were too small to make it attractive enough as a commercial product. Almost forty years later, in 1891, the Englishman John Anderton filed a patent covering the use of polarized light for stereoscopic projections. However, he did not develop any market applications for lack of polarizing filters usable outside the laboratory. So, by the end of the century all the basic ingredients for one of the most popular stereoscopic display methods were already present.

Animated Stereoscopy

Many scientists took an interest in 3D vision in the wake of Wheatstone and Brewster's discoveries, and many variations of the stereoscope appeared, making it possible to look at a series of images that gave the illusion of movement. The general principle was to present successive images for viewing that were located on the walls of a cylinder. Slits in another moving part of the device served as a shutter so as to reveal a different image to each eye only when the image was in the right position. We see here the foundations of the movie projector's mechanics.

We cannot yet talk about motion pictures *per se,* but it took less than five years for inventions that displayed slightly different images in rapid succession to come on the scene. A host of devices with exotic names were produced: the stereophoroscope (Czermak, 1855), stereotrope, bioscope, stereophantascope, and kinematoscope (Sellers, 1861), and many more. They all let one view very short animated sequences, such as a full step sequence, a jump over an obstacle, a juggler throwing a ball, and so on. The cycle's length depended on the number of views that could be inserted in the cylinder or disk, the rotation of which gave the impression of movement. Sellers's kinematoscope used a chain of photographic plates or stereoscopic pairs that passed in front of the viewing cylinder like the rolls of a player piano. This increased the number of views that could be used without increasing the device's size too much. 3D animation existed thirty years before movies were invented!

Stereoscopic Camera

Stereoscopy and photography were discovered just a year apart. No more time was needed between Edison's invention of the first movie camera in 1888 and that of its stereoscopic counterpart by the Englishman William Friese-Greene in 1889. The device was full of technical flaws and thus never developed commercially. However, Edison and his patent factory never lost a chance to patent a camera model. The first patent application went nowhere, but the second one was filed by Dickson, Edison's assistant at the time, in the UK in 1893. Edison had invented the film camera, which he christened "Kinematographic Camera," and had had the forethought to provide a two-lens variation for 3D filming. However, in the absence of a projector, no more than one viewer at a time could watch the film using a variation of the stereoscope dubbed the kinetoscope. Not until 1895 and the Lumière brothers did cinematographic projection become a reality, and even more time went by before the first 3D movie projection took place.

First Decline

Public interest in stereoscopy gradually declined from 1860 to the end of the century, despite technical advances, with a concomitant shift to photography, which was more within the technical reach of everyone and, above all, appreciated as an object of socialization. Indeed, the stereoscope's major flaw was that it could be used by only one person at a time, whereas photo albums could be leafed through by a group of people and become the topic of conversation.

Some of the masses' lack of interest in 3D imagery at the end of this first "3D wave" stemmed from the poor quality of most of the stereoscopic prints. It was effectively even easier in the 19th century than today to make poor 3D photographs than good ones. The brightness differences between the two prints, vertical shifts, and other misalignments were much harder to correct at the time than today and the constancy of quality in the mass production of stereoscopic cards was a far cry from that achieved by our modern digital printers.

Twentieth Century

3D Photography

The availability of ready-to-use industrial photographic plates and the mass production of cameras of reasonable dimensions enabled photography to spread through the public at large from the start of the 20th century on. In 1902 *The Amateur Photographer* published an article explaining how to make a stereoscopic card using a single camera by shifting the camera's position laterally between two shots.

Rolleidoscop of 1931

© Henri Peyre

Industrial manufacturers proposed various stereoscopic cameras based on their standard ranges up until World War II. The formats tended to become standardized around the popular 6 x 6 - cm format, which for stereoscopy led to 6 x 13 - cm double plates. The glass plates were replaced little by little by rolls of film, which accounted for the bulk of the market in the late 1930s. The quality of this type of camera peaked with the Rolleidoscop in 1930. The Rolleidoscop gave two images on 6 × 13 cm film frames. It had three lenses: two side lenses for taking the pictures and a central reflex lens for focusing the shots.

Jim the Penman
The first public projection of a stereoscopic film (*Jim the Penman*, by Edwin S. Porter, produced by Paramount Pictures Corporation) was done using this technique in New York's Astor Theater in June 1915.

Anaglyph Cinema

As we have seen above, colored filters were invented at the end of the 19th century, but were not used commercially until the start of the 20th century. In 1897 the Frenchman Claude Grivolas adapted the colored filters technique for use in a movie camera that exposed two reels of film through the filters. The resulting images were then projected onto a screen simultaneously and looked at through red and green glasses. A first way to make 3D films was taking shape and becoming a topic of discussion. The only thing missing was a simplification phase to make the technique accessible to the public.

Funnily enough, the first public projection using anaglyphs did not take place in a movie theater, but during a shadow puppet show in the United States in 1918. An itinerant dance troop performed behind a semi-opaque screen lit by red and green lights. The audience watched the dancers' red and green shadows through anaglyph glasses and, lo and behold, saw them dancing before the screen, just in front of their noses!

A first wave of public showings unfurled in the 1920s under the aegis of the producer Jacob Leventhal. The first commercial 3D film, Harry K. Fairall's *The Power of Love,* was shown in Los Angeles in September 1922. It was followed by a short film, *Plastigrams,* in 1923, and then the four films in the Stereoscopiks series, *i.e., A Runaway Taxi*, *Zowie*, *Ouch* and *Luna-cy!,* in 1925. For these first attempts two black-and-white films had to be projected by two mechanically synchronized projectors, each fitted with a different colored filter. In December 1922 William Van Doren Kelley invented an anaglyph film covered with two colored emulsions, one on each side. This film, which he named Plasticon, made it possible to project 3D films using a single projector! Using this process, William Van Doren Kelley showed two films in New York, one of them being the short film *Movies of the Future.*

Just a few years later, in 1927 to be precise, Abel Gance in France shot his famous *Napoléon* in an extra-wide format called "Polyvision." Actually, because of the complexity of synchronizing the projectors, only the last reel was in a wide format. This format called for the use of three movie cameras mounted side by side to give a total aspect ratio of 4:1. This ancestor of Cinerama was expensive to set up in a projection booth, so much so that the wide format was used in Paris and London only. In the rest of the world, audiences were treated to the film's central panel only. Some scenes were shot in 3D and viewed in anaglyph format in private projections. Unfortunately, the 3D segments were cut from the final film version, probably because this silent film's box-office success was no longer guaranteed, given competition from the talkies, which were just starting to be shown that same year.

The processes of the time were too rudimentary to eliminate the drawbacks of anaglyph movies, such as the differences in brightness between the two views and the impossibility of showing true colors. The problems that certain viewers had perceiving the depth and the eye strain and headaches experienced by many more viewers limited the commercial success of the 1920s' 3D movies. We can speak of a more or less anecdotal wave of interest in 3D movies, but not mass infatuation.

The anaglyph process had a new wave of popularity starting in 1935 thanks to MGM, which commissioned a first 3D film from Leventhal. Called *Audioscopiks*, it was a succession of short talkies demonstrating 3D effects.

The anaglyph movie craze ended with the film *Third Dimensional Murder*, produced by the same Leventhal in 1940. Louis Lumière fiddled with anaglyphs that same year, but never launched any commercial activity in that direction. The discovery of polarized projection in 1932 sealed the anaglyph movie's fate. Despite all this, the technique subsists today for printing, for it alone lets one print 3D images on ordinary paper.

Active Glasses 3D Cinema

Some people might believe that active glasses, which make it possible to watch films composed of left- and right-eye images projected in rapid succession on the same screen, belong to the digital era. Nothing is farther from the truth: The alternating projection system was invented by the Frenchman C. Dupuis in 1903. Dupuis took the example of the dual projectors that had been used for fifty years already to use cross-fades or dissolves in slide shows.

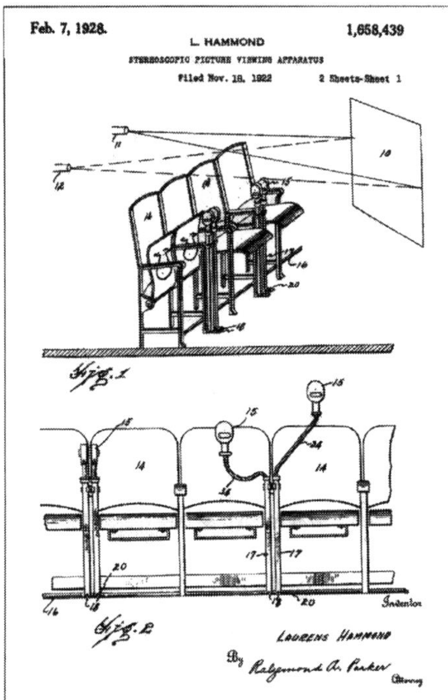

L. Hammond's Eclipse system, 1928

The system was used commercially for the first time by Laurens Hammond and William Cassidy in the Selwyn Theatre, New York, in 1932, under the moniker "Teleview." Two films were shot with two separate 54-mm cameras. The films were then projected by two electrically synchronized projectors with a half-frame shift between them. The images for the left and right eyes were thus projected on the screen in alternation. To hide the view not intended for each of the eyes in turn, the viewers sat behind a Televiewer, a sort of wooden housing with "rotary interrupting shutters." The two shutters for each seat were activated by synchronous motors working at 25 rotations per second and powered by the same alternating current as the projectors. In this way, 3D viewing was not only possible, but deemed highly convincing. The system was patented under the trade name "Eclipse" (three patents, one in 1922, the second in 1924, and the last one in 1928).

Unfortunately for the two inventors, the system's exorbitant installation cost made it unprofitable. What is more, despite its advantages, the Eclipse system still had a major flaw: Its 25 Hz switching rate was too slow to eliminate bothersome flicker. Finally, since the images were shot simultaneously but projected with a half-frame shift, viewers saw quick horizontal movements as parallaxes. This produced undesirable variations in objects' distances that gave the impression of "gelatinous vehicles" that the critics of the time were quick to pick up.

What is more, the reviews of the first show were disastrous because of the poor quality of *The Man from Mars*'s script. Indeed, the movie's sole merit was that it was the second film in the world to be shown in "stereoscopic relief." More than sixty years would go by before the idea of active glasses in a more marketable form cropped up again.

Autostereoscopy

The inter-war period was one of many technological developments in Russia. 3D movies were one of these bubbling fields. Various articles on synchronized shutters, anaglyph projection, and polarization were published. However, works in the area of autostereoscopy were what brought Russian research to the fore. Three-dimensional shooting and display methods, called "parallax stereograms," were developed by Berthier in France in 1896, but never got beyond the research stage to true commercialization.

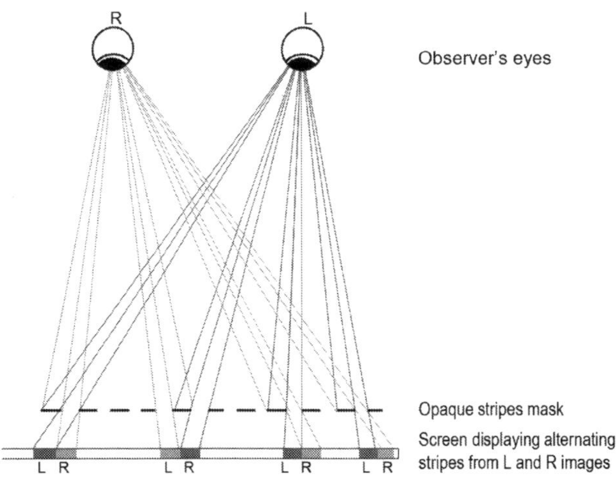

Berthier's principle of lined network 3D photography, 1896

Parallax Stereogram

A parallax stereogram is produced by putting a screen composed of narrow vertical banks or grooves in a glass plate a few millimeters in front of the surface of the photographic film in a two-lens camera so as to mask part of the film's surface when it is seen through the left lens and another part when it is seen through the right lens. After exposure, the film is thus covered with narrow alternating bands belonging to two (or more) stereoscopic images. To see the picture in three dimensions, one must look at it through a grid of lines similar to the one used for the shoot. No special device is necessary, neither glasses nor stereoscope. On the other hand, the depth effect is visible only if the eyes are in a very precise location opposite the image and its grid.

First the American Frederic Ives, working in 1902, and then the Frenchman Eugène Estanave, in 1908, tested various variations of the procedure. Estanave's autostereoscopic plate had the photographic emulsion on one side and the grid on the other, thereby ensuring the right spatial match between the two. Besides 3D images, Estanave used the same method to produce what he called "*images changeantes*" or changing pictures, where the two views were not stereoscopic pairs but variations of the same image. In this one, he could have a character open and close his eyes by changing the point of view slightly. In 1912 Louis Chéron filed a patent application for a process in which the number of images increased greatly, but at the expense of brightness.

Relièphographie

Still in France, the inventor Maurice Bonnet refined the idea even more and created an eleven-lens camera in 1934 and then a thirty-three-lens version in 1937. The images that they produced were sold directly as part of a light box that protected the image and gave it a uniform brightness. He christened his invention "Relièphographie."

Lenticular Photography

In 1940 Bonnet invented the lenticular network, *i.e.*, a plastic film composed of longitudinal lenses instead of alternating opaque and transparent lines. The number of possible views and brightness of

3D Photography

Here is Maurice Bonnet's definition:

"A 3D photograph consists of two parts: the optical selector and the recorded image. The optical selector takes the form of a transparent plate, one surface of which bears fine vertical grooves that play the role of tiny lenses. The photographic film or plate is put in contact with the optical selector's surface. A lens mounted with a diaphragm projects the image of an object onto the selector, which breaks it down into much finer linear images than the grooves' width. This makes it possible to record a large number of images of the same object seen from different angles by moving the picture-taking lens.

When you look at a photograph recorded in this way, your right and left eyes see two aspects of the same object, corresponding to two different points of view, separately but simultaneously. This gives the impression of depth. What is more, if you move in front of the photograph, you have the impression that you are walking around the photographed object."

the images increased greatly. The lenticular picture was born and became commercializable. In fact, in 1942 Bonnet even opened a commercial studio on the Champs-Élysées that became very successful.

Autostereoscopic Cinema

Meanwhile, various researchers in Russia, one of whom was Nikolai Valyus, were frantically trying to apply this technology to moving pictures. The Russian engineer Semyon Pavlovich Ivanov perfected the idea and filed a patent on it in 1935. He gave a first public demonstration of his Stereokino system in 1937.

In 1940 he placed a grid of more than 150 kilometers of wires in front of 4 x 6-meter projection film in the Moskva movie house in Moscow. The film had been shot by exposing the two pictures side by side on the same film. A grid of vertical wires in the projector separated the two images, which were projected onto the screen from behind. A second grid of wires placed in front of the screen enabled each of the 200 viewers to see two separate images. The seating had been very carefully planned, because 3D viewing was not possible from every point in the theater. What is more, the viewers had to find just the right spot to see in 3D and then avoid moving their heads on pain of losing the 3D effect.

Two films, *Concerto* (1940) and *A Day in Moscow* (1941), were shot and projected in this movie theater. After the war (in 1947 to be exact), another film using this technique, *Robinson Kruzo (Robinson Crusoe)*, was presented on a three-meter-wide screen in Moscow's Vostok movie theater. The Stereokino technique spread little by little and by 1955 there were a dozen movie theaters equipped for it in the Soviet Union and about just as many films using the technique had been shot.

A few experiences with autostereoscopic movies occurred outside Russia, in France, thanks to the cyclostereoscope that François Savoye developed from 1940 to 1950. The large cyclostereoscope's screen was seven meters wide and, set up in the Clichy-Palace movie theater in Paris in 1953, enabled 200 viewers to watch the action in 3D without glasses. A system was invented whereby the grid was mounted at the periphery of a quickly rotating cylinder inside which was the screen. The rotating leaves passed behind the screen and interrupted the projectors' light beams at the same time as the leaves that passed in front of the screen cut off the viewers' left or right view. As always with the lined network technique, the viewer's position was crucial: Once the viewer

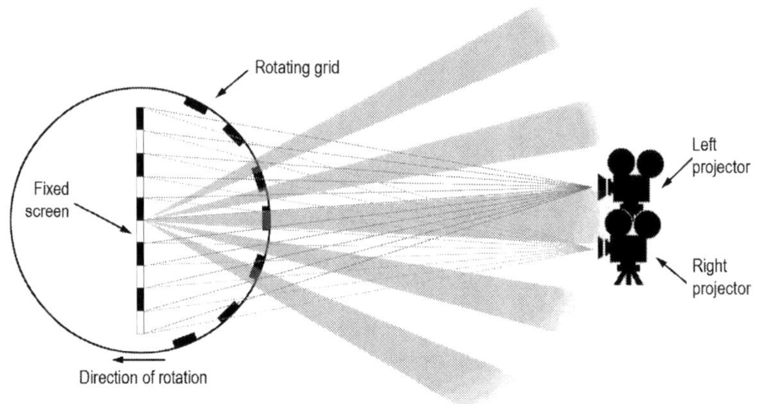

Diagram of François Savoye's cyclostereoscope, 1940

found the spot where the 3D effect was present, s/he had to keep her/his head perfectly still: More than a few millimeters' shift in position, and the effect was gone!

The complexity and above all great difficulty of finding acceptable points of vision led to the lined networks' being abandoned. Despite its limitations, however, this branch of the search for 3D effects was not a dead end. It laid the foundations for the technique of lenticular films printed on postcards and posters as well as autostereoscopic television sets, which were pioneered by the Frenchman Pierre Allio, who showed his first prototype in 1987. The company Alioscopy continues to invest in this promising field. Indeed, the glasses-free 3D television based on lenticular films is perhaps the future of home 3D viewing!

View-Master

The View-Master, which displayed seven transparent stereoscopic pairs of images mounted on a disk, was released on the market during the New York Expo of 1939. The first Model A viewer was patented in 1940 by two Americans, Gruber, who was an amateur photographer, and Graves, who produced postcards. Their aim was to take over the postcard market and replace postcards with their 3D images as souvenir illustrations for tourists. However, the US Army immediately recognized the View-Master's possibilities for training its troops and ordered 100,000 units that very same year. In just five years the army also bought 6 million disks from the two inventors. After the War, in 1950, View-Master built its

© www.3dimages.co.uk

A View-Master and its disk of 7 stereoscopic views

first factory. It then bought the rights to disseminate stereoscopic images from Disney Studios in 1951. The View-Master did a lot to popularize and dispel the mystery surrounding stereoscopy for the public at large. The company's ownership changed many times, and View-Master is now part of the toy industry giant Fisher-Price's corporate portfolio. To date, the View-Master has been produced in twenty-five different versions and more than 1.5 billion of disks have been sold.

Polarized Cinema

Research on applications for polarized light in image projection continued between the two world wars. In Europe, Boehner from the company Zeiss-Ikon developed polarizing filters and a process for shooting 3D movies on a single film on which the two images were exposed side by side so as to avoid all risks of desynchronization. The first projection with a polarizing filters system took place in Italy, in late 1936. The film was Gaultiero Gaulterotti's *Nozze Vagabonde,* a black-and-white film shot with a movie camera that was developed by Gaulterotti but very similar to Boehner's.

The first color photographic films date back to 1935 (Agfa had just invented the two-color film), making it possible to produce movies in 3D and in color. The first projection of a 3D color film took place in the *Haus der Teknik* in Berlin in June 1936 and the first commercial showing, also in Berlin, followed in December 1937. The film was called *Zum Greifen Nah* ("Close Enough to Touch") and was shot at a fairground.

Meanwhile in the United States the American Edwin H. Land and his company, the future Polaroid Corporation, continued the work of Herapath and ultimately patented, in 1928, easy-to-produce polarizing filters that were more effective and cheaper than those developed by Zeiss. The method consisted in depositing a thin layer of herapathite crystals, all aligned parallel to each other, on a transparent film or glass plate. These films/plates could then be fashioned into polarization filters for projectors and lenses for glasses.

Land did not give a public demonstration of his invention until 1935, however. He nevertheless focused a lot more on applying polarization to glasses, car headlights, and anti-reflecting windshields than to 3D movie projection. The first commercial stereoscopic 3D movie showing using a system of polarized glasses to be held in the US took place during the 1939 New York World's Fair. The film, *Motor Rhythm*, was a Chrysler assembly-line documentary (Chrysler financed the film). Both the shoot and the projection used side-by-side arrangements of standard equipment, *i.e.*, cameras and projectors. The simple addition of orthogonal Polaroid filters in front of the projectors and the corresponding Polaroid glasses for the viewers completed the projection system. By the end of 1941, more than three million Americans had attended 3D movie projections during exhibitions and various other public events. However, the first 3D movie theater did not open until after the war. It was the Telecinema, the first stereoscopic movie theater using the polarized system, which opened in London in 1951.

Vectograph

Edwin H. Land's vectograph is another interesting invention of the 1940s, despite the fact that it was never marketed. The vectograph did away with the need to modify the movie projector; polarized glasses alone were needed. The invention worked as follows: Ordinary film was replaced by a two-sided film, with one of the two images of each stereoscopic pair on each side. The images were deposited in the form of a more or less polarizing emulsion, rather than a more or less opaque emulsion, with, of course, orthogonal directions for the two images. In this way the projection's luminosity was not reduced by mirrors, filters, and other more or less complex accessories: The film itself played

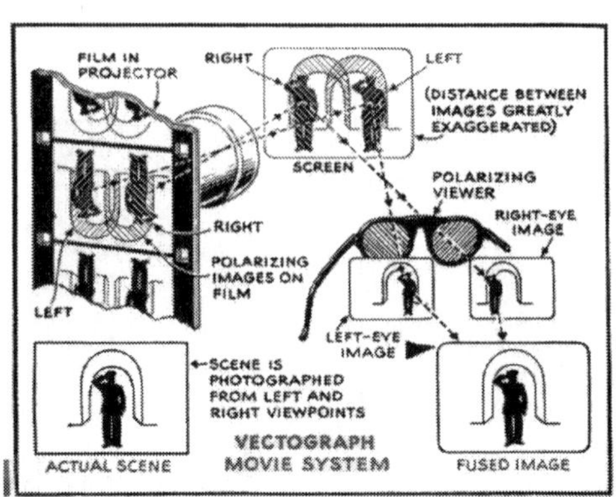

Principle of the vectograph explained in the March 1943 issue of Popular Science

© Richard Bolt

8 × 12 - cm advertising vectograph of 1938. A large-size version of this vectograph was exhibited in Grand Central Station's Grand Concourse, in New York, in 1939.

the role of the polarizing filter. What is more, the two images' synchronization was guaranteed, since they were on the two sides of a single vector.

The first vectographs, in black and white, were developed for the needs of the US Army, which used them for aerial reconnaissance photography during World War II. At Warner Bros.'s request, Land eventually developed a color version of the vectograph, but too late, unfortunately, to reap any commercial benefits from the wave of interest in 3D movies that swelled in 1952. Consequently, the vectograph was never used for commercial movies.

Hollywood and the Fifties: the Second Wave

Interest in 3D images between the two world wars was more scientific than from the public at large, and the interest that the latter did show was mere curiosity that could not be considered a great success. However, things began changing as of the end of World War II.

All industrial activities were relaunched after the war, and photography and cinematography gradually embraced color with the invention of Kodachrome film. 3D photography surged in 1954. Photography increasingly used the 35-mm film used in the movie industry, and more than 250,000 stereoscopic cameras are thought to have been in use in the second half of the 20th century. The best known such camera on the market was the

© www.pacificrimcamera.com

The Stereo-Realist still camera

American-made Stereo-Realist, of which more than 100,000 units were sold. As for the View-Master, its success as a household item was confirmed in Europe as well as in the United States.

1950 was the year that television came on the scene. The direct consequence was a 50% drop in Hollywood movie box-office earnings in 1952, and the "witch hunt" that the US government launched at the same time (under McCarthy's impetus) cut huge swathes through the forest of creative minds responsible for the US movie industry's great successes. Motivated by the obvious concern to boost their flagging box-office receipts, the majors threw themselves body and soul into innovation, the major one of which was the invention of Eastman Color film, which superimposed the three emulsions on the same celluloid strip. Color already existed in filmmaking, but required triple movie cameras, developed by Technicolor, that were extremely heavy and bulky. Doubling these huge triple cameras for stereoscopic shoots was out of the question. Using one alone was already at the limit of what was feasible. Color film was thus a true "open sesame" for stereoscopic 3D movies.

With the advent of Eastman Color film, any movie camera could become a color movie camera, and the age of black-and-white movies was definitely over. This was followed in 1952 and 1953 by the development of Cinerama and then CinemaScope, which was popularized by Walt Disney's 1955 animated film *Lady and the Tramp*, and finally Natural Vision stereoscopic projection with two projectors.

The New Formats of the 1950s

CinemaScope

CinemaScope is a widescreen movie format with a 2.55:1 aspect ratio. The image is recorded on a single film by means of a cylindrical lens with a vertical axis, called an "anamorphoser," that squashes the horizontal dimension on the film. A similar lens is used during projection to restore the image's natural format. The cylindrical lens is also called a "Hypergonar," the name that its inventor, Henri Chrétien, gave it in 1926. Starting in 1957 the dimensions of the CinemaScope projection window were standardized to an aspect ratio of 1.1734:1 before decompression and 2.35:1 after de-anamorphosis.

CinemaScope premiered with Henry Koster's movie *The Robe* in 1953.

Cinerama

Cinerama is an immersive movie process on an ultrawide curved screen that uses three cameras and three synchronized projectors mounted side by side to cover a 146° visual field. The first showing of *This is Cinerama* took place in New York in September 1952. Only ten films were shot and shown using this format between 1952 and 1962. The last one was *How the West Was Won*, with John Wayne.

Natural Vision

The Natural Vision process was developed by the screenwriter Milton Gunzburg, oculist Julian Gunzburg, and cameraman Friend Baker in 1952. Having failed to sell their invention, they decided to use it themselves and produced the film *Bwana Devil*, which popularized the technique. The Natural Vision process used two movie cameras that were placed at 90° to the left and right of the viewing axis and looked at the scene in mirrors. The interaxial distance was set at 3.5 inches (9 cm) and the variable convergence could be adjusted by changing the mirrors' angles.

Bwana Devil

The first movie shown by the Hollywood industry in 3D and color was Arch Oboler's *Bwana Devil*. Shot using the Natural Vision process and shown in a single movie theater in Los Angeles, it was an immediate hit as of its release in 1952, raking in 100,000 dollars the first week. As the Natural Vision process used two projectors with polarized filters, projection time was limited to one hour because of the problems of synchronizing the two projectors. Full-length feature films thus had to be shown with an intermission to allow the projectionist to change the reels on the two projectors.

This resounding success by an independent producer woke up the major studios, which had already been caught up short by the transition to the talkies. This time they were determined not to miss the boat and immediately threw all of their forces into the battle. In 1953 already no fewer than 45 stereoscopic films (or 15% of the majors' productions) came out of Hollywood. Another score of stereoscopic films were added in 1954.

Decline

The blockbusters of the Golden Age
The blockbusters of 1953-54 included *Dial M for Murder*, *Creature from the Black Lagoon*, and *House of Wax*, which was the first 3D film with stereophonic sound. The last 3D film of the Golden Age was *Revenge of the Creature*, released in February 1955. Ironically, this 3D release was highly acclaimed by the critics and was relatively successful.

This golden age of 3D movies did not, however, go beyond the stage of curiosity and fell like a soufflé after less than a year. The public's disenchantment was an almost inevitable consequence of the studios' poor management of their success: An overdose of bad taste and an arms race – to see who could release the film with even more impressive pop-out effects than the previous one – sped up the decline in screenplay and production quality from every angle. Numerous technical flaws made it impossible to achieve sufficient quality, and the lack of experienced stereographers meant that these flaws were not always even detected before the films' theater releases. The pairing of the geometry and luminosity of the two cameras and two projectors alone, although far from the only problems at stake, was a real problem at this time, when digital processing was still unknown. Just one degree of difference in the two development baths' temperatures was enough for a stereoscopic film to become a source of headaches.

The theater operators' shortsighted keenness to save money, which led them to skimp on the quality of their silver screens and polarized glasses, and the projectionists' lack of training exacerbated these problems. What is more, the theater operators let the viewers sit at the far sides of the rows, where the images rendered by the silver screens that polarization required were much too dark. In an attempt to stem the tide of disappointments, the Polaroid Company conducted a survey of some 100 projection booths in late 1953 and found that a quarter of them presented serious projector synchronization failures.

The other causes of the decline of this second wave of interest in stereoscopy include CinemaScope's success and the advent of television: The first live television broadcast in the US occurred in 1951 when President Harry Truman's speech in San Francisco was transmitted to broadcast stations; the first televized live event in the UK was the coronation of Her Majesty Elisabeth II in 1953.

The negative news that spread through the grapevine about the quality of 3D films and the headaches that resulted was such that the last films shot in 3D in 1954 were not even projected in 3D, on pain of being boycotted by the public. Audiences much preferred another novelty of the moment, CinemaScope, which was extremely impressive with its extra-wide screen and did not give viewers headaches!

In Europe, too

The Telecinema movie theater opened in London as part of the Festival of Britain in 1951, just slightly before this wave of Hollywood 3D blockbusters. It could seat 400 and was run by Raymond Spottiswoode, one of the greatest promoters of 3D movies in Europe and effectively the person who first used the term "3D" to refer to depth photography.

With his brother Nigel he built a double camera very similar to the Natural Vision camera. Four films were shot and shown at the Telecinema that year, including *The Distant Thames*, which was shot in color with a pair of enormous three-film Technicolor cameras. More than 1.5 million people attended the 3D showings at the Telecinema between 1951 and 1953. The Spottiswoode brothers, who had seen *Bwana Devil*, berated Hollywood for overusing the cameras' convergence to get pop-out effects in front of the screen. They claimed – probably rightly so – that the excessive use of special effects created headaches and audiences would eventually get tired of the "trick." It should be pointed out that *Bwana Devil*'s director, Arch Oboler, said more or less the same thing in the United States.

The Seventies Burst: the Third Wave

The Mask, The Bubble, The Stewardesses

The Mask

Stereoscopy remained a marginal technique in the 1960s and 1970s. A few productions nevertheless made a name for themselves. One was *The Mask* in 1961. Warner disseminated red-and-green anaglyph 3D version of the film. Viewers were asked to put on masks with red and green glasses when a voice told the hero, "Put the mask on now!" This occurred four times, for just a few minutes each time. In such relatively special circumstances, the drawbacks of anaglyphs remained relatively tolerable. This also explains why the film was such a hit during its many television broadcasts in the US in the 1980s.

The Bubble

Arch Oboler, the man who launched the 3D movie wave in the 1950s, was also the man behind *The Bubble*. This film, which was released in late 1966, used "Space-Vision 3D," a technique in which the two images were placed one on top of the other on the

Gala's Christ (S. Dalí, 1978)

same celluloid film in a standard academic frame (22 × 16 mm, or 1.37:1). A lens on the projector rectified the two images' positions. This technique did away with the need for two projectors and all the associated problems of synchronization and geometric adjustment. For projection, the polarized system, which had now become the standard, was used. People flocked to see *The Bubble*, despite rather unfavorable reviews.

The Stewardesses

Allan Silliphant's *The Stewardesses* kicked off a new wave of interest among the studios in 1969, especially because of its extraordinary ROI. Produced at a cost of US$100,000, it brought in US$26 million at the box office, although it was shown in just 800 theaters. Its distribution was not hobbled by technical problems, for the film, which was produced in side-by-side format, could be shown on just one projector thanks to the Stereovision process developed for the occasion. A special anamorphic projection lens had to be adapted, a silver screen set up, and disposable glasses distributed to the audience, but synchronization was no longer a problem.

A few 3D films arose once again in commercial movie theaters in the period from 1970 to 1985, thanks to this side-by-side format. Thirty-six films, such as *Amityville 3D,* were shot using Allan Silliphant's Stereovision process between 1969 and 1984. However, stereoscopy seemed to have been relegated once and for all to the commercial niche of amusement parks. Huge resources were effectively made available for 3D films in theme parks, for they were a major drawing point for visitors. Theme park managers thus had no qualms about paying whatever was necessary to get a top-quality result. To immerse viewers in the action even more, the parks equipped facilities especially for 3D viewing. So, Disney World installed a special system in its futuristic Epcot Center that used two high-luminosity 70-mm projectors in tandem and an extra-wide (2.45:1) screen with an 18-m base for spectacularly convincing results.

Dial M for Murder

The 1954 hit *Dial M for Murder* was re-released in 1980 and was an excellent box-office hit. Most of the 3D hits of this period were also transferred to Video Discs, a now forgotten format that had tried to compete with VHS cassettes. They were distributed with active-shutter glasses coupled with the interlacing of TV images.

Dalí and 3D

Salvador Dalí, who was a keen science and technology watcher, took an interest in stereoscopy. The illusion of depth fascinated him and gave him food for artistic thought. He met Dennis Gabor, the inventor of holography, in the 1970s and began painting stereoscopic canvases: *Gala's Foot* in 1974 and many other works, such as *Gala's Christ* en 1978.

IMAX 3D

In 1985, six years after the first IMAX movie theaters were set up, the Canadian company began producing 3D movies by juxtaposing two of its enormous 70-mm projectors to project the films with the polarized process. Disney and Universal Studios set up IMAX 3D theaters in their amusement parks in 1986. Films of excellent quality, such as *Captain Eo*¸ starring Michael Jackson, were shown in these facilities. Starting in 1990 IMAX also set up 3D movie theaters with the active-glasses technique, notably because projecting the films on dome-shaped screens proved incompatible with polarization.

IMAX 3D

The IMAX system is the perfect example of the immersive movie. The huge screen is 22 m wide and 16 m high and covers a viewing angle greater than that of the human being. As a result, viewers no longer truly see the edges of the picture and are immersed in the story taking place on the screen. The images were recorded on 70-mm, 15-perforation films. IMAX reels weight up to 250 kg and range from 120 to 180 cm in diameter. An IMAX 3D camera weighs more than 100 kg (220 lb). For 3D projection, either linear polarization glasses or active-shutter glasses are used. In the latter case, a single projector running at double speed is used.

The IMAX format compared with conventional 35-mm film

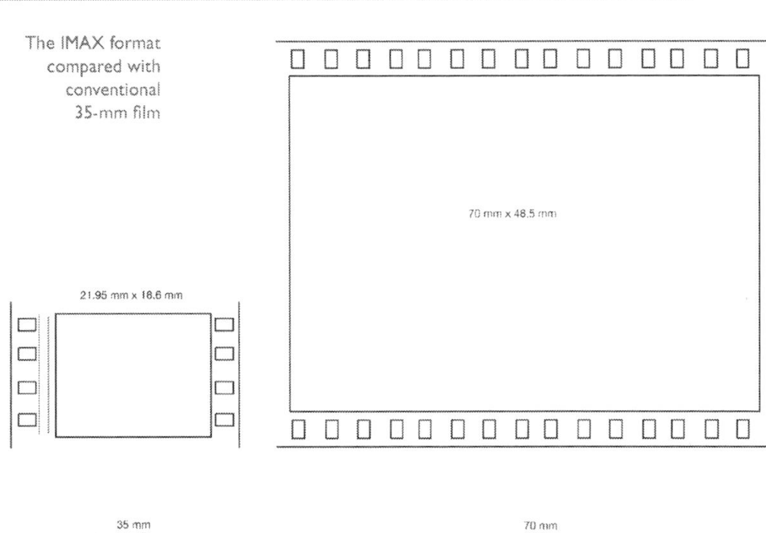

The first IMAX 3D fiction film was Jean-Jacques Annaud's famous *Les Ailes du courage* (*Wings of Courage*), released in 1996. More than half of the 300 IMAX theaters in the world are compatible with 3D projection, and close to thirty films were shot specifically for the IMAX 3D format between 1985 and 2005. Various 3D films - for example, *Monsters vs. Aliens* and *Avatar* - have also been postproduced in IMAX 3D format.

Autostereoscopic TV

The Frenchman Pierre Allio adapted the principle of lenticular images to video films in 1987 and began producing autostereoscopic screens. He created the brand Alioscopy to market 3D screens that could be watched without glasses. The main characteristic of the Alioscopy screen is to reconstruct eight different points of view so as to enable each viewer to see two consecutive points of view and thus the depth effect.

The Digital Age

Announced Death of Film

Since the start of the 21st century the entire image production chain has turned resolutely towards full digital. There is no lack of arguments in favor of digital: lower production and postproduction costs, less chemical pollution, ease of correction, instantaneous viewing possibilities, constant quality, and so on. The advantages of digital filming for stereoscopy further include the possibility of precise automatic complex initial settings.

Photography

The public at large has had access to ready-to-view stereoscopic photography since 2009 thanks to the first digital 3D cameras, the Fuji Finepix Real 3D W1 and W3. Both models work with two 10-megapixel sensors separated by a fixed interaxial distance of 77 mm and have an autostereoscopic rear monitoring screen. With its fully automatic mode, the Fuji is as easy to use as any ordinary camera. The pictures it takes can be viewed on a home PC or printed as anaglyphs using software that requires no special training.

In addition, many amateur stereographers build their own "3D rigs" composed of pairs of digital cameras and get excellent results with technology that is likewise available to everyone.

Cinematography

The movie industry had the most to gain from the computerization of image processing. The number of images handled in making a movie is very high and they must all undergo strictly identical processing. It was thus natural to convert movie camera images to digital files and then, after editing the film and adding special effects, to put the edited movie back on film. This intermediate step is called "DI" for Digital Intermediate.

The rest of the movie production chain gradually converted to digital technology in the first years of this century. On the one hand, the quality of digital movie cameras caught up with that of cameras that use film and the advantages of immediate viewing tipped the scales. At the other end of the chain, digital distribution became possible with the marketing, by Texas Instruments and its license holders, of DMD (Digital Micro-mirror Device) projectors, the release of which was quickly followed by a specific standard, known as the DCI specification, which all the majors have adopted. It is important to note that the stereoscopic formats were integrated into the DCI specification from the start, which facilitated their distribution enormously.

Digital postproduction offered a leap in quality and possibilities for pairing the left and right images of a stereoscopic pair that were not imaginable in the era of optical transfers. The first decisive successes were achieved in the first years of the century with *Spy Kids 3D* in 2003, *The Polar Express* in 2004, and *Chicken Little* in 2005.

The public at large was somewhat skeptical at first, but quickly developed confidence in the quality of the "new 3D movie" that digital technology made possible. The 3D films that followed each garnered a little more success than the last. So, in the course of 2009, *My Bloody Valentine*, and then *Monsters vs. Aliens* paved the way for *Avatar*, James Cameron's planetary hit. *Avatar* definitely marked a major turning point in the history of movies, proving to the whole world that 3D movies could rival the greatest hits in the history of film. The movie reaped more than 3 billion dollars for an estimated production cost of 460 million.

Digital Projection

The point of no return was reached in 2005 with the start of digital projection in commercial theaters. Walt Disney Studios Entertainment effectively showed *Chicken Little* in 3D digital format in the Chinese Theatre, Hollywood. Since then, the world's movie theaters have been converting to digital technology. Digital's penetration rate is growing each year and the possibility of showing 3D films for a marginal cost increase is the main driver of the theaters' conversion to digital technology.

2007: the First 100% Digital 3D Movie

Scar 3D is the movie that truly marked the transition. Released in 2007, it was the first 3D film that was produced digitally from start to finish, from the camera to the movie theater's projector. National Geographic and 3ality Digital released *U2 3D*, the first movie of a concert shot live with digital 3D cameras, in 2008, not even twelve months after *Scar 3D*'s premiere.

The Future of 3D

After Talkies and Color, 3D is Coming

There are many reasons for 3D's resurgence, all of them linked to the advent of digital technologies. After a slow but steady decline in movie theater audiences, the industry saw 3D as a way to attract people once again to the big screen. 3D effectively works best in large theaters, and neither HDTV nor video on demand, nor even Internet can meet the expectations of a public that yearns for strong emotions and full immersion.

According to the Motion Picture Association of America (MPAA), US box office receipts rose 10% between 2008 and 2009 under the impetus of 3D movies, which accounted for more than 10% of the tickets. What is more, pirate copies of films made with under-the-coat cameras automatically disappear with 3D movies, which means smaller losses for the studios. Finally, the advent of 3DTV and Blu-ray disk players offer the studios the promise of substantial earnings on the home movie market. Following the example of Fox's *Avatar*, all the studios are once again banking on 3D, as they did in the Golden Age of 1953.

It is hard not to draw parallels between the two periods. In 1953, television was the cause of declining movie-going that 3D movies tried to combat. At this start of the 21st century, Hollywood is trying to counter the great popularity of Internet and immediate

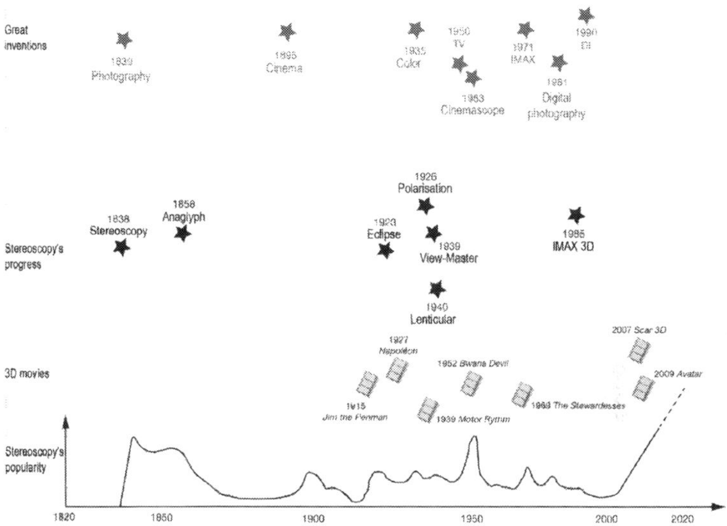

Stereoscopy's rough ride down the time line...

Interactive 3D Movies

The depth and high definition of digital images contribute to the feeling of immersion. As audiences become more and more closely entwined in the action, they will logically want to be part of and interact with it more and more as well. The high degree of public satisfaction with "4D" shows in amusement parks has already been ascertained. In such theaters, the seats move with the action and the animation of other elements of the set, such as rainmakers and wind-blowers, strengthens the feeling of immersion. The next challenge will be to add interaction to immersion and enable viewers to influence the storyline by means of their reactions, even to become part of the fiction presented on the screen. Practical applications along such lines already exist in amusement parks.

What is more, live 3D broadcasts of sports and other events also already exist. It remains for us to develop live interactive shows to blur the line between immersion and participation even more. In actual fact, the well-defined boundaries between video games, amusement parks, and movies could dissolve and, in so doing, challenge a great many of today's story-writing rules in general.

Long Term

Public enthusiasm for stereoscopy does not seem to be weakening and the reason most often given for this is the enhanced feeling of immersion in the pictures. The enhanced realism that stereoscopy provides compared with conventional movies is yet one more step along the road to perfect realism, but it is not the last step. Presenting the viewer with two views is not enough. If the scene were perfectly realistic, it would be possible to walk around the characters and objects and see them from every angle. There is nothing natural about watching a film through special glasses while sitting in a theater and eating popcorn! To get a realistic scene from many points of view and not just one, you have to generate and project many views. That means going from stereoscopy to omniscopy. This generation of many images is necessary to enable the viewer to have the two views required for 3D viewing regardless of his/her position. That is the necessary condition for creating quality autostereoscopic screens. Glasses-free immersive 3D movies are perhaps the next step on the 21st century's technological highway.

Narrative Grammar

More than a century of filmmaking has produced a very precise narrative grammar that all directors and viewers take for granted. The actors' positions, depth of field, choices of focal length, and subdivision into shots are all so many tools used to give the impression of depth with a "flat" medium.

A real 3D movie cannot exist without two basic elements: a good story that lends itself to 3D, and a good story-telling manner. Now, to date, our combined experience in screenwriting for 3D movies is only some tenths, even some hundredths of the experience amassed in conventional movie screenwriting. One of the things that experience has taught most directors – but not all of them! – is that 3D must serve the story, and not be a show itself. Overusing pop-out effects is what destroyed 3D filmmaking at the height of its success, in 1953-1954, and James Cameron's restraint in this regard is probably what made *Avatar* such a hit.

The way the story is told must change: The great depth of field that 3D requires means that you cannot focus attention on a character as in conventional movies and other methods will have to be found to attract the viewer's attention to the main characters on the set. Action scenes can no longer be cut up at as quick a pace as they used to, and depth changes between scenes will have to be planned most carefully. In a nutshell, the entire grammar of cinematography will have to change. Writing a new chapter will not be enough. We shall also have to force ourselves consciously to cast out some of the chapters of the past. The next few years will be crucial for the development of this new grammar.

Holographic devices producing views that can be seen from all directions currently do exist, but they suffer from extremely severe limitations: small size, lack of opacity, poor contrast, and so on.

Is This the Fourth 3D Wave?

Are we in a fourth wave of interest in 3D, a wave that, like the three previous waves, will break and recede? Or are we at the start of a new era? Only the future will tell, but there are grounds for believing that digital technology has given photographs and 3D movies the tools that they lacked to join the arsenal of modern media forever. It is now up to us to use the right technical, story-telling, and artistic elements to turn them into methods of expression that are useful, pleasing, and fun to use!

Stereopsis and Physiological Aspects of 3D

1

Why do we see in 3D?

How does a one-eyed person judge distance?

Does everyone see 3D?

Why do 3D films sometimes make people nauseous?

What is the Pulfrich effect?

Must the cameras converge?

3D Perception

Our eyes are sensors that are very similar to cameras in many respects. They are connected to the central nervous system and brain by the optic nerves, which are high-speed information channels. The brain transforms the received images into mental representations of the surrounding world with all its features, including depth. The entire set of processes for perceiving the world in three dimensions is called "stereopsis" from the Greek *stereos* meaning "solid" and *opsis* meaning "vision." Stereopsis was first described by Charles Wheatstone in 1838, but depth perception has always fascinated scholars over the centuries.

Notice that stereopsis does not exist to the same degree in all creatures. It is found primarily in predators and birds. Most large animals do not perceive depth as well as we do. Human beings have an approximately 120° field of binocular vision, whereas our panoramic vision is limited to about 180°. Prey animals can spot their predators at great distances but do not perceive details at close range. In contrast, predators – a group that includes humans – perceive things very accurately in their zones of action. Stereopsis helps them catch their prey with a high rate of success.

Although only a small fraction of the population is visually impaired, a larger fraction – between 3 and 15%, depending on the source – does not have binocular depth perception. This can be due to temporary visual disorders that affect certain individuals in the first months of life, when the nerve processes for perceiving the environment form. The causes of this hypostereopsis are still not well known. Luckily for such individuals, they can use many other information sources to create their own representations of the world in three dimensions. What is more, stereopsis is useful only for the portion of the world that is located no more than twenty or so meters away from us. Our perceptions of the distances and volumes of things are also based on other clues called "monocular perception factors."

Monocular Perception Factors

We use many monocular perception factors of depth. The term "monocular" comes from the fact that they come from the images that are given by a single eye or, more precisely, because they are carried even by images that are devoid of depth. Traditional movies give viewers all these factors, so that the world that we see on the big screen looks realistic. Only the binocular factors that stereoscopy will bring are missing.

Accommodation

In the human eye, the lens contracts or dilates to focus the image of a nearby or distant object on the retina. The resulting muscular feedback tells the brain if you are focusing your gaze on a very close object (less than a yard away). Since the ability to modify

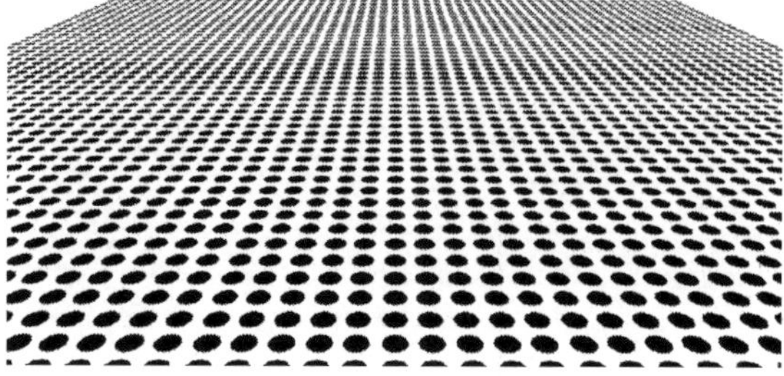

Optical density gradient: The relative sizes of identical objects decrease with increasing distance.

this accommodation decreases with age, this factor becomes less important for people over fifty. In practice, it is useful for very short distances only.

Size

An object's size, or rather the size that an object occupies once it is projected on the retina, gives the brain a very precise idea of its distance. That is all the more true if the object in question is known and already filed in the brain's database. Even in the case of unknown objects, if one sees two identical objects that appear to be of different sizes, the brain will assume that the one that appears to be larger is closer than the other one. This effect is also called the "optical density gradient." Perspective, which is explained below, is moreover merely a consequence of this effect.

Perspective

Distant objects appear smaller than close objects. Parallel lines stretching towards infinity converge with distance. This effect can be ascertained by looking at a straight line or a building's façade from its foundations. Note that perspective exists even when no actual line stretches away from the observer. So, a line of telephone poles that become shorter with distance constitutes a vanishing line that gives very precise distance information.

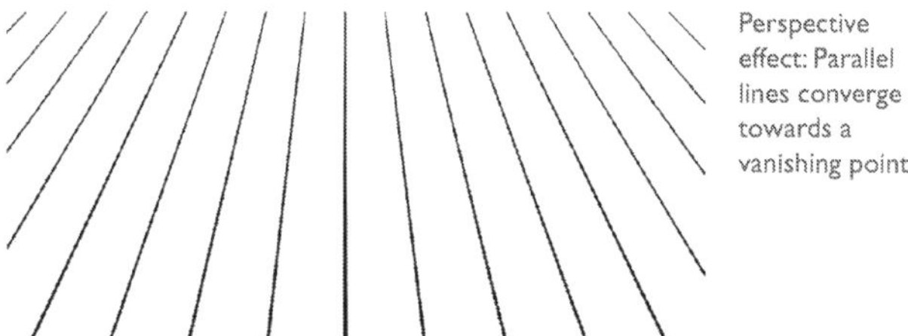

Perspective effect: Parallel lines converge towards a vanishing point.

Occlusion

Close objects hide more distant objects from view. From this we deduce which objects are closer and which are farther from us. Our internal representation of the world memorizes objects' relative positions much more than their absolute distances. So, at a great distance and without occlusion it is often difficult for us to say which of two trees of apparently identical size is taller. In contrast, seeing a player in front of another player during a soccer match is a sure indication that the first player is the closer of the two.

Shadows and light

The play of light on the surface of objects gives us additional clues as to the positions of the elements of this surface, for hollows and bumps will be clearly differentiated if the lighting is sufficiently harsh. As a rule, soft lighting and weak contrast reduce the preciseness of our perception of objects' volumes. What is more, shadows thrown on the ground and on other objects give us information about the objects' relative positions with regard to each other.

Motion Parallax

The changes in distant objects' relative positions that occur when we move about tell us about their distance. The classic example of watching the landscape through the window of a moving train makes the respective distances of trees, buildings, poles, and vehicles very obvious, even at distances well beyond those at which stereoscopy suffices to give us information. The higher the speed of the scene's lateral displacement, the greater our ability to gauge distance. This comes from the increase in the parallax between the two successive visual representations just before the brain analyzes the scene.

Texture Gradient

Many surfaces, especially manmade ones, have repetitive elements in their textures. The cobblestones of a street, bricks on a façade, and lampposts along a road are examples of repetitive textures. The interval between two such elements is perceived to get smaller and smaller with distance. When combined with the effect of perspective, this visual clue gives us information about the relative distances of characters and objects positioned near these textures. Trompe-l'oeil effects, on the contrary, are textures in which the pitch is deliberately changed in order to give a false impression of distance. So, paving a castle's driveway with smaller and smaller stones as it approaches the building reinforces the visitor's impression of a long approach. If the viewpoint allows it, the human brain is easily taken in by such illusions.

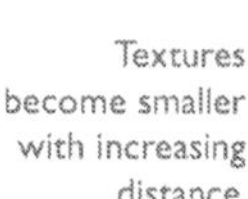
Textures become smaller with increasing distance.

© B. Michel

Atmospheric haze: Contrast diminishes with distance and colors become bluish.

Atmospheric haze

Air is not absolutely transparent. At large distances it reduces contrast and attenuates objects' colors. From this we can determine which peak in a mountain chain dozens of miles away is closest.

Geometric and Cognitive Factors

Our representations of the world depend mightily on our experience. The first time you see an object or person you examine it more or less unconsciously and add its characteristics to your mental database. On the second meeting, an even very partial vision of a single eye in the shadows of a portion of the scene enables you to identify it. The brain then calls up the representation of the object or person that it memorized.

Memory of Shapes and Sizes

The knowledge that we have of an object's exact shape and dimensions, when combined with its perceived size, enables us to gauge its distance. So, even if you look at a building with only one eye, you can assess its distance as a function of the heights of its commonest elements, such as its windows and doors. When you see a cathedral for the first time from afar, it is not necessarily very impressive, for the large door and stained-glass windows will make you underestimate its size. On the other hand, when you draw closer, you see other details on the human scale that you can identify, and then stereopsis kicks in: You correct your assessment accordingly, and only then are you surprised by the edifice's size. We use such clues very often, especially when we look at photographs.

Binocular Factors

All of the visual distance cues that we have just run through are provided to us, whether a scene is seen by one or both eyes. They give us exactly the same information, whether we are looking at snapshots, paintings, television screens, or the big screen. Artists have been using all of these methods – most of the time unconsciously – to make their works realistic for centuries. There is, however, one major difference: Our eyes focus on the surface of the photograph or screen, not on the point where the object represented in the picture is assumed to be. We are used to looking at photographs, pictures, and screens from an early age, so our brains became used to dissociating an image from the three-dimensional reality that it represents very early on. Watching 3D films forces us to reverse this operation: We have to continue to focus our gazes on a screen at a set distance while converging our eyes on each of the virtual objects projected on it in turn. Luckily, this exercise is not too stressful for the brain, which manages to adjust after sixty seconds or so.

The binocular factors of depth perception are very important for looking at our close surroundings, up to some ten, even twenty, meters. Horizontal disparity is the most important of these factors and is the cornerstone of the 3D movie industry and, more generally, 3D images.

Convergence

To get a single image of an object the two eyes must look at a point located at the same distance as the object. The set of points in space that are seen as one when the eyes converge on a given spot is called the "horopter." In other words, when we look at a close-range object, the eye muscles get our eyes to converge so that the horopter is moved to the same distance as the observed object. This muscular effort is detected by the nervous system the same way as that of accommodation is and used by the brain as a distance cue. Why does the brain have the eyes converge on the point of interest? We are entitled to wonder since, even without this convergence, the two eyes already see the object. The answer becomes obvious when we study the eye's anatomy. The eye does not have the same sensitivity everywhere. A central area, the fovea, is supplied with many more sensitive cells and thus gives a much better image than the more peripheral areas. The brain thus automatically tends to have each eye pivot so that the object being observed is centered on the fovea. In so doing, it forces the eyes to converge on this object. The result of convergence is thus doubly advantageous: We see the object of interest as a single object and with the finest resolution possible.

Binocular Vision Exercise

Hold your arm out straight in front of you at eye level and focus on the tip of your index finger. Slowly move your finger towards your nose while continuing to focus on it. You will become aware of the muscular effort required to make the two eyes converge. Now look at a spot on the horizon without moving your finger. Your finger will appear double as the horizon comes back into focus. The eye muscles' efforts and changes in the visual field's sharpness form a set of clues that confirms the distances at which the objects you see are located.

Unlike accommodation, convergence is not affected by age. This distance clue is extremely important for short distances, typically those that are under a meter. Convergence, and thus our ability to move the horopter back and forth, is what enables us to have a sharp and accurate view of the things that interest us in our environments. Convergence is thus closely linked to the interocular parallax, which is the most important of the depth perception clues for our purposes.

Interocular Parallax

The images captured by the two eyes are processed by the brain's visual cortex and combined into a single representation of the world – which we often wrongly call an "image" of the world – that includes the notions of distance, shape, and relative position. Information about the volume and distance of each object is deduced from the horizontal disparity, or parallax, between the objects in the stereoscopic pair. In the real world, this difference depends on only three parameters, namely, interocular distance (IOD), convergence, and, of course, the object's distance from the observer.

The parallax determines the object's perceived distance. A parallax greater than the interocular distance (IOD) must be avoided, as it does not exist in nature.

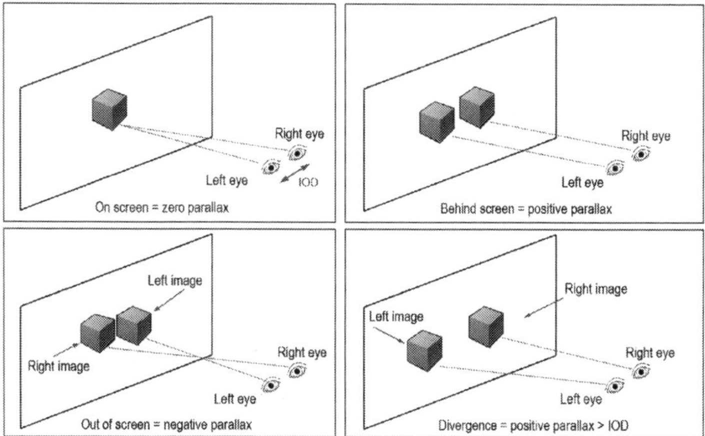

If an observed point has zero parallax, that means that the point is at the same distance as the point on which the eyes are converging. The set composed of all these points is called the "screen plane," because if one displays on a screen an image in which all the points seem to be in the same place for both eyes, then the scene is perceived as being flat and situated in this plane. Looking at a 2D film with a 3D system gives that effect. Attention should be called to the fact that, for the human virtual system, this so-called "screen plane" actually is not flat, but curved both vertically and horizontally. That is why the large screens used in amusement parks are often curved: They give a better sense of immersion in the 3D world than a flat screen does.

For all parallax values not equal to zero the point observed is either behind or in front of the screen. We speak of positive parallax when the point is behind the screen and negative parallax for objects that are closer to the audience.

Stereoscopic representation systems, such as 3D photographs and movies, try to give each eye images that are identical to what it would see if it looked at the same scene in the real world.

Proprioception and Sense of Depth

Proprioception is self-perception, whether conscious or unconscious, that is, people's awareness of the positions of the various parts of their bodies with regard to gravity. People are aware of their positions in the world, which enables their brains to insert their own bodies into their three-dimensional representations of the world, most of which comes from stereopsis, or at least for short distances. The fusion of these two abilities, proprioception and stereopsis, is what enabled human beings to invent tools and all the complex actions that they are constantly performing. Peeling an apple without cutting oneself and typing a text on a computer would not be possible without the mental fusion of proprioception and stereopsis.

Stereoscopic cinema, video, and photography are all aimed at immersing the viewer in the reality of the presented image. Once this illusion is accepted, the brain does not suspend this impression of immersion as long as the consistency of the factors of perception remains good. Unfortunately, perfect consistency is almost never possible. So, when the camera follows a flying dragon our representation of the world tends to adjust to the 3D image, including when it makes a sudden dive. But proprioception tells us that we are still seated and motionless in a movie theater! The inner ear has not detected the diving movement seen on the

© Pictorvision

A Cameron-Pace Group 3D rig mounted on the gyrostabilized Pictorvision Eclipse system guarantees perfect horizontality when shooting in 3D.

screen and the illusion of immersion is likely to be shattered. These breaches of consistency must thus be avoided or, in any event, minimized if we want to keep viewers in the illusion. Tests carried out by Discovery Channel information documentary film makers during an ocean fishing season led to the finding that it was impossible to mount a 3D camera on a trawler's deck in rough seas and keep the viewers in the illusion. Seeing the horizon constantly rising and falling away should greatly upset the viewer's sense of balance, but the viewers' inner ears did not receive the same stimuli. The inconsistency between the viewer's proprioceptive information and what the film images showed was found to cause nausea even more surely than if the viewer were on the trawler. That is why one should always keep the horizon scrupulously horizontal when shooting in 3D.

To correct this problem to a certain extent, many amusement park theaters are equipped with mobile seats that move in sync with the action. In these "4D" theaters the audience's proprioceptive sensations jive with the stereoscopic image and the sensation of immersion is considerably enhanced.

Pulfrich Effect and Other Strange Phenomena

The Pulfrich effect is a psycho-optical phenomenon that was documented by Carl Pulfrich in the early 20th century. It is due to the difference in the speed of perception according to an object's luminosity. The Pulfrich effect is thus perceptible only in the case of moving objects. For example, look at the movement of a pendulum swinging from left to right. If you cover one eye with a piece of dark glass you will see the lateral swing seconded by a movement in depth, as if the pendulum were oscillating in a circle rather than a plane. The darker the filter, the more pronounced the effect becomes.

If the filter is over the left eye, the moving object seems to be in front of the plane of the screen if it moves to the left and behind the plane if it moves to the right. The explanation is simple: The presence of a dark filter over one eye produces a lag in the perception of the scene coming from this eye. The left eye thus sees the pendulum a split-second later, and that causes a horizontal disparity with respect to the right eye. The brain interprets this difference as an interocular parallax and deduces a depth proportionate to the speed and direction of the moving object. Motionless objects are not affected.

The Pulfrich Effect on the Web

As surprising as it may seem, some amateurs have shot films based solely on the Pulfrich effect (google "Pulfrich video" on the Web). To watch a "Pulfrich effect" film, the viewer must obviously wear an appropriate pair of glasses, such as a pair of sunglasses with one of the lenses removed.

It is possible to make whole short video sequences in "Pulfrich mode" with a good 3D effect. However, the constraints are very severe: All of the objects in front of the screen plane must constantly be moving in one direction and all of the elements behind the screen plane must move in the other direction. You can imagine a sequence filmed from a moving train window or occurring on a merry-go-round, but otherwise, in practice, this huge problem makes the Pulfrich effect unusable in film-making. On the other hand, it is important to know about it, for it can be the cause of errors in gauging distances in a 3D movie if one of the images in the stereoscopic pair is too dark, which can happen in theaters that use two separate cameras for the stereoscopy. The same problem can crop up on a control monitor during shooting with a mirror rig, which can darken one of the images more than the other one. Note that taking shots in "Pulfrich" mode is easy: An ordinary movie camera suffices, for the only things that count are the staging and lateral speeds of the filmed objects.

Stereoscopic Vision

All animals have two eyes, but predators – to which we belong – have better stereoscopic vision in order to locate their prey. Less aggressive animals, such as cows, have more extensive peripheral vision but not-as-good distance perception. They are thus better armed to detect a predator and flee as quickly as possible. A few examples: The human's field of vision is about 180-190°, of which 110° is binocular. The horse has a field of vision of 350°, of which only 60° or so is binocular vision. The rabbit, with eyes situated on the sides of its head, covers all 360°, but has binocular vision in only 20°. The chameleon and many fish have complete panoramic vision but no binocular vision, given that their eyes are situated so far apart, on either side of the head.

The Eye and Visual System

Vision has always seemed natural to us. It enables us to perceive a little less than half of the environment, from very close distances of the order of a few centimeters to infinity. All of this is made possible by two identical organs, the eyes. For a photographer, the human eye's resemblance to a camera is striking: The lens of the eye projects an image on a sensitive surface covered with sensors and the iris limits the amount of light to adapt to the lighting conditions. A few dimensions: a diameter of 25°mm, an opening that ranges from f/2.8 to f/22, and a focal length of 25°mm. The main difference is that the light-sensitive surface, the retina, is curved, and has higher resolution in the center than at its edges. The retina has two kinds of sensor, cones and rods, distributed unevenly over its surface. The cones are sensitive to brilliance only, whereas the rods perceive color but are less sensitive to low light levels. There are three subgroups of rods, the maximum sensitivities of which correspond to blue, green, and red.

The central zone is called the fovea or macula. It contains cones only and thus has more acute vision than the rest of the retina. A small, slightly off-center region, the optic disk or papilla, is blind because it is devoid of sensitive cells. It corresponds to the spot where the optic nerve is connected to the eye. The brain automatically fills in the missing part of the image with a copy of the corresponding image from the other eye, so that this gap is never consciously perceptible.

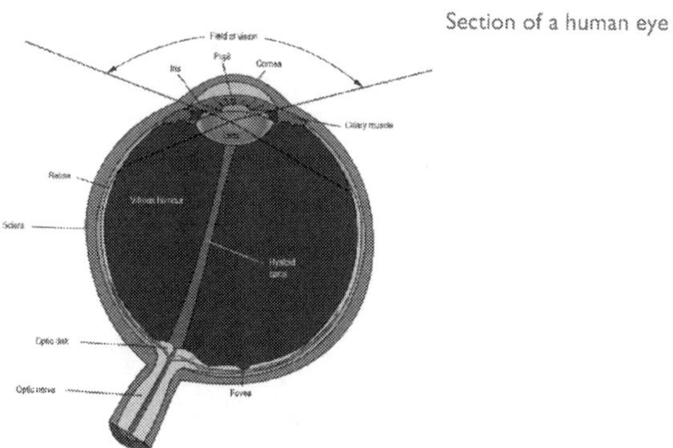

Section of a human eye

The human visual system is far from perfect. Glasses make it possible to correct focusing problems (near-sightedness, far-sightedness) in many people, as well as convergence problems (strabismus), to a certain extent. About 10% of the population has differential visual acuity, that is, one eye is better than the other one and contributes more than half of the brain's perception of details. In such people, depth perception through binocular vision is impaired, sometimes to such an extent that stereoscopy cannot give them any useful information (this is called amblyopia). If the flaw is not corrected soon after it develops – usually between the ages of 6 and 10 – the brain tends to ignore the bothersome image and bases its interpretation of the person's vision on one eye only. The amblyopia becomes permanent and the subject loses her/his ability to use binocular vision. An estimated 6-10% of the population sees little or no depth through binocular vision.

Deliberate Squinting and Autostereograms

Bela Julesz, a specialist of perception and the visual system, was working for Bell Labs in 1959 when he invented random dot stereograms, known by the acronym SIRDS for "Single Image Random Dot Stereogram". Twenty years later he and his student Christopher Tyler came up with autostereograms. They had discovered that if one looked at a repetitive texture it was possible to force one's gaze to focus far behind or in front of the picture's plane so that the brain would fuse the images of two different iterations of the texture. If tiny horizontal offsets were introduced into this texture, the brain would interpret them as a parallax and a three-dimensional shape would appear in the image.

A autostereogram

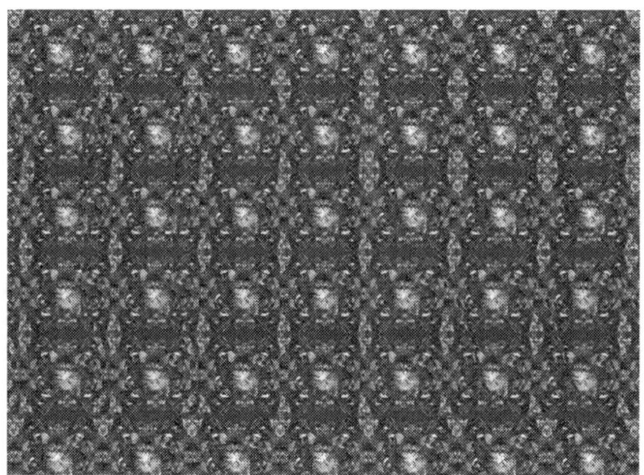

How to See the Image Hidden in a Stereogram:

Set the image to the right scale: The pattern must repeat two or three times in the space of the interocular distance, *i.e.*, 6.5 cm.

Come very close to the screen or printed image, with your nose just a few centimeters (an inch) from the image, and look towards infinity, behind the screen, without trying to focus. As your vision will be blurry, don't try to see anything. Next, move away from the image little by little, still without trying to focus on anything. Once you've reached a distance of 50°cm (20 inches), stop moving and wait for the image to appear.

This will be difficult to do the first time. With practice, you will be able to see subsequent stereograms more and more easily.

To see the shape hidden in an autostereogram you have to make a conscious unnatural focusing effort, *i.e.*, deliberate squinting. Strabismus (or squinting) is a flaw found in the visual system that provokes an erroneous convergence of the two eyes and induces double vision of the observed object (diplopia). In the case of autostereograms, the image that forms is not perceived as being double because the repetition of the texture's pattern prompts the brain to fuse two different images that resemble each other enough to be perceived as identical.

There are two ways of looking at an autostereogram: either keeping your eyes relatively parallel or making them converge more than normal. In both cases the brain will fuse the views of two different patterns, each seen by one of the two eyes, into a single image. This unnatural situation obviously differs from the usual conditions in which both eyes look at the same pattern and the brain uses this information to forge a flat representation situated in the plane of the paper. Most autostereograms are designed for viewing by the first method, that is, by gazing at the horizon.

How Does One Create an Autostereogram?

Various software applications exist for turning a horizontally repeated texture into an autostereogram, but the principle is simple enough to be applied with image-processing software such as Photoshop. Many free software applications able to generate autostereograms also exist. One example is Stereogram Creator.

Stereoscopic Errors, Headaches, and Other Unpleasant Effects

Newcomers to stereoscopic 3D imagery are sometimes surprised by the depth effects, which can be exaggerated, unexpected, unpleasant, sometimes even literally nauseating. All of these adjectives pop up regularly in audience reactions to imperfect stereoscopic presentations. It is also possible that some of the equipment used is faulty, poorly set, or poorly used.

If you are in one of these cases, it may be due to poor convergence settings or too much depth. Unfortunately, it is not always easy to set the intensity of your depth. In movie theaters, sitting half as far away from the screen will suffice to halve the depth effect. With a 3D Blu-ray disc displayed on a 3DTV, it is not always possible to change the intensity of depth. On the other hand, most multimedia drives and games on PCs or consoles allow you to modulate this intensity (the maximum parallax value) as well as changing the focus (having the scene move forward or backward with respect to the screen plane).

Warning
If discomfort or headaches persist, consult an ophthalmologist. It is possible that an ordinarily minor problem of vision may be revealed by the additional efforts required to view stereoscopic films.

For players who are just starting with 3D, it is advisable to set the depth to 15% of its maximum and the focus to its maximum in the direction that pushes the scene behind the screen. The scene can then be brought closer and the depth increased little by little until values that they deem comfortable are reached.

With habit, people gradually adapt to watching stereoscopic productions: Their eye muscles gradually get used to controlling convergence and focus separately and they learn the subtleties of 3D imagery unconsciously. The viewers' lifestyles also influence their adaptability: A programmer who spends all his time in front of a computer screen will take longer to adapt than an Alpine shepherd who is constantly shifting between close-range and remote points of interest.

Convergence

When we look at a close object, each of our eyes focuses on the observed spot, that is, the lens dilates or contracts so that the image it transmits will be projected right on the retina. At the same time, the two eyes converge on this same spot. These movements are unconscious and often so well synchronized that it is hard for us to do them separately. Feedback from the eye muscles tells the brain at what distance the observed object is located. When you look at a picture, whether stereoscopic or not, your eyes converge on one point in the picture. If it is a photograph or the screen of a portable device the two eyes are oriented at just the right angle to converge a few dozen centimeters from the surface. If you are in a movie theater, on the other hand, looking at a screen several meters – or even dozens of meters – away, the point of convergence is so far away that the two eyes are actually parallel. The visual effects, whether perceptive or muscular, linked to convergence disappear or are negligible for all distances of more than a few meters.

The size and distance of the observed image have major influences on the way we see 3D. During a film the eye muscles will remain more relaxed if we are focusing on a large, distant screen.

Accommodation/Convergence Conflict

A movie or photograph must present maximal consistency among the various depth perception factors. Still, however much care is given to the image, it will not be possible to achieve realistic consistency between convergence and accommodation, since the image is presented on a plane at a set distance from the viewer while trying to make the viewer believe that the image consists of various objects at different distances. The eyes will thus be able to converge to various extents to observe the details of the scene at various distances but will have to remain focused on the picture's plane. That is not much of a problem, for the brain

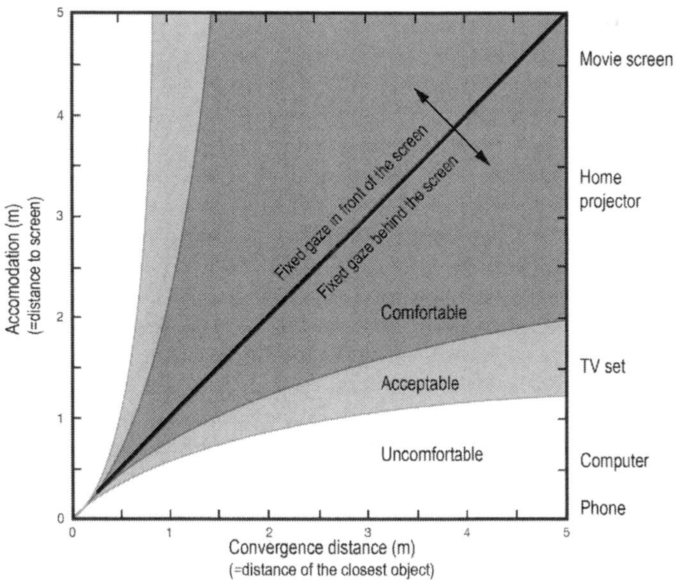

Accommodation/convergence comfort diagram

can accommodate to that easily, at least as long as the screen is more than a few meters away. So, in a movie theater, the viewers are sufficiently far from the screen to be able to accommodate their sight to the screen (and a dozen meters can be assimilated with infinity from this point of view). During the film their eyes will usually converge on objects that are at least several meters away and the absence of correlation between convergence and accommodation will remain negligible. That is one of the reasons why watching a 3D film is always less tiring in a movie theater than on TV.

It is easy to understand from the above accommodation/ convergence comfort diagram that the closer the object of interest, the smaller the degree of tolerance of the link between these two parameters. The relationship between the two along the diagonal is perfect and discomfort is minimal; this corresponds to objects displayed in 3D at exactly the same distance as the screen. Above the diagonal line, the gaze converges on a point closer than the screen plane, to which the eyes accommodate. If no object is presented at less than a third of the distance to the screen the viewer will almost never experience any discomfort, regardless of her/his distance from the screen. Below the diagonal line, the gaze converges behind the screen plane. A true feeling of infinite distance will be truly acceptable only if accommodation takes place at least three meters from the screen, which is always the case in a movie theater. For small and medium-sized screens, however, the screen's size must be taken into account to set the maximum parallax for distant objects.

As soon as the eyes accommodate far enough, typically at more than five meters, there will be practically no conflict between convergence and accommodation. This case corresponds to the upper edge of the diagram and is found in movie theaters.

A short accommodation distance corresponds to the case of TV sets at home, which are usually placed 2 to 4 meters from the viewer. The space at the bottom of the diagram thus corresponds to the case of very close TV sets, computer screens, and portable devices, which are usually less than a meter from one's eyes. In all these cases, objects that are too close, such as out-of-screen objects, cause discomfort.

The space along the left edge of the diagram corresponds to strong in-your-face objects. So we see that such spectacular 3D effects must clearly be used with caution.

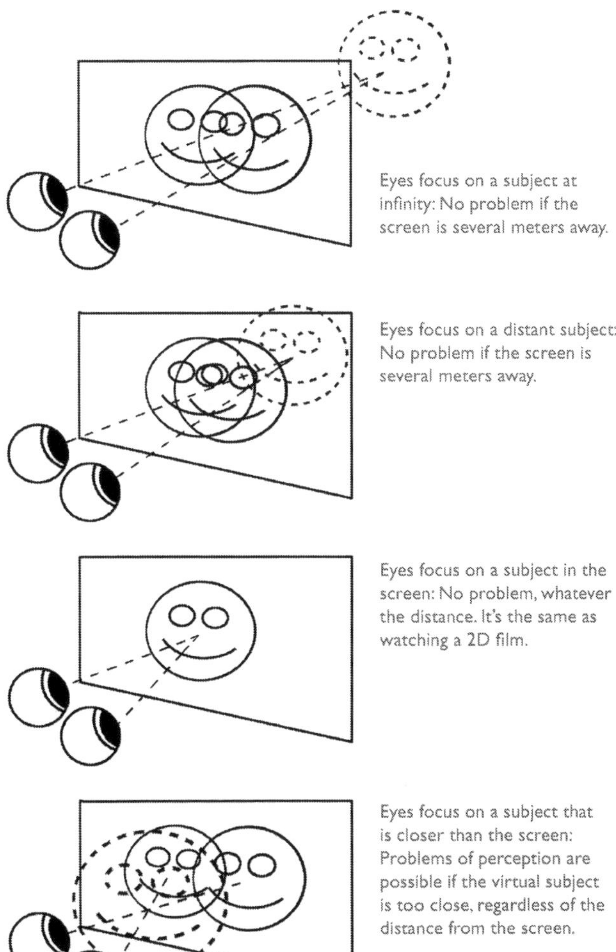

Eyes focus on a subject at infinity: No problem if the screen is several meters away.
Eyes focus on a distant subject: No problem if the screen is several meters away.
Eyes focus on a subject in the screen: No problem, whatever the distance. It's the same as watching a 2D film.
Eyes focus on a subject that is closer than the screen: Problems of perception are possible if the virtual subject is too close, regardless of the distance from the screen.

Visual and Stereoscopic Acuity

The human eye's angular resolution is on the order of a 60-second arc, which corresponds to 20,000 separate points to sweep the horizon. However, this resolution is obtained only in the central zone of vision, the fovea, and decreases towards the sides. To determine an object's distance, the brain uses the horizontal differences between the two views and is capable, under the best conditions, of detecting a parallax difference that is less than the optical resolution; ten times less, according to some sources.

To get the most out of an image projected on a screen there is thus a maximum distance that must not be exceeded, on pain of losing details and decreasing the impression of immersion and feeling of depth.

The standard SMPTE EG-18-1994 recommends a minimal field of vision of 30° for movie theaters and is widely adopted for home installations (*www.smpte.org*). Shorter distances enhance immersion and reduce eye strain. THX (*www.thx.com*) has also published a standard with which THX-labeled movie theaters comply. This standard requires a minimal field of vision of 26° for the seats farthest from the screen and an optimal value of 36°. For home cinema installations, THX recommends a slightly larger angle: 40°.

The following table gives the maximum distances from the screen for optimal viewing of 3DTV (16:9, 1920 × 1080 pixels) based solely on visual acuity or following the SMPTE (30°) and THX (26° to 36°) standards.

Interocular Distance

In the human species, the mean distance between the two eyes varies little. The interocular distance in the average adult is 65 mm for men and 63 mm for women. In children this distance is logically smaller, and an interocular distance of 48 mm is generally used.

Optimal Viewing Distance According to Screen Size

Width of the 16:9 Screen (+ diagonal in inches for HDTV)	Max Distance According to Visual Acuity (m)	Max Distance According to SMPTE (m)	Max Distance According to THX (m)	Recommended Distance According to THX (m)
70 cm (32")	1.3	1.3	1.5	1.0
100 cm (46")	1.8	1.9	2.2	1.7
110 cm (50")	2.0	2.1	2.4	1.5
143 cm (65")	2.6	2.7	3.1	2.0
3 meters	5.4	5.7	6.6	4.8
4 meters	7.2	7.6	8.8	6.3
8 meters	14.4	15.2	17.6	12.7
12 meters	21.6	22.9	26.4	19
20 meters	36.0	38.1	44.0	31.7

Orthostereoscopy

"Orthostereoscopy" is the depth effect obtained with an image acquisition and reproduction system that complies with a reference interaxial distance of 65 mm. The depth effect is deemed natural and does not trigger negative reactions because the scene presented to the viewer's brain is identical to what would be seen if the viewer were looking at the original scene.

There is no one strictly orthostereoscopic position for viewers looking at a stereoscopic image. The position is determined by the angle at which the scene is intercepted. This angle must be identical to the camera angle at the time of filming. If the viewer is farther or closer than the camera was, the scene's perspective will be wrong, with either exaggerated or foreshortened depth. Similarly, if the viewer is not exactly in the viewing axis, the scene's volume will be distorted trapezoidally.

Nevertheless, if these deformations are not exaggerated, the brain will tolerate them quite well. This is because the brain effectively recreates its vision of the world in three dimensions by assessing the various objects in a scene's relationships with each other. So, if all the objects undergo similar deformation, the scene's mental reconstruction will not be hampered.

Hypostereoscopy

When shots are closer to each other than the standard interocular distance, we speak of "hypostereoscopy." Hyperstereoscopy is sometimes indispensable. So, in macrophotography of an ant, if you want to get a correctly framed view and realistic depth, it is obviously impossible to use two cameras set 65 mm apart. While the framing will be correct for one of the cameras, the other one will not even see the scene, which will be outside its field of vision. The two shots will thus have to be separated by a very small distance. This distance can vary according to the size of the object to film and even go below a few fractions of a millimeter for shots taken with a binocular microscope. Sony presented a prototype 3D camera in 2009 that was designed for taking left and right close shots with sensors that were so close to each other that they could share a single lens. The interaxial distance was small, but large enough to give an impression of depth nonetheless.

Reproducing a hypostereoscopic scene on a standard viewing system designed for human beings with an interocular distance of 65 mm is possible and the result may even be pleasant to see. However, the brain will interpret the projected images as if they had been taken normally. It will thus interpret the depth clues wrongly and exaggerate the scene's dimensions, as if it were seen by a miniature human being. This effect is called "gigantism." Logically enough, this effect is not very disturbing if the final viewing screen itself is very small, such as that of a mobile phone.

Hyperstereoscopy

The opposite of the preceding case, when the points of view are farther apart than the standard interocular distance, is "hyperstereoscopy." Shots that are very far apart give a parallax between images for very distant objects. The obvious drawback of hyperstereoscopy is that this artificial increase in the parallax is very large for close objects. You thus have to avoid them completely, on pain of creating major visual discomfort. Hyperstereoscopy exaggerates the observed depth. Contrary to hypostereoscopy, hyperstereoscopy forces the brain to perceive a scene as if it were seen by a giant with eyes that are logically very far apart. It is thus perceived in miniature and the resulting effect is called "dwarfism."

Hyper-, Ortho-, or Hypostereoscopy?

Reconstruction of Depth	Interaxial Distance	Effect on the Viewer	Use
Hypostereoscopy	<< 65 mm	Gigantism	Macrophotography
Orthostereoscopy	65 mm	Realism	Conventional realistic imagery
Hyperstereoscopy	>> 65 mm	Dwarfism	Distant landscapes

Zoom lens and "cardboarding" effect. One artificial effect sometimes occurs in poorly framed 3D scenes: The viewer does indeed perceive depth, but the objects and characters seem to be flat and situated in parallel planes that are strongly separated from each other. This cardboarding effect is due to a difference in the camera and projection rooms' fields of vision. So, a camera with a zoom lens covers a field of vision of 10° or even less, whereas the field of vision of a person seated in one of the first rows of a movie theater is on the order of 50°. Zoom lenses are suitable only for simulating viewing at a distance, such as looking through binoculars. When we look at a distant scene through binoculars, the "cardboarding" effect does not strike us as abnormal. If you decide to use this type of view in an immersive 3D film, however, you should prepare the viewer for what is coming, for example by embedding the image in an oval mask to simulate the use of binoculars.

To avoid cardboarding you must first of all avoid long focal lengths and prefer long shots. As this is not always possible, the most widely used remedy remains that of moving the cameras apart. This will increase the horizontal parallaxes and the various planes in the picture will thicken and objects regain some volume. The closest objects will recover more acceptable volume and depth. In contrast, since these same close objects will be farther apart in the two views, the risks of discomfort will rise rapidly. This solution thus cannot be applied unless great care is taken to avoid close foregrounds. Characters that are situated farther away from the camera will not gain much thickness, even if their distances seem to increase. What is more, increasing the interaxial distance will have another effect, that of miniaturization. The brain, which is used to processing images taken 65 mm apart, will work on the same assumption. So, if a village square was filmed with cameras stationed 650 mm apart, the brain will interpret the scene to be a miniature village seen by a normal person.

In actual fact, it is difficult to get around the undesirable effects of long focal lengths. It is thus better to organize the shoot differently and bring the camera closer to the set. In cases where a zoom lens remains the only solution, you will have to compose the scene so that the foregrounds remain as close as possible to the backgrounds and find a good compromise between a large zoom and large interaxial distance in order to produce a satisfactory 3D effect.

Cardboarding is not generated by large zooms alone. It can also occur with wide angles when the scene being filmed is shallow or close to the foreground and the interaxial distance has been increased exaggeratedly in order to restore volume to the scene. So, characters leaning against a living room's back wall and filmed with a wide-angle lens may be detached from the wall but lack volume. The subtlest 3D will always be rendered best with a small interaxial distance and wide-angle lens. Unfortunately, these conditions require proximity to the camera, which is not always possible.

Reproducing the Stereoscopic Image

Stereoscopic takes tend to imitate the human visual system, with two sensors separated by the same distance as the human eyes. But what about their reproduction? As a rule, the left and right images of the stereoscopic pair are projected on the same spot on the screen. They are superimposed, not offset by 65 mm. Is this compatible with the filming system? Luckily, the answer is, "Yes!"

Taking the Shot: to converge or not to converge?

There are various ways to superimpose the left and right images: by having the cameras converge or by leaving them parallel and bringing the images closer to each other. In the first case, you just have the cameras turn around their vertical axes. In the second case, two courses of action are possible: either displace the sensors laterally inside the camera or offset the images in postproduction.

Convergence

The cameras' convergence raises a few problems: The horopter, that's to say, the plane of points with zero parallax, is a cylinder with a vertical axis. All of the objects situated on the horopter will thus be reconstructed in the plane of the screen during projection. The objects situated on the horopter, such as the two squares and small circle in the figure below, will be perceived in the plane of the screen. And certain objects situated on the

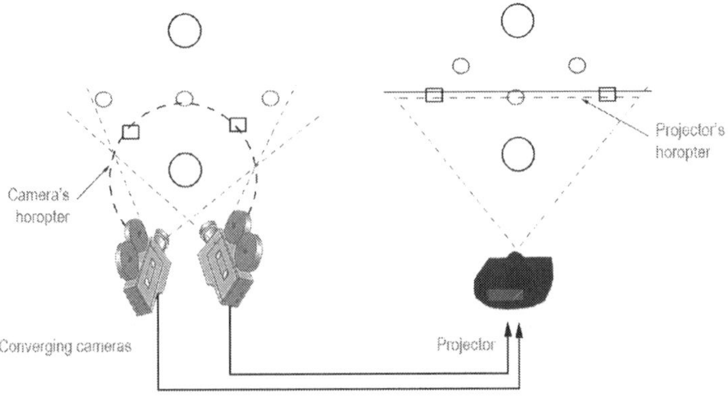

Converging cameras: Spatial curvature is deformed during projection. The objects and characters become rounder (the angles in this diagram are greatly exaggerated).

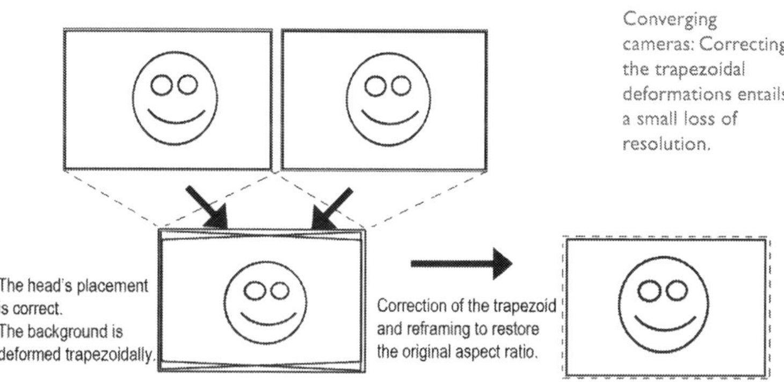

Converging cameras: Correcting the trapezoidal deformations entails a small loss of resolution.

same line parallel to the cameras' plane, such as the three small circles in the figure below, will not be returned to the same plane. We have the impression that the volume of the scene is curved rather unnaturally. Another drawback is that the images are rotated between the take and display, which produces trapezoidal deformation. Of course, with the very small angles that are usually used, these deformations remain minimal and barely perceptible.

Another good reason for using a moderate angle of convergence is that when it becomes large, the horizontal disparities in remote objects become large. Consequently, strong convergence means you must not photograph backgrounds that are too remote.

Remark
In the 3D business, this horizontal shift is known as HIT or Horizontal Image Translation.

Convergence is often used for artistic reasons. It gives characters and objects close to the camera's axis more roundness and lets you have better control over the scene's depth. Skill and experience are necessary to manage convergence effectively without being subjected to its drawbacks.

Parallel Cameras with HIT in Postproduction

With two parallel views, the two resulting images are offset the distance between the two lenses. Once these images are superimposed on each other on the screen, the entire scene is perceived as being in front of the screen. To move it back, you need only shift the left image with respect to the right one so as to superimpose a specific point in the scene in both views. This point will then be perceived as being in the plane of the screen.

This operation is very easy to do in postproduction, but has the disadvantage of losing a certain amount of the image's width, since the pixels on the images' left and right sides will no longer

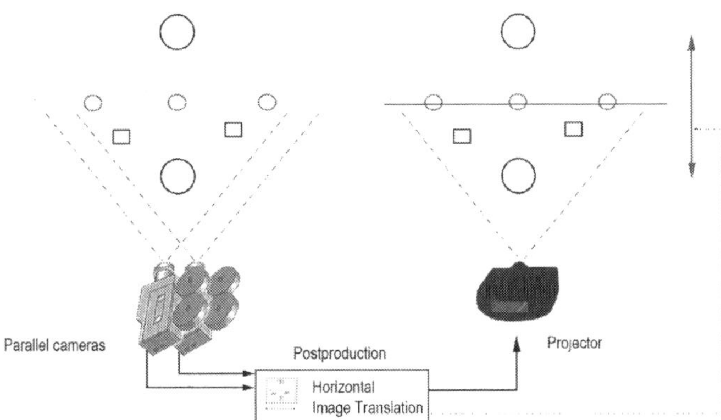

Parallel cameras: Spatial curvature is not deformed. The scene's depth with respect to the screen is set in postproduction. This reframing reduces the picture's width.

match up. If, in addition, you want to keep the image's format, you will have to reframe it completely and thus sacrifice a few lines at the top and bottom of the picture. It is obviously advisable to keep this loss in mind when shooting and to frame a bit too wide on purpose.

Parallel Cameras with Offset Sensors

A last solution is made possible by the fact that cameras are manmade and do not necessarily reproduce every aspect of the human eye. So, in cameras such as the Panasonic AG-3DA1, the two sensors film the scene in parallel, but a mechanical device shifts the sensors sideways while keeping them in the same

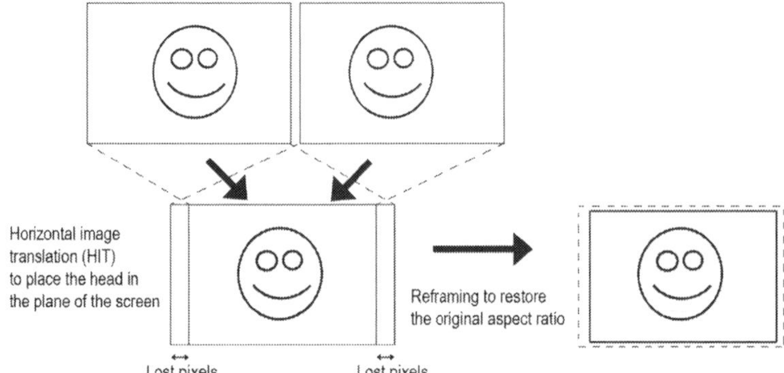

Parallel cameras with HIT: loss of resolution due to the lateral reframing

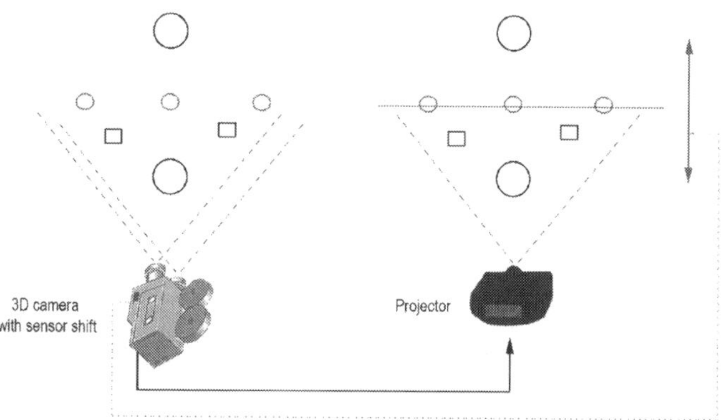

3D camera with offset sensors: Spatial curvature is not deformed. The scene's depth with respect to the screen is set in the camera.

plane. This optical shift is actually a special case of convergence that neither deforms the picture trapezoidally nor reduces the picture's usable width.

Parallel or Convergent?

The human visual system uses convergence and accommodation to obtain the most suitable pair of images for fine interpretation of the part of the world that it is analyzing. What we call the "gaze's natural position" is a combination of accommodation and convergence. Still, beyond a few meters, the angle of convergence can be mistaken for parallelism, as the angle between the two eyes' lines of sight becomes negligible.

Mirror Rigs

When the cameras' interaxial distance is smaller than the camera's width, simply mounting the cameras side by side becomes impossible. A mirror rig is then used. As a rule, a single semi-transparent mirror is placed at 45° in front of one of the cameras. Part of the image of the scene is sent to the camera located behind the mirror and the other part towards another camera mounted perpendicularly above or below the mirror. This arrangement makes it possible to have interaxials as small as 0 mm with stepless control. The rig's optical and mechanical quality is primordial.

The mirror's quality is capital: In theory, the color balance, reflections, and brilliance must be identical for the two cameras. Any asymmetries in these effects must be compensated for in the camera's settings or in postproduction. What is more, the picture filmed in the mirror will have to be turned around, since the mirror will have turned it upside down.

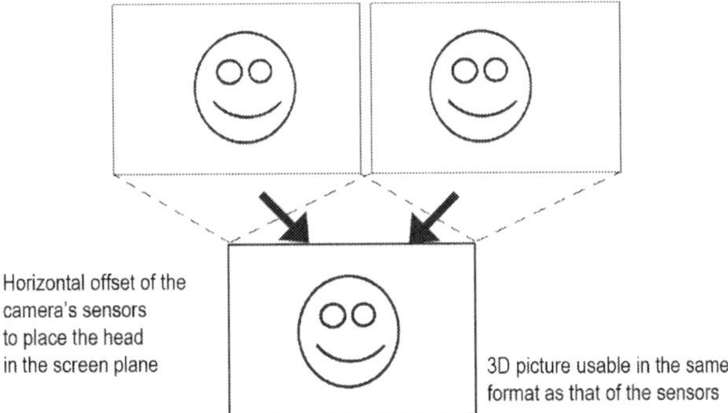

3D camera with sensor translation: No losses due to reframing

To converge or not to converge? The debate is still raging among stereographers. The advocates of convergence, such as Pace and 3ality in the US and Binocle in France, claim that the quality of their images is better, camera control is easier, and the number of pixels lost in postproduction is smaller. The "everything parallel" clan also brings together some great names in the field, such as Sony Pictures Imageworks and Dreamworks. It stresses the absence of deformation of perspective, the greater depth leeway when it comes to the background, and the freedom to set the depth in postproduction. In their view, the fact that a small percentage of resolution is lost is not important, especially if the images are computer generated or the cameras have high resolution. 3D imaging greenhorns will always prefer to learn the ropes using a parallel assembly, since it creates fewer problems and ones that are easier to correct afterwards.

Correcting the Flaws in the System

In 3D images, all disparities other than parallax must be avoided. As we have seen above, it is not always possible to avoid introducing disparities, such as the vertical offsets due to convergence, into a picture. What is more, the equipment may not be perfect and there may be differences in rotation, vertical position, zoom factors, and colorimetry in the filmed shots, to mention only the most frequent ones. These flaws must be removed for comfortable viewing. Luckily, there are many solutions on the market to do just this, from freeware such as StereoPhoto Maker and StereoMovie Maker to real-time processors such as Sony's MPE-200 SIP.

Advantages and Drawbacks of 3D Filming Methods

Method	Advantages	Drawbacks
Convergence	• Increased roundness of characters and objects; may be used to artistic advantage • Easy depth control on the set • In some cases it is not necessary to use a mirror rig, because the cameras' angulation avoids having to bring them too close to each other	• More complex, expensive rig that goes out of order quicker • Slight deformation of spatial curvature, which can be problematic for architecture • Risk of exaggerated parallax in backgrounds • Obligation to correct trapezoidal deformation • Slight loss of resolution due to correction of trapezoid • Impossible to correct excessive background parallax in postproduction
Parallel without HIT	• Very easy to use • Cheap • Robust	• Horizon will be in the screen plane • Window violations cannot be avoided • For these two reasons, this technique is not suitable for IMAX giant screens
Parallel + HIT	• Very easy to use • Cheap • Robust • Correct previewing requires a monitor that allows horizontal offsets	• Previewing on the set requires expensive equipment • Obligatory reframing in postproduction • Major reframing may be necessary if the background is close • Loss of resolution due to HIT (typically, 2-3% is lost) • Mirror rig needed for close scenes
Parallel + Camera Offset	• Easy previewing on the set • Very easy to use • Easy depth control on the set • No loss of resolution • Content usable without subsequent corrections; parallax can always be adjusted in postproduction	• Impossible to change lenses. You have to make do with a single zoom lens • Invariable interaxial distance

Starting with the simple horizontal or vertical translation of one image with respect to another image, the corrective operation entails recomputing the image's pixels and thus a certain deterioration of quality. It is nevertheless preferable to have a slightly softened corrected image than to keep unaligned images.

Screen Size and Depth Perception

Unlike flat images, 3D images are not perceived the same when the screen size changes. There are several reasons for this, namely, divergence, depth deformation, and the convergence/ accommodation conflict.

Divergence

Presenting an object on the screen with a positive parallax greater than the interocular distance (typically 65 mm) forces the eyes to diverge, something that never happens in the real world, where objects "beyond infinity" never occur in one's field of vision! "You just have to be careful when shooting the film," you might think, but things are not that simple. Imagine a producer who is doing a televised report and has projection on a screen of no more than 5 meters in mind. He sets his cameras to present the horizon with a parallax of 65 mm. Next, given the documentary's success, a distributor decides to show it in a large venue with a 20-meter-wide screen. The image's size will be quadrupled, but so will the parallax! And viewers in the first row will have to make huge efforts to focus their eyes on points that are 26 cm apart, which will be painful, if not impossible.

Divergence for Children

According to MacLachlan, the interocular distance in human beings can start at 43 mm at birth and increase by about 1.2 mm a year until it reaches 65 mm by adulthood. Fortunately, children adapt easily to less-than-ideal conditions, but smaller maximum parallaxes must be planned when shooting children's films, on pain of causing audience discomfort or headaches.

Depth Deformation

Let's return to the hypothetical television documentary for a five-meter screen. If the filmmaker wants a character's arm to pop out of the screen, he will give it a negative parallax, causing the character's screen to advance 10% of the distance between the viewer and the screen, for a realistic effect. If the film is projected on a 20-meter-wide screen, with viewers seated four times as far from the screen, the arm will still pop out by a factor of 10%, but this 10% will correspond to four times the distance that the filmmaker intended: The depth of the entire volume of the scene will be stretched, with remote planes that will move farther back and foregrounds that will be too far forward of the main character. The human brain can adapt to a certain extent, but too much deformation will ultimately break the thread with the story in which the viewers are immersed, bringing them back to the reality of a movie theater in which the 3D is no longer believable.

Convergence/Accommodation Conflict

All 3D projection systems in which the screen is located less than a few meters from the viewers have the potential to cause viewer discomfort. On a living room TV screen, if the positive parallaxes for editing the horizon achieve 65 mm, there will be no divergence. Everything would thus be perfect if, at the same times as the eyes focused on some remote mountains, they were not obliged to accommodate to the screen plane located two to three meters away. The resulting combination of eye muscle positions is unnatural and tiring. So, filmmakers will not be able to give free rein to their desire for large parallaxes if the images are to be viewed on television or computer screens. As for the viewers, if the horizons of a 3D newsreel give them headaches, the only remedy is to sit farther from the screen to attenuate the convergence/accommodation conflict. But in so doing, they will get an exaggerated perception of depth.

Accommodation and small screens
The convergence/accommodation conflict becomes the main criterion for accepting 3D on very small screens, *i.e.*, those of cell phones, portable video players, and so on. Experience shows that it is preferable to use small parallax values for images intended for small screens. Small parallaxes effectively limit the scale of the accommodation/convergence conflict and thereby minimize viewer discomfort.

Ideal Screen

The fullest immersive experience will be achieved in projection rooms where the screen is at least 8 meters from the closest viewers. And since the maximum parallax introduced in a 3D film depends directly on the screen's size, each film must be designed with the maximum screen size in mind right from the start. Projecting a film on a larger screen than foreseen will make it difficult to look at the horizon comfortably and cause discomfort, even headaches. Showing this same film on a smaller-than-intended screen will remain comfortable, albeit with slightly subdued 3D.

A film intended for movie theaters thus will not be shown to best advantage on television, but will not pose a problem. The reverse is not true: Content designed for even the largest flat-screen televisions on the market is likely to be uncomfortable to watch on the big screen. Made-for-television or big-screen movies will give an acceptable viewing experience on portable devices, with less pronounced 3D. However, that corresponds fairly well to the miniaturization effect that is to be expected from a screen that is just a few inches across. Finally, if a child finds the 3D too strong to watch, it is wise to advise her/him to move farther away from the screen to reduce the problems of divergence.

Position Opposite the Screen

Horizontal Positioning

All stereoscopic display systems are optimized for an image-centered viewing position. As a rule, a 30° viewing angle is considered the limit not to exceed, especially in the case of systems that require a silver projection screen. Outside this comfort zone the viewers may see not only differences in brilliance, but, above all, deformation of the images due to their viewing perspective. This deformation is already unpleasant when one watches a 2D film, but in the case of 3D it distorts volume. The brain must thus compensate for such distortion. In most cases it does this quite well, but the additional effort can cause unusual general fatigue, even localized eye muscle fatigue due to the unnatural positions that the brain forces the eyes to adopt.

Vertical Positioning

The ideal position for looking at a 3D picture is always in the axis of the screen. Besides the trapezoidal deformations that occur when one is too high or low with respect to the screen, the other problems that crop up include differences in brilliance on highly reflective projection screens and more or less marked loss of vision on the flat screens of certain television sets. The rule is to stay within plus or minus 15° of a horizontal line going through the center of the screen. In the case of a wall-mounted TV, the screen has to be tilted downward if it is high up.

Depth Budget and Comfort Zone

Given the physiological criteria and technical limitations that we reviewed above, both positive and negative horizontal disparities clearly must be limited to a range of values that is called the "comfort zone." You must avoid venturing outside this zone on pain of creating unpleasant effects for viewers.

The positive parallax must never exceed the interocular distance. As for the negative parallax, which causes objects to pop out of the screen, higher values may be reached occasionally, but for a few brief instants only. That is because the eye muscle fatigue that is caused by accommodation/convergence conflicts or rapid focal point changes is not felt immediately. Brief passages of strong "in-your-face" effects are not upsetting for audiences if they are sufficiently well spaced out, typically once every fifteen minutes.

In addition to these physiological limits, the psychological element of having an object intrude into the viewer's personal space also limits their proximity: Such objects may not go beyond a fraction of the viewer-screen distance. Since negative parallaxes correspond to percentages of the viewer-screen distance rather than absolute distances, the usual recommendation is never to have things pop out more than 20% of the viewer-screen distance, even for small screens. Watch out for exaggerated deformations of depth, too, for the perceived distance increases faster and faster when the negative parallax increases. The following distance/ parallax diagram shows well the nonlinearity of this relationship.

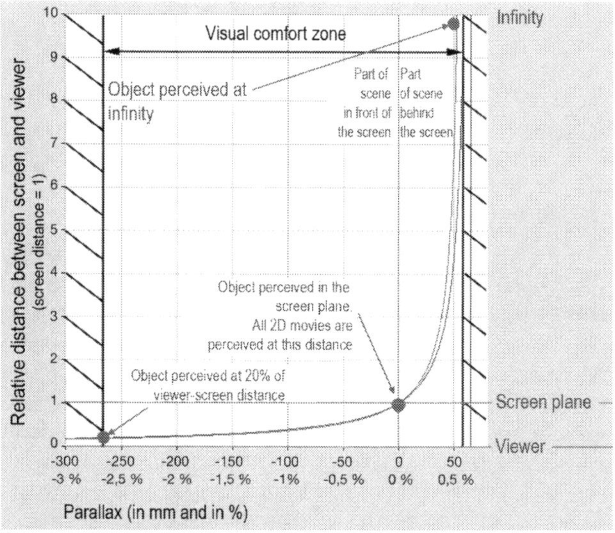

Relationship between parallax and perceived distance for a 10-meter-wide screen

The 3% Rule

Given the variety of screen sizes and viewer-screen distances with which we must deal, the rule of thumb is to keep the range of parallax values to 3% of the image's width. For extremely wide screens, such as IMAX screens, the range drops to 1.5%, whereas if no screen larger than a television screen is foreseen, you can go up to 4%, even 5%.

Maximum Parallax Range Using the 3% Rule

Image Format	Screen Width (in pixels)	Maximum Parallax (in pixels)
DV	720	24
HDV	1440	48
Full HD	1920	64
2K D-Cinema	2048	68
4K D-Cinema	4096	136

The proportion of this depth budget that may be used behind the screen's plane depends on the screen's size. The smaller the screen, the more pixels can be used for the scene's depth behind the screen for these pixels represent fewer millimeters of width and you thus are less likely to hit the limiting value of 65 mm at which ocular divergence occurs.

If we consider the values in the table opposite, it is obviously possible to design an entire film with 3D pushed behind the screen only, provided that it is to be projected on one-meter-wide TV sets only. However, immersion in the action will be hard to achieve, as the perceived distance will be too great. It would probably be more reasonable to place most of the scenes just behind the screen (even if the horizon is visible, this remains possible) without reaching a positive parallax of 3%. In this case, the horizon will be closer than in reality, but the effect will remain acceptable and the risks of discomfort on the largest screens will be limited. And of course, action in front of the screen will be allowed when artistically necessary, but in such instances great depth will be limited.

In another extreme case, a film designed for a 20-meter-wide screen can devote 6 pixels from its 64-pixel parallax budget to the space between the screen plane and infinity and assign the remaining 58 pixels to elements located between the screen and the viewer. Since pop-out effects should be used with moderation, this will produce a film with much smaller parallaxes and

Relationship Between Parallax and Screen Size

Full HD 1920 pixel Image	Maximum PositiveParallax According to the Largest Screen Size			
Screen's Width (in meters)	Max Positive Parallax To Avoid Divergence (in mm)	Max Positive Parallax (in % width)	Max Positive Parallax Corrected for Accommo-dation/ Convergence Conflict	Max Positive Parallax (in pixels)
1	65	6.5%	4.5%	86
2	65	3.2%	3.2%	62
3	65	2.1%	2.1%	42
5	65	1.3%	1.3%	25
10	65	0.6%	0.6%	13
20	65	0.3%	0.3%	6

pleasant 3D, with the entire depth budget being used only for some rare, carefully planned, moments.

The two examples given above show how important the screen's size is in the well-thought-out use of stereoscopy. Films for which the depth budgets allow it will be processed differently in post-production to push the scenes back for television broadcasting and Blu-ray discs or DVDs. The parallax budget will gradually be reduced if the content is to be disseminated on very large screens, such as on IMAX screens, which can attain widths of 30 meters.

Window Violation (WV)

A "window violation" refers to letting the sides of the frame partly mask foreground elements. Such elements attract attention and, since they cause great retinal rivalry, are extremely disturbing if not corrected by a floating window.

Floating Window

The projection screen, computer monitor, and border of a printed photograph are all natural edges of pictures. We find it normal for the scenes in an ordinary snapshot or movie to be cut off by the frame, as if it were a window through which we watched the scene. Things are more complicated in stereoscopy because the edges of the frame can very well cut off an object that is perceived to be in front of it, which is not at all natural. The illusion of being immersed in a 3D world will be shattered and retinal rivalries become very disturbing: One part of an object is visible to one but not the other eye and the brain refuses to fuse the two images. If it is a fast-moving object exiting the scene, the problem may escape notice, but if it's a massive solid-oak wardrobe, something will have to be done. The question is, what? Changing the point of view, widening the field, or changing the decor can solve the problem temporarily, but then another object appears at the edges of the scene and creates another window violation.

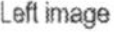

Left image

Right image

Stereoscopic photograph with retinal rivalry on the right

Object in front of the screen on right and invisible in the left image

Left image unchanged

Right image with right edge masked

Stereoscopic photograph corrected with floating window

A stereoscopic pair with retinal rivalry (above) is corrected by a floating window (below).

© B. Michel

Scan the QR code in the figure and look at the color picture on the web. You will see the red-cyan anaglyph of the preceding stereoscopic pair with the floating window correction.

The solution is actually simpler than it appears. You simply have to remove the bit of object that is not seen by one eye from the visual field of the eye that does see it. In this way, the two eyes will perceive the same reality at all times. It is as if the objects disappeared behind the edges of a virtual window between the viewer and the real window of the screen. This method is known as a "floating window." To create a floating window in front of the screen, you simply have to mask left edge of the left image and right edge of the right image. The edges do not have to be masked symmetrically. In the example given below, only the right edge of the right image was corrected. It is even possible to mask the edge obliquely to simulate a tilted window. Nothing forces you to have the virtual window parallel with the screen.

If the depth budget changes with time, as during a camera movement, the floating window can be animated and its edges can change dynamically over time to protect the scene from window violations. This dynamic floating window technique was used for the first time by Brian Gardner in *Meet the Robinsons* (2007), but is now common practice.

Window Violation without Retinal Rivalry

Foreground elements that are partly occluded by upper or lower edges do not cause retinal rivalry because the two views' corresponding pixels appear and disappear at the same time.

Nevertheless, nothing is more unnatural than to see the screen scalp an actor situated in front of the screen. Such a window violation prevents the brain from reconstructing the scene correctly and destroys the feeling of immersion. This flaw cannot be corrected by a simple floating window. It is thus important to avoid such inconsistencies altogether.

The same goes for the bottom edge of the frame, with a slightly mitigating circumstance: In movie theaters, the bottom of the screen is often masked by the preceding row of people. Only the first rows in the audience are truly affected by such window violations. The others can imagine that the scene is hidden simply by the seat backs of the rows in front of them instead of by the edge of the screen. Still, if you want to make a perfect film for the ideally placed viewer, bottom window violations are to be avoided as much as top window violations.

Stereoscopic Shooting

3

What is a 3D rig?

How does one synchronize and set up a pair of cameras?

What workflow should be put in place to shoot in 3D?

How are 3D CGI movies created?

How is a stereoscopic slideshow assembled?

Technical Parameters

Resolution, image frequency, and focal length are the three main parameters to consider when shooting in 3D. In classic movies, lots of compromises on those three parameters are acceptable, but in 3D they are not tolerated. As seen in the previous chapter, good depth effects sometimes hinge on a parallax of a fraction of a pixel; approximations are no longer possible if top-quality results are expected.

Resolution and Aspect Ratio

You can forget everything not HD if you want to achieve a decent feeling of depth on one-meter-wide screens and above! Lower resolutions are acceptable on portable devices with screens only a few inches across and where depth is also much less pronounced. At the other end of the spectrum, large movie screens, and especially those of IMAX theaters, have the highest resolutions allowed by modern technology and offer the best, must immersive 3D images in the world. In the overwhelming majority of cases, 3D images are shot with 2K resolution. Big-budget films can afford to use 4K or more cameras, but their output is ultimately reduced to 2000 pixels in width for distribution, after benefiting from the extra quality at all intermediate steps. 720p resolution (1280 x 720 pixels) is the lowest resolution format bearing the "HD profile" name. It compensates for its slightly low pixel count with its higher frame rate, typically 60 frames per second. The most popular format on the 3D television landscape is 1080p or 1920 × 1080 pixels at 25 or 30 frames per second. As for 3D cinema, it is currently distributed in 2K, with some extremely rare exceptions, meaning a width of 2048 pixels (the height varies with the chosen aspect ratio).

The format/brigthness compromise
One more technical reason pushes the balance in favor of aspect ratios close to the native format of cinema projectors (2048 x 1080 pixels), namely, brightness. It is well known that all 3D display systems devour a large share of the projector's light, leaving the viewer with a noticeably dimmer movie than the 2D version. Using the whole available area of the projection lens guarantees the best possible brightness on the screen. With CinemaScope, 15% of the total available brightness would be lost compared with 16:9.

What is the ideal aspect ratio for 3D? Artistic and aesthetic choices are always subject to discussion, but since the advent of 3D, many favor less elongated formats than before. Thus, James Cameron decided to use two different aspect ratios for the 2D and 3D versions of *Avatar*. The 2D version is in CinemaScope format (2.39:1 or 2048 × 858 pixels in theaters), while the 3D version is consistent with the format used at shooting time, or 16:9 (1.78:1, or 1920 × 1080 pixels indoors). Thus, James Cameron decided to use two different aspect ratios for the 2D and 3D versions of *Avatar*. The 2D version is in CinemaScope format (2.39:1 or 2048 × 858 pixels in theaters), while the 3D version is consistent with the format used at shooting time, or 16:9 (1.78:1, or 1920 × 1080 pixels in theaters). Like many directors, James Cameron uses CinemaScope for 2D because it heightens at the same time the feeling of openness procured by wide open spaces and the oppressive effect of confined spaces. However for the 3D version – his obvious main target – he achieved the same feeling by playing with the depth budget, increasing it in outdoor scenes and compressing it strongly indoors.

The larger relative height of the 16:9 format also helps to keep the viewer focused on the main action. In 3D, very large image sizes must present a large number of details located away from the main action because one cannot afford to use shallow depths of field to refocus the viewer on the main action.

Image Frequency

The standard cinema image rate has been 24 frames per second since 1927. This value was chosen because it was the lowest acceptable value giving a reasonable illusion of smooth motion. The broadcasting industry understood as early as the 1950s that 50 or 60 images per second were required to render fast motion, such as found in all sporting events. Now that the digital cinema revolution has reached all steps in the production workflow, from camera to projector, nothing is left to stop filming at higher speeds. Movies shot at 48, or even 60, frames per second are not yet mainstream mainly for cost reasons, in addition to the Hollywood industry's great inertia.

However, the advent of 3D completely changes the situation: Stereographers refine scene depth by playing with horizontal parallax of a few pixels and adjust some disparities down to a twentieth of a pixel. Besides this, the low frame rate of the screen display induces large horizontal offsets between two successive images. Let's do a quick calculation to locate the problem: During a pan performed at 15° per second, or a quarter turn in six seconds, shooting at 24 fps in 2K with a 50-mm focal length causes a motion parallax of 0.6° per second, or about 1.5% the width of the lens field, or 27 pixels. The magnitude of this shift is comparable to the entire positive parallax budget of a commercial movie, which sometimes is as low as half this value!

Now let's add a technical element: Active glasses systems display two images in rapid succession, the left image a split second before the right one. In this case, our brains tend to interpret part of the motion parallax as a parallax distance; we imagine objects moving in one direction as being more distant than those moving in the opposite direction. The excessively slow rate of 24 frames per second can really degrade the depth quality in such cases.

The 48 Hz and 60 Hz frequencies are now part of the DCI Digital Cinema standard, following strong demand from various stereographers, including Kommer Kleijn from Imago and Steve Schklair from 3ality Technica. It is no coincidence that stereographers are the most insistent advocates of high-speed shooting: They know how damaging the huge horizontal disparity induced by the use of 24 frames per second is. Several great directors have already understood this, as the pioneers of 48 Hz shooting are none other than James Cameron with the follow-up of *Avatar* and Peter Jackson with *The Hobbit*. Peter Jackson himself acknowledged during the first 48 Hz tests that depth perception was much more comfortable and reduced eyestrain, even likening the difference to the one between vinyl records and audio CDs.

Focal Length

When it comes time to choose the lenses for shooting, the two big questions are "wide-angle or tele?" and "zoom or prime?" In both 2D and 3D, the reasons favoring one or the other choice are artistic, physiological, and technical.

Artistic Choice

Long focal lengths mean detachment; short focals rhyme with presence and intimacy. Those well-known rules remain true when shooting in 3D.

To observe a reality in which we are not involved, we will use only medium or long focal lengths. This will be the case when reporting on team sport events: To gauge an attack strategy during a football match, you have to have some ten players in the frame, and that will be possible only with a remote camera. In contrast, a dialog or a duel scene, where the viewer identifies easily with one of the protagonists, will require a short focal length and great proximity to the action.

Physiological Choice

The feeling of immersion is reinforced when the image on the screen is similar to that provided by the eye. Given the human visual field, short focal lengths are better: The "right" focal values are in a range from 20 to 50 mm. But, as long as the technical constraints such as the minimum and maximum parallax values are met, using longer focal lengths remains possible. The price to pay will be a less pleasant, less "round" look, and a cardboard effect that separates the scene depth in a succession of planes.

Wide-angle = proximity
The corollary of choosing short focal lengths is the obligation to remain close to the subject. For a character to fill the frame, s/he cannot be located more than a few meters from the camera. Under these conditions, the human visual system converges on and accommodates to the subject; the depth feeling is strong.
These two elements combined plead in favor of short focal lengths.

Technical Choice

The technical choices stem from three constraint categories: size of the action area to film, available equipment, and quality optimization.

The action area is defined by the frame and the depth budget, i.e., the distance between foreground and background. A dialog at the top of a mountain imposes a very large depth budget; the same dialog shot at a restaurant table will need only a few meters of depth. In the first case, we will shoot the scene with medium focal lengths and parallel cameras; in the second case, short focal lengths and converged cameras are preferable.

The equipment is often limited by the available budget. With large production resources, we will have a choice of several zoom and prime lenses. In many other cases, we will be forced to deal with the integrated zoom of monobloc 3D cameras that will provide restricted opportunities. In this case, the fixed interaxial distance will prohibit extreme depth values, in long telephoto as well as in wide-angle conditions.

Short focal length lenses are expensive, but their main disadvantage is their bulk, which almost inevitably forces one to use a mirror rig. Consequently, the choice of short focal lengths will have budgetary implications. The best lenses are fixed focals, also known as primes, and if the shooting of a sequence requires multiple lens changes, we should ask ourselves whether a zoom should be used instead. This will obviously be much less time consuming, as each lens change requires readjustment of the rig. However, these time savings between shots will be offset by increased rig complexity, as the off-centered shift adjustment of each zoom must be possible in order to match up the two lenses perfectly.

Varying the zoom factor during a shot is risky: The human visual system does not zoom, so the resulting shot will look unnatural. In addition, the mechanical limitations of zoom lenses disrupt the perfect match between both left and right frames so that the rigs must compensate for those mechanical imperfections with complex electronics. In addition to this drawback, the focal length changes are often accompanied by an interaxial distance change, which adds even more complexity to the system. High-end rigs offer real-time control of all of these parameters and leave the door open to all possible artistic decisions without increasing setup duration. Unfortunately, the price tags and daily rental fees for such rigs mean that they are out of reach for all but the biggest productions.

Summary
We can optimize the quality versus shooting cost ratio with a schedule that will minimize lens changes, which can mean limiting oneself to one, two, or three focal lengths per shooting day. Most of the time, zooms will be used to facilitate framing and minimize handling, while most of the takes will be filmed with a fixed focus.

Depth of Field

Textured and in-focus objects are necessary for the brain to be able to correlate the two views of the scene and deduce a distance. This imposes omnipresent image sharpness and a maximum depth of field. We must therefore encourage small openings and high sensitivity to prevent blur. Good stereoscopy requires short focal lengths and close subjects. The risk of creating overly large disparities in the background is thus real. It then becomes very tempting to reduce the depth of field in order to blur the background and thus lead the viewer's gaze forward. This may be an acceptable method, but it is better to avoid it completely and play with other effects such as modulating the scene's lighting to refocus the viewer's attention away from areas with a lot of high divergence.

Shallow depths of field cleverly modulated to highlight one character after the other during a dialog must be avoided in 3D movies. The viewer's gaze must be able to wander on foreground and background objects to estimate their relative distances. This unconscious process is mandatory to map the virtual world and to create a satisfactory mental representation in volume. Motion blur is also to be avoided: Even a vehicle crossing the field at speed should remain in focus; otherwise, its parallax will be difficult for the brain to interpret.

Lighting for 3D

This demand for a large depth of field leads us quite naturally to mention lighting. To open the lens's iris minimally, we need a very well lit scene and highly sensitive camera sensors; a bit of noise induced by the sensor's high sensitivity is better than a blur caused by too much opening. All objects in the scene must be in focus, so they all need light. Even the darkest corners in night scenes need proper lighting! It is easier to push the colorimetry towards dark blue tones in postproduction than to correct an inadequate depth of field.

All planes in focus ▶ Lighting of all planes is required.

The use of mirror rigs means that the brightness reaching the sensor in each camera is halved. How can we compensate for that? By doubling the lighting power, of course!

Mirror rig ▶ even more lighting required!

Stereoscopic projection systems use various active and passive glasses. The best of these systems swallows more than 60% of the brightness of the projector. Parts of images that you feel are dark on a control monitor will very often be even much darker when seen in a room with a wheezy little projector.

We must therefore ensure that the scene's lighting is well distributed. Strong contrasts between two parts of a screen do not work well in 3D. Soft, even lighting of less attractive areas is essential; it will reduce retinal rivalry, and thus ghosting, in the less interesting parts of the scene, and avoid offering unnecessary large dark areas with indecipherable depth values to the viewer's gaze. Let's not forget that a patch of black sky does not offer any eye-catching parallax. The viewer will perceive it therefore erroneously in the screen plane, which will destroy your depth staging of the scene.

Stereoscopy ▶ lighten dark areas.

Window Violation and Lighting

Light! More light! With 3D, nothing must be left in the shadow.

Sometimes you have to use lighting in a creative way to compensate for inevitable mistakes. Over-the-shoulder shots are common in classical cinema but often go wrong in 3D, since in such shots, almost by definition, a character is seen from behind, half cut off by the edge of the frame. If this character is dimly lit and has low contrast, his figure will be perceived with little or no depth and in the plane of the screen, which will save the situation. This lighting effect is also often reinforced in postproduction.

Flares and Front Lighting

Filming the headlights of an approaching car may give interesting effects in a movie. In 3D, the sun or other bright light sources present in the field will cause lens flares due to light reflections inside the lens. These effects can be used for artistic purposes in 2D but ruin the picture in 3D because they are identical in both left and right images, and thus perceived at infinity. We then have the impression that this reflection "digs a hole" in the foreground or actors, which is not the goal!

Lens flares are unwanted in 3D: Avoid front-facing light sources inside the frame.

Hard or Soft Lighting

Strong light is necessary to magnify the depth of the main action. As this takes place most probably close to the screen plane, the parallaxes will be small, as is the risk of ghosting (ghosting may be present when one eye receives (part of the) images intended for the other eye). Rather harsh lighting is thus possible. Still, you may have to push the lighting a bit into shadowy areas to avoid creating flat zones without any visible texture, as otherwise the eye would not catch enough details to determine depth in this part of the scene. In contrast, backgrounds and close foreground objects will have extremely high parallax values and therefore may cause harmful ghosts at projection time. They will therefore require softer lighting, leading to less contrast. The same rule applies to the side edges of the frame when using a floating window: Soft lighting will reduce retinal rivalry.

Cameras for 3D Shooting

The choice of cameras and their associated lenses will determine the final result's quality, but it is mostly a compromise between the director and his director of photography's preferences, trade agreements, budget, and technical constraints. Knowing that, at one end of the scale, professional integrated 3D cameras are available for less than $25,000 and, at the other end of the scale, the price of digital cinema cameras with their set of associated prime lenses may often exceed $200,000, the compromises one makes will obviously depend on the available budget!

Choosing Cameras for 3D Rigs

In addition to the considerations just mentioned, stereoscopy imposes its own technical limitations. First, the cameras must be accurate and as well matched as possible. Otherwise, unintended image disparities may render all the efforts made elsewhere completely useless. Therefore, rigid cameras, stiff rigs and accessories, and easy-to-block mechanical settings are required. If each camera is available in two separate modules (sensors and control electronics), you can save space on the rig carrying them. If they are small and accept small-diameter lenses, you can usually make do with a parallel rig, which will be cheaper and less complex than a mirror rig. Digital cameras fall into two families depending on the type of sensor used: CMOS and triple CCD.

Cameras with Tri-CCD Sensors

Triple CCD cameras have a separate sensor for each R, G, and B light component. The shooting is simultaneous for all pixels. The optical complexity required to separate the beam into three parts can generate aberrations and requires a large camera head. The tri-CCD has the theoretical advantage of better brightness and resolution as the area of three sensors is greater than that of a single one. On the other hand, the RGB prism separator adds bulk and complexity to the camera.

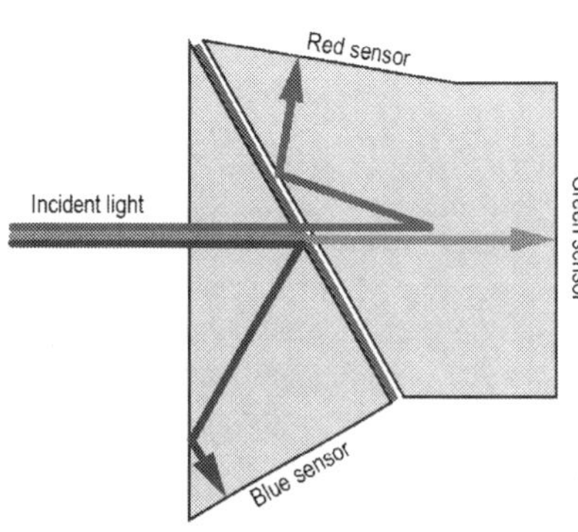

Principle of color separation inside a tri-CCD camera head

Numerous HD broadcast cameras belong to the tri-CCD type. You will find cameras with 2/3" sensors that are small enough to be shoulder mounted or put on a Steadicam, such as the Sony HDC-P1 (*www.sony.com*) or the Panasonic AK-HC1500G (*www.panasonic.com*).

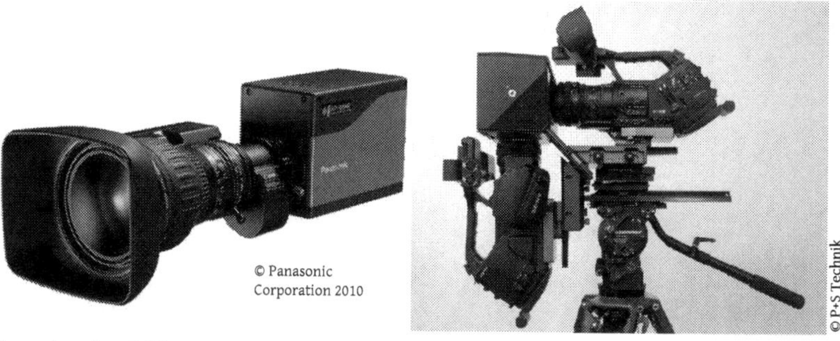

Examples of tri-CCD cameras usable on a 3D rig: on the left, the Panasonic AK-HC1500G; on the right, two Sony PMW-EX3 mounted on a P+S Technik Neutron mirror rig that can be easily shoulder mounted.

Cameras with CMOS Sensors

A single sensor records the three components by interleaving the R, G, and B pixels on the same matrix fitted with a Bayer filter. CCD cameras must conduct "demosaicing" of the raw image (RAW) to transform it into a conventional RGB image. This can be done directly in the camera for direct use of the RGB image, but is usually done remotely, in an OB van or later in postproduction. Sophisticated algorithms then exploit the large dynamic range and extensive spatial resolution of the sensor to interpolate colors intelligently and render a high quality picture, even in very high or very low light areas. Some "debayering" methods are better suited to natural scenes and some others are better with more contrasted and artificial images. If the RAW conversion is done in post, it will be possible to optimize the codec parameters according to the type of scene and adjust the colorimetry at the same time. In many cinema workflows, the RAW format is kept throughout the production; the associated metadata recorded while shooting and editing images make it possible to apply the correct color grading only at the display step. That way, the possibility of adjusting the image from the original data is maintained throughout the production chain. The disadvantage of CMOS sensors is that they must be read line by line, causing a curtain effect (rolling shutter) that the tri-CCD cameras do not have.

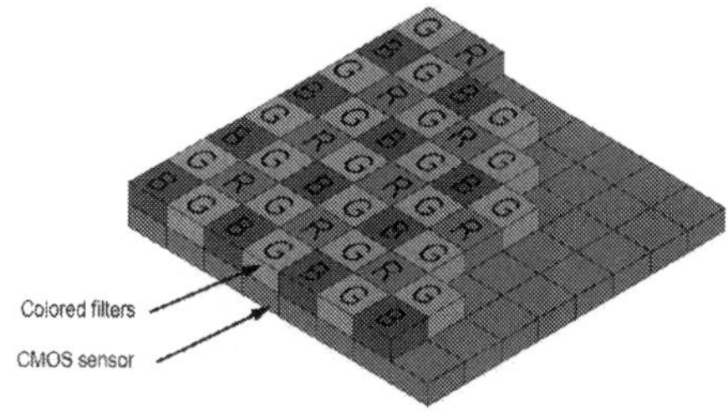

Principle of color acquisition with the Bayer matrix filter used by CMOS sensors

The RAW Format

CMOS sensors pixels are covered by R, G, and B color filter elements arranged in a matrix pattern called a "Bayer matrix" containing one red, one blue, and two green pixels for each "RGB pixel" of the final image. The higher number of green pixels is due to the fact that the human eye has more dynamic sensitivity in this part of the visible spectrum. The RAW format directly records the brightness values of all pixels as transmitted by the sensor. They are coded with a variable number of bits, usually 8 to 16, covering a wide brightness range that a specific codec will transform into a more classic image with R, G, and B pixel values. Uncompressed RAW data from large CMOS sensors can be extremely bulky. Therefore, codec compression methods are often a trade-off between the quality and quantity of data to store or transmit.

To match up both images from a 3D rig perfectly, we must be sure that the shutters are activated in the same direction, which is not always guaranteed with mirror rigs, which reverse one of the two images. Among the benefits, note that CMOS sensors are now reaching impressive sizes and resolutions; see, as an example, the digital SLR cameras and their gigantic sensors, which are used more and more for video recording. The most famous CMOS camera is probably the Red One (*www.red.com*), together with its big sister the Red Epic, selected by Peter Jackson to shoot *The Hobbit*.

Digital Sensors' Sizes in Inches

To determine the size of a three-CCD or CMOS sensor expressed in inches, converting inches to millimeters is not enough! For historical reasons, the dimensions in inches are always larger than the sensors'real sizes. In the 1950s, the Vidicon TV cameras used glass tubes on the tips of which was a sensitive rectangle: the sensor. At that time, the dimensions in inches did not refer to the sensor size but to the diameter of the glass tube supporting the sensor. However, no camera manufacturer dared to change the naming convention for fear that its sensors would be considered smaller than its competitor's... and the situation continues after more than fifty years!

Pixel Count and Sensor Size

The number of pixels of a sensor can be a misleading marketing argument. For example, a tri-CCD Full HD 3D camera such as the Panasonic AG3D-1A has six sensors with 2 million pixels (1920 × 1080) each, three for each left and right view, but a 3D camera with a single 4000 x 2000-pixel CMOS sensor for each view does not necessarily give better results. Everything will depend more on the quality of the sensors, lenses, and associated electronics.

Correspondence between Commercialand Real Sizes ofVideo Sensors

Name	Aspect ratio	Sensor Width (in mm)	Sensor height (in mm)
1/3.6"	4:3	4.000	3.000
1/3.2"	4:3	4.536	3.416
1/3"	4:3	4.800	3.600
1/2/7"	4:3	5.371	4.035
1/2.5"	4:3	5.760	4.290
1/2.3"	4:3	6.160	4.620
1/2"	4:3	6.400	4.800
1/1.8"	4:3	7.176	5.319
1/1.7"	4:3	7.600	5.700
2/3"	4:3	8.800	6.600
1"	4:3	12.800	9.600
4/3"	4:3	18.000	13.500
35 mm	3:2	36.000	24.000

The sensors' physical dimensions are very important. Large sensors catch more light and are therefore less sensitive to noise. However, they will require larger lenses and therefore bulkier rigs. Some sensors are the same size as the traditional 35 mm film; so we can use existing lenses of the usual dimensions from the analogous film world. This is the case, for example, of the Arri Alexa CMOS cameras (*www.arridigital.com*).

First in the list of mandatory criteria is the possibility of synchronizing (genlocking) two identical cameras, followed by the ability to orient the shutters in the same direction (for CMOS cameras). Finally, one must make sure that the weights and volumes, lenses included, are suitable for the shooting location and the selected 3D rigs.

© BandPro and Silicon Imaging © Vision Research © TSF.BE

CMOS cameras: on the left, an SI-2K Mini camera from Silicon Imaging fitted with a Zeiss DigiPrime lens; in the center, a Phantom 65 from Vision Research with its impressive 65 mm sensor shooting at 140 frames/second; on the right, the Red Epic.

Integrated 3D Cameras

Uncomplicated shots avoiding extreme close-ups, distant views, and scenes with complex lighting can be filmed with a parallel rig. Cameras like the integrated Panasonic AG3D-1A or Sony PMW-TD300 belong to the "parallel offset camera" category, where convergence is adjustable by laterally shifting both sensors relative to their lenses. This type of camera, provided it is well made, is the ideal choice for shooting a first 3D movie: sync between the two views is always guaranteed, mechanical and optical matching are preset, and as the two cameras' settings are natively linked, manipulation errors are rare. The biggest disadvantage of this type of camera is its fixed interocular distance, which limits the range of possible distances to the subject. For the PMW-TD300 and its 45 mm interaxial, Sony announces a minimum shooting distance of 1.2 meters. As these cameras belong to the professional range, they are fitted with dual HD-SDI output as well as memory cards. They can therefore be used to capture and transmit live events if necessary.

The first integrated 3D camera: Panasonic AG3D-1A, dual tri-CCD sensor, Full HD resolution, adjustable convergence and zoom

Note also the advent of high-end integrated cameras that offer opportunities for interaxial adjustment. The first to appear on the market is the Mk1 from Meduza Systems (UK). Its maximum resolution is 3288 x 4284 pixels for each of its two CMOS sensors. Its specially designed optics are matched and inserted into sleeves only 38 mm in diameter, allowing for relatively small interaxial values. Depending on configurations, the Mk1 can record 160 images per second in HD, 100 per second in 2K, and 60 per second in 4K.

© Meduza Systems

A high-end integrated camera: the Meduza MkI. Adjustable convergence and zoom, adjustable interaxial distance (from 38 to 110 mm).

Many consumer 3D cameras began to appear in 2011. They all have more or less similar features, but many of them use an anamorphic lens to record directly an image that is compressed by half horizontally in a "compatible frame side by side" format. They therefore do not meet the full HD 3D specification. It is nevertheless possible to find some that are truly Full HD 3D at affordable prices for amateurs, such as the Panasonic Z10000 or the Sony HDR-TD10E.

Choosing a Rig

Parallel Rigs

Parallel rigs cover a wide range of products from the most rustic to the most professional. Among the professional rigs, many are reconfigurable in "parallel" mode or in "mirror" mode. The rigs are generally manufactured by specialized companies and accept almost all existing cameras. The choice is based on their weight, stiffness, and mainly on their ease of setup. Some are motorized and allow remote adjustment of the interaxial, convergence, and usual camera features such as zoom, iris, and focus. In some cases, the controller stores the corrections applied during the focal length change internally in order to keep both zoom lenses perfectly matched at all times.

In some specific applications where the scenes to be shot are always similar, you can make do with a rig with a fixed interaxial distance. It is then natural to integrate the rig with its support or its enclosure into a single sturdy functional unit. A typical example is the small fixed-lens sports camera integrated in a waterproof housing.

© Redrock Micro

The Redrock Micro, a simple, sturdy, mechanical parallel rig with adjustable interaxial distance (up to 400 mm), here fitted with two Canon XF100 cameras.

Sports Cameras. Under difficult conditions where portability and ruggedness outweigh setup flexibility, sports cameras can take spectacular shots in places where no standard camera rig would survive. The best known is the 3D sports camera kit GoPro 3D HERO (*www.gopro.com*). This waterproof housing has a fixed interaxial distance of 33 mm and contains two small fixed-lens cameras with a 132° aperture angle capable of shooting in 1080p at 25 or 30 fps or in 720p at 60 fps. GoPro cameras are widely used in capturing sports – surfing, skiing, motocross, rallies, etc. – including by the BBC.

Mirror Rigs

The workhorse of shooting with separate cameras is the mirror rig. This is the only technique that offers complete adjustment freedom for all parameters, making it the versatile tool of choice. Suppliers of quality rigs are few. The best known – because of

© StereoscopyNews

The GoPro 3D HERO dual camera in its waterproof housing. The red part (in grey, in the center of the right image) on the back of each camera is the synchronization module. Each camera records a 1080p stream on an SD or SDHC memory card.

their strong involvement in Hollywood blockbusters – are 3ality Technica and Cameron-Pace Group. The fully automated versions of high-end mirror rigs (usually so complex that they are available for rental only) allow very fast automatic calibration and setting, saving valuable time on shoots where minutes are worth thousands of dollars. However, we should not underestimate numerous manufacturers of high-performance rigs, each of them offering specific advantages. We include in the non-exhaustive list, in alphabetical order, 21st Century 3D, 3D Film Factory, Binocle, GenusTech, P+S Technik, S3D Technologies, Screenplane, and Stereotec.

Although this is not an absolute rule, the manufacturers put their rigs in three main categories:

- Studio: Size, weight, and price are not the main parameters. Extreme rigidity and stability, features, and ease of use are, however, the key words;
- Pro: Such rigs offer a compromise between mobility, robustness, and functionality for indoor and outdoor shooting;
- Mini: Lightweight, portable, and suitable for Steadicam mounting. Versatility comes second to ease of handling.

Mirror rigs' prices depend on their sophistication. The range goes from $9,000 to $150,000, not counting the price of the cameras and their lenses. The mirrors alone can reach very high prices, in line with their performance.

© Binocle.com

A parallel rig and a mirror rig (the mirror itself not yet in place) on the European Rugby Final shooting in the Stade de France, Paris. Cameras are Grass Valley LDK800s.

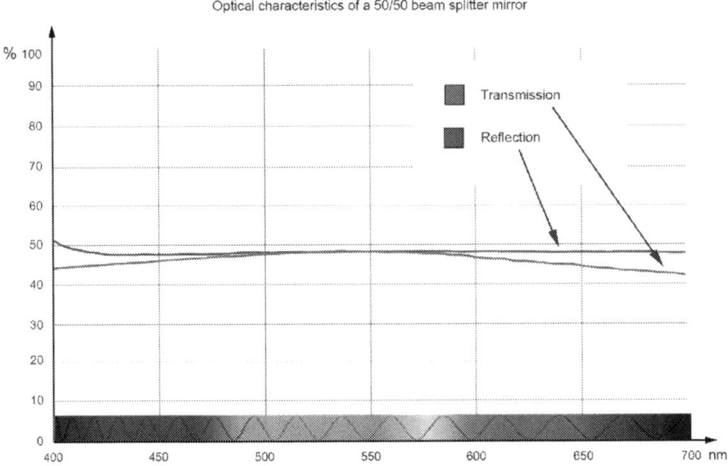

An ideal semi-transparent mirror reflects 50% and transmits 50% of the incident light. A good mirror in the real world tries to come close to this ideal.

Mirror Choice. The mirror is the most crucial component of a rig. Semi-transparent mirrors should ideally let 50% of the light pass through in a straight line and reflect the remaining 50% at 90° to the input axis; all this without introducing any geometric or color distortion and with perfect uniformity. It must of course be strong, not vibrate, and withstand shocks and scratches.

The reflective layer of the mirror is made by metal evaporation and deposition in a vacuum. To ensure that exactly 50% of the light will be transmitted, the manufacturer measures the mirror's reflectivity continuously during the process. Once the exact 50% figure is reached, the metallization stops.

3D rig requirements are contradictory and choosing the right mirror is always a compromise. For example, a thick mirror is resistant to vibration but distorts the image's geometry more; a front-coated mirror will avoid double reflections but will be much more sensitive to scratching.

Protective Glass. In the case of outdoor shooting in bad weather, exposing an expensive high-quality mirror to the elements is always unpleasant. There is a solution: Protect the front of the optical housing with a protective glass. For example, P+S TECHNIK and Schneider Optics (*www.schneideroptics.com*) offer a front-facing protective glass including a quarterwave filter (also known as a lambda/4 waveplate or incorrectly as a "depolarizer"). This added

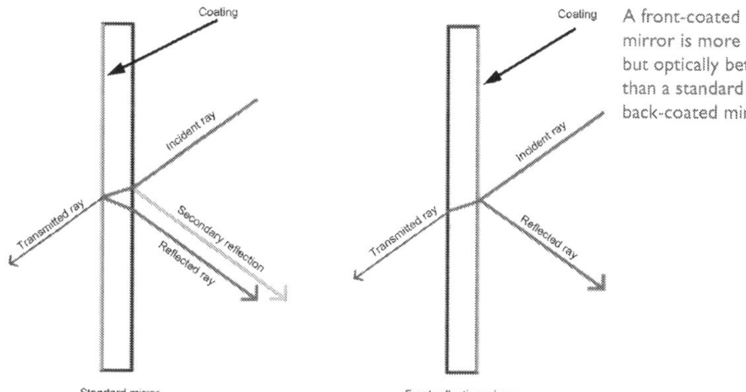

A front-coated mirror is more fragile but optically better than a standard back-coated mirror.

layer changes the angle of polarization, resulting in a severe reduction of polarization artifacts. The filter causes minimal loss of light, so the exposure stays practically the same as without the filter. And since the quarterwave filter is laminated between two layers of optical glass, it is protected from scratches and gives the protective glass better shock resistance than normal glass, similar to the performance of the safety glass used in car windows.

Above or Below? Two configurations are possible with a mirror rig: the front-facing camera is always in the direct viewing line, but the reflected camera can be installed above or below. Depending on the circumstances, one or the other configuration will be chosen: The top-mounted position is more often used for cinema shots and the bottom-mounted position for TV, but this is not an absolute rule.

© P+S Technik

© Element Technica

Setup with camera above: on the left, P+S Technik rig with Red cameras; on the right, Element Technica Atom rig with Red Epic cameras; the mirror is removed so we can see the front-facing camera.

Setup with camera under: on the left, shoulder-mounted rig; on the right, Steadicam mounted rig.

Advantages of camera on top:

- possibility of using very bulky cameras;
- better access to the various settings;
- more backward center of gravity for easier panning;
- better freedom of movement for low-angle shots;
- possibility of positioning very close to the ground.

Advantages of camera below:

- less view blocking for the audience located behind the camera;
- lower center of gravity;
- easier handling when mounted on the shoulder or on a Steadicam;
- better protection of the reflecting mirror from dust and scratches;
- fewer parasitic reflections in the mirror.

Camera Synchronization

Cameras synchronization is of utmost importance for depth quality. For example, a car traveling at 100 km/hr travels 3 cm in a millisecond, and 3 cm seen at a distance of 10 meters with an average focal length accounts for about 0.3% of the field of vision. If the left camera films the car a millisecond before the right one, the motion parallax created by the sync error can be five or six pixels, which will be interpreted as a depth difference of several meters! Without synchronization, we can always realign the two recorded streams with the closest image. Therefore, the maximum timing error will be half an image, or 21 milliseconds if shooting at 24 frames per second. Those images will be acceptable for very slow movements only and any fast action will cause disruptive effects: incorrect depth, distortion, duplicate objects, etc.

Synchronization with Genlock

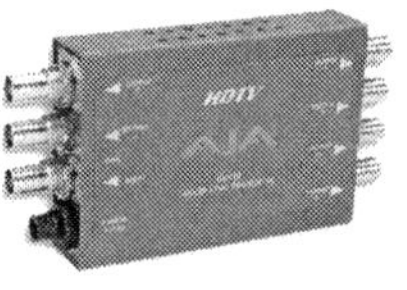

© Aja

An HD sync generator suitable for 3D: The AJA Gen 10 is fitted with 6 BNC outputs.

Professional cameras have a genlock input accepting a synchronization signal from a reference source. The most precise genlock signal is called tri-level sync, because it synchronizes all connected devices on the level of the image, line, and even the pixel in the line.

A tri-level sync generator is neither expensive nor bulky (see photo below). It usually has multiple outputs, so we can use it to synchronize two or more cameras, a SIP, a recorder, etc. A word of caution about connectivity: Not all cameras use the BNC connector and a much smaller DIN 1.2/2.3 adapter will sometimes be required.

© B. Michel

The BNC to Din 1.0/2.3 adapter required by many cameras (RED, etc.).

The sync generator will be configured for the chosen frame rate; the usual values are 23.98, 24.00, 25.00, or 29.97 frames per second, more rarely 48, 59.98 or 60. The frame rate in each camera's configuration menu will have to be adjusted to the same value as the "genlock" signal. If the frequencies do not match, the camera will ignore the "genlock" and will not be synchronized! In general, a padlock icon appears in the viewfinder to confirm that sync is active.

Synchronization without Genlock

Shooting without "genlock" is still possible, thanks to one important detail: The cameras are shooting all the time, as soon as they are energized. When you press the Rec button, they begin to store images from the first scan following the record command, but the instant when the electronics was started is also when the camera's internal clock started. So to synchronize two cameras all you need to do is to start them at exactly the same time.

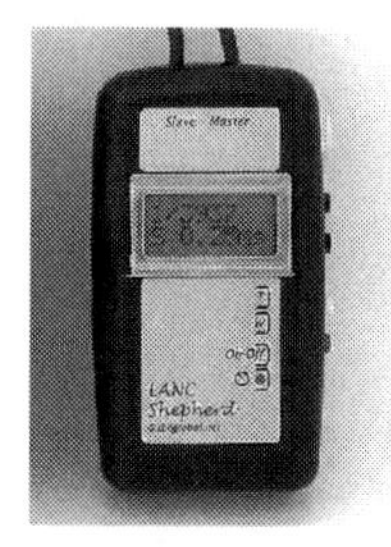

© Berezin Stereo

The LANC Shepherd electronic sync generator from Berezin Stereo

If your cameras are Sonys or Canons, you can use a LANC Shepherd controller (*www.berezin.com/3d/lanc/index.html*). The LANC is not a "genlock," it is simply a way to start the two cameras at the exact same moment and then check the gap between them afterwards. Depending on the match quality of the cameras, you can usually hope to keep acceptable synchronization for about 15 to 20 minutes. To resynchronize the cameras after a take, they must be shut down and the procedure has to be restarted.

Still and video cameras from Canon usually have a LANC. On Sony cameras, the LANC connector is referred to as "ACC" or "Control-L." The LANC uses a communication protocol specific to Sony that allows a remote controller to synchronize not only the start of two cameras, but also to control focus, shutter, and zoom. The LANC Shepherd display shows the offset between two cameras in milliseconds. If the offset is too large, the box is used to stop and restart the camera. In general, a first switch generates a relatively large timing offset, but once the electronics reaches its working temperature – which takes only a few seconds – the following restarts offer a sync that is good enough for 3D shooting (less than a millisecond).

What is a good enough synchronization? In photography, it is commonly accepted that the timing must be accurate to one tenth of the exposure time (0.1 milliseconds for a shot at 1/1,000). In video, an offset of 0.02 ms will produce only one hundredth of the statistical offset obtained without using LANC at 25 frames per second. Fortunately, the LANC Shepherd regularly achieves 0.01 ms.

Non-Synchronized Cameras

What can be done if the cameras are incompatible with the LANC protocol?

If the cameras are not synchronized at all, the best you can do is to start the two cameras simultaneously. The traditional method is to couple the power sources and synchronize them. We use the external power connectors to power the cameras from a single source, battery or mains, and then we power them on with a unique switch common to both cameras. In this way, we can hope that their internal clocks will be synchronized. Tests have shown a success rate of 75% on Panasonic HVX200 cameras and on some JVCs. An empirical way to verify the sync is to film a rubber ball bouncing quickly vertically. It is quite easy to compare the left and right images visually and then to calculate the offset in milliseconds by interpolating the distance traveled by the ball between two successive images and comparing the left-right offset. A more accurate, but more cumbersome, method is described below (see later in this chapter, "Synchronization Measurement").

The 3D Slate application for iPad or iPhone. This application displays in a fast sequence all information about the actual shot, i.e., interaxial distance, convergence, foreground and background distances, and various notes, to ensure that it is transmitted to postproduction.

Even with internal clocks synchronized, care must be taken to trigger the shooting at the same time on both cameras. Again, the simplest empirical solution is to use an infrared remote control. Most cameras come with a remote control that should trigger the two cameras at the same time. Synchronizing the two streams with the image manually will of course be required in postproduction. If the cameras are recording the sound, a clap at the beginning of the shot will serve as a benchmark. Several low-end editing software packages (e.g., Magix Video Deluxe) incorporate such audio sync functions in their 3D-enabled versions. In all cases, using a visual and audio clap is a way to check the sync, even when using a "genlock." Various clap applications for the iPad and other tablets are available on the web at very low prices. The time display in this type of application may be used to synchronize the clips with accuracy to the image and even to assess the offset between the left and right recorded streams. Those applications cost typically less than a simple mechanical clap.

Synchronization Measurement

If you have any doubts about the sync's reliability, there's nothing better than to check its accuracy. To do this, you will need a PC, testing software, and a good old CRT monitor (CRT). As scanning in a CRT monitor is progressive, it can serve as a yardstick for measuring time lags, even very short ones. Camera Sync Tester software is freely available on Peter Wimmer's website (*www.3dtv.at*); Peter Wimmer is the author of Stereoscopic Player. Once the characteristics of the monitor are known, the software displays a numbered vertical scale. Whatever the method you used to sync the cameras, after taking a synchronized view of the screen, simply note the difference between the last displayed line numbers and multiply them by the time needed to display a line on the monitor.

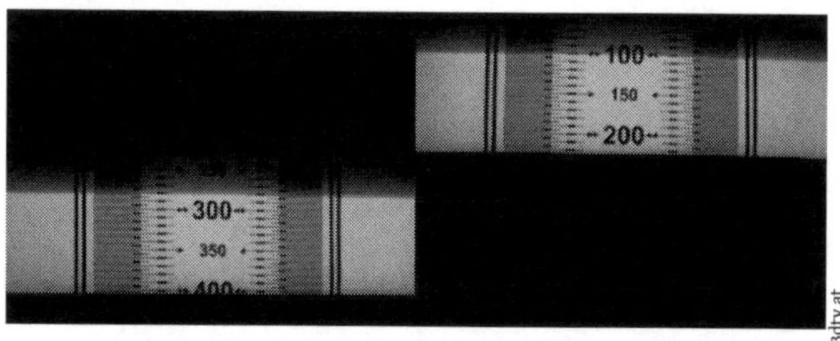

© 3dtv.at

Two screen shots of the Camera Sync Tester software running on a CRT monitor recorded by two slightly out-of-sync cameras: The last displayed line is #403 on the left and #223 on the right. With 31 microseconds needed to display one line, we determine a sync error of 5.7 ms.

Stereoscopic Calculators

The 1/30 rule, according to which "the interaxial distance should be equal to one thirtieth of the distance to the nearest object," was already used empirically a century ago. It is still used now, but is insufficient to set up a rig properly for a given 3D scene. The interaxial distance's adjustment depends on various factors, the main ones being, of course, the minimum and maximum distances between the camera and scene elements and the maximal acceptable positive parallax. This parallax value depends on the size of the screen where the image will be viewed: The parallax will be a smaller percentage of the screen size if the screen is very large. So, on a thirty-meter-wide IMAX screen, the positive parallax (behind the screen) is so small that virtually the entire scene is in front of the screen plane. The focal length and the size of the camera's sensor are also involved in the calculation, as well as the width of the image in pixels. All lengths are given in meters to maintain consistency in the formulas, so a 16 mm focal length is given as 0.016 m.

The two main parameters to determine are the interaxial distance and convergence angle. Next is a table summarizing the parameters that will determine the setup of a 3D rig with their usual value ranges. To illustrate the values with a concrete example, we use an interior scene with a background located nine meters away from the 3D rig, fitted with 2/3" cameras.

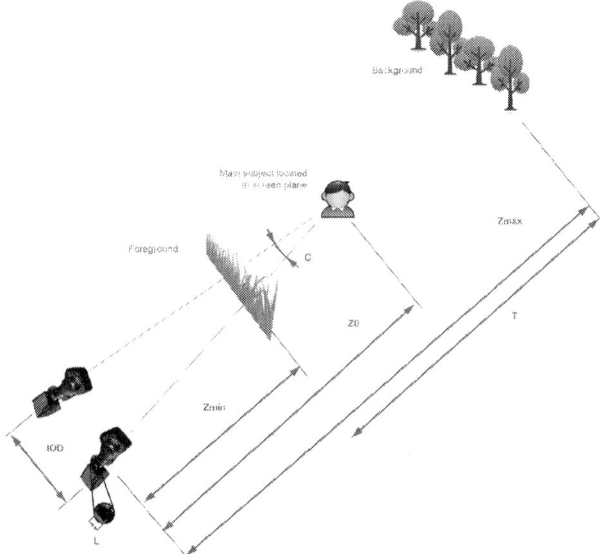

Main parameters used to determine interaxial distance and convergence

Characteristic Values of 3D Rig Parameters, with Examples

Parameters	Category	Unit	Symbol	Typical Values	Example
Foreground distance	Scene	m	Zmin	1 to 50*	1.8 m
Subject distance (zero parallax)	Scene	m	Z0	1 to 50*	2.0 m
Background distance	Scene	m	Zmax	1 to 50*	5.0 m
Depth weight (arbitrary value)	Scene	dimension-less	PR	0.1 to 10	0.9
Focal length	Camera	m	F	0.001 to 1.0	0.005 m
Sensor width	Camera	m	L	0.001 to 0.1	0.0088 m
Digital image width	Camera	pixels	W	500 to 8000	1920 pixels
Vision screen width	Vision	m	S	0.05 to 30	10 m
Max positive paral-lax allowed on screen	Vision	m	PP	0 to 80	0.060 m
Max out-of-screen effect	Vision	dimension-less	J	0 to 0.80**	0.75
Interaxial distance	Vision	m	IOD***	0 to 0.5	0.030 m
Convergence	Vision	angle degrees	C	0 to 5.0****	0.76°

* Beyond 50 meters, we assume no more depth perception; infinity is assimilated with 50 meters. This has practically no influence on the calculation's results.

** 75 %: With a viewing distance of 10 m, the object will pop out to a distance of 2.5 m from the viewer.

*** The acronym IOD *(InterOcular Distance)* is not strictly correct as we should use "interaxial" and not "interocular," but that is what is used in the literature.

**** The eyes converge 5° easily. With efforts that are hard to maintain for a long period of time, they can converge up to 12° maximum (thus converging around 30 cm from the nose).

In addition to the above parameters, a number of derived parameters are discussed in the literature, but in general they are derived by a simple formula from the above values.

Derived Parameters of a 3D Rig, with Formulas and Examples

Derived Parameters	Category	Unit	Symbol	Formula	Example
Scene depth	Scene	m	T	Zmax - Zmin	3.2 m
35-mm equivalent focal length	Camera	m	F35	F * 35/L	0.020 m (or 20 mm)
Max positive parallax on sensor	Camera	m	MAOFD*	L * PP/S	0.000053 m
Magnification ratio	Vision	dimension-less	M	S/L	1136
Max positive parallax in %	Vision	dimension-less	Prel	PP/S	0.006 (or 0.6%)
Max positive parallax in pixels	Vision	pixels	Pmax	W * PP/S	11.5 pixels

**The MAOFD acronym stands for "Maximum Acceptable On Film Deviation."*

Mathematical Formulas

Interaxial distance and convergence can be determined relatively independently. First the interaxial distance is determined based on the foreground and background distances; then the convergence angle that will bring the main subject in the zero parallax plane is determined. In this way, the second calculation does not invalidate the results of the first one. However, there are other setup methods than this one, which is considered the official method since its popularization by Samuelson in 1994.

If you do not wish to use convergence and prefer to work in parallel, bringing the main subject in the zero parallax plane will be done by HIT (Horizontal Image Translation) in postproduction, at the cost of losing a few pixels at the edges of the image. The maximum possible shift corresponds to Pmax (in pixels). Therefore, you should frame your scene a little wider and accept losing some resolution (or, budget allowing, shoot in 4K instead of 2K). With this reframing in mind, you must take the chosen aspect ratio into account: If you shoot in 16:9 and sacrifice 12 pixels of width, you will also lose 12 × (9/16) = 7 pixels vertically during the cropping.

Interocular Distance Computation

The general formula for calculating an accurate interaxial distance is Bercovitz's formula. Here is a simplified version, assuming that the distance from the camera to the scene is large relative to the focal length. This formula is therefore suitable for cinema or television, but not for macrophotography.

The calculation is performed in two steps: First we compute the maximum positive parallax on the camera sensor by dividing the maximum acceptable parallax on the screen by the magnification ratio between the sensor and the screen:

$$MAOFD = L \times PP/S$$

Then a trigonometric relationship gives us the maximum allowed interocular distance:

$$IOD = (MAOFD / F) \times (Zmax \times Zmin)/(Zmax - Zmin)$$

Convergence Angle Computation

After imposing the distance between both cameras, one parameter is still adjustable within some limits, namely, convergence. The cameras' angle may be changed inwards only (convergence), never outwards (divergence). The angle of convergence is a few degrees maximum; 5° is generally considered an absolute maximum.

The convergence effect drags the entire scene volume back. Indeed, a "parallel" setup will create a scene representation completely in front of the screen, which will be the horizon. It is therefore possible to increase the convergence angle until the background (objects at the distance Zmax from the camera) gives the maximum positive parallax allowed on the screen (Pmax value, when measured in pixels). At that time, the maximum convergence for the set is reached.

Note that pushing convergence to the max is not always the best choice, and far from it! If you want the scene to be orthostereoscopic, and thus to appear exactly as the original scene, the parallax must be reached by the horizon and not by the farthest objects in the scene (unless they are combined). A typical scene offers backgrounds that are not necessarily on the horizon (indoors or outdoors but under trees, for example); it could be pushed back beyond its theoretical position without creating any divergence, but this is more an artistic choice than the result of applying a mathematical formula.

Like the depth weighting factor, which lets you exaggerate or minimize the 3D effect in the interaxial calculation, convergence may be more or less strong for artistic or aesthetic reasons: strong convergence distorts the observed volume, gives it a more shaped and round look, thereby strengthening the sympathetic side of a character, but can become annoying when you are shooting a cube.

Convergence computation formula:

C = arctangent (IOD/Z0)

Remark
The convergence angle is expressed in degrees. The two cameras should be rotated inwards symmetrically at half this angle.

Calculation example. Using the above table values, we get:

MAOFD (in m) = 0.0000528 (or 52.8 microns)

IOD (in m) = (0.0000528/0.005) × (5.0 × 1.8)/(5.0 – 1.8) = 0.0297 (or 29.7 mm)

Having decided for artistic reasons to reduce the depth effect to 90% for the scene, we reduce the IOD value by 10% before computing the convergence angle:

True IOD = IOD × PR = 26.7 mm

Z0 = 2 m

C (in degrees) = arctangent (0.0267/2.0) = 0.7649° (or 0°46")

Connection between Angle and Convergence Distance

The trigonometric relationship connecting angle and distance of convergence is highly nonlinear, which is why this angle must be determined very precisely. The table below illustrates this nonlinearity well.

Convergence for IOD = 65 mm

Distance (meters)	Angle (degrees)
0.3	12.2251
1.0	3.7189
2.0	1.8614
3.0	1.2412
5.0	0.7448
10.0	0.3724
20.0	0.1862
30.0	0.1241
50.0	0.0745
100.0	0.0372
1 000.0	0.0037

Stereoscopic Calculators

Stereographers never do the above calculations without the help of a computer. There are lots of applications for streamlining the calculation process on all available platforms: PC, Windows CE, iPhone, Android, etc. You must choose the most appropriate one for the device or mobile that you have on hand.

In Practice

Adjusting the stereoscopy is just one step in a film crew's daily practice, and not necessarily the first one. As on a classic 2D shooting stage, you begin by setting up the set, characters, camera, and lighting. Then, you adjust, at least roughly, the focal length to reach the desired framing. After that you check the foreground and background distances, and finally, experience and sometimes a 3D calculator will help to adjust the interaxial and convergence settings.

If you are shooting with parallel cameras, after setting the interaxial distance, you adjust the HIT (horizontal offset of the images) on the control monitor to bring the scene to the desired depth range. If the background is distant and part of the scene reaches the screen plane or comes even closer, HIT will push the

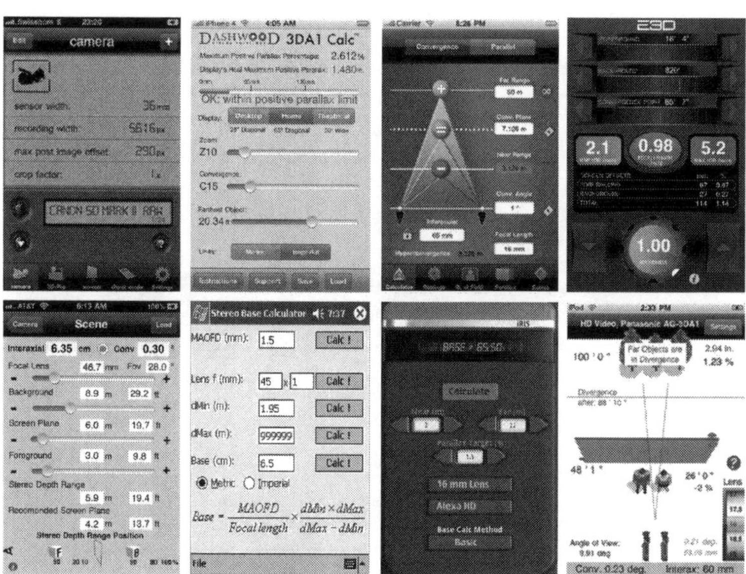

Examples of 3D calculators for mobile devices. Some record parameters to document the shot, others add a "clap" function (perfect on tablets such as the iPhone. Prices vary from free to over $100 according to the features offered.

Parallax Budget for Broadcasting

3D shooting for 3D TV is still in its infancy, so there are actually few rules firmly accepted by all TV stations. The few that have been published are thus even more worthwhile considering. So let's look at the British channel Sky 3D's recommendations to its content providers:

The majority of Positive Parallax (into the screen) shots should not exceed 2%. Negative disparity (out of the screen) at close points should be used with care and not exceed 1%. These guidelines are intended to deliver managed and comfortable stereoscopic viewing and can be exceeded for specific editorial needs such as graphic content or short-term visual impact. Such instances should be constrained to 4% Positive and 2.5% Negative.

background to infinity (HIT = Pmax). If the whole scene is near, its farthest plane does not even reach the screen before translation. In this case, HIT will be even greater than the maximum positive parallax value (HIT> Pmax). Obviously, care must be taken not to push the scene further than necessary and the positive parallax of the background must be checked to ensure that it will never exceed Pmax. As HIT may request a large frame crop, you must check that you don't "bite" into the frame more than expected. You must not forget to note, for postproduction use, the HIT value used on the control monitor.

If you are shooting with angled cameras, after setting the interaxial distance we must test on the control screen the positive parallax of the background, while increasing gradually the convergence angle up to the point where the background parallax reaches its maximum allowed value. From this moment on be careful, for any interaxial decrease will cause the maximum positive parallax to "'explode" and force you to reduce the convergence to keep it under control.

In both cases, you must finally make sure that negative parallaxes in the foreground do not exceed the limit set for the parallax budget. For example, if you decided to stay under 3% maximum total parallax and have already used up 0.5% positive parallax, you are left with 2.5% to use for out-of-screen effects. If the control monitor shows that you have exceeded this value, you must repeat the procedure again with a lower interaxial.

After this first round of adjustments, you then move to the large display screen that you must have taken care to install near the shooting stage for a finer check of the 3D. Only on a large size screen can the visual details of the scene be validated. You must check the overall compliance with technical requirements and artistic wishes: depth budget, characters roundness, background distance, window violation, foreground lighting, and so on. Obviously, if something goes wrong, the rig's position or settings will be adjusted and the result checked again on the big screen.

Dynamic Modulation of Interocular Distance and Convergence

The two basic settings of a rig can change during a take, but in practice this will occur mainly when shooting live, and only the convergence will evolve during the shot. Indeed, as soon as off-line editing is provided, it will be possible to reconverge images by applying a HIT at edit time. The convergence puller assisting the cameraman will modify convergence gradually (and preferably by mechanical lock) between the preset limits. With some high-end rigs, focus and convergence synchronization is available, so that one assistant can handle the roles of both focus puller and convergence puller.

When action moves away or comes significantly closer, it may nevertheless be necessary to adjust the interocular, but almost always in synchronism with the zooming. This dynamic tracking is not common because it modifies the apparent size of the scene, but such a situation can occur when following some outdoor sporting events such as golf. Tracking a golf ball in flight is a delicate exercise and with 3D only experts will take a go at it. But fortunately for them, tracking the ball in the middle of its course is usually done against the sky or a distant horizon as the backdrop, so at a time when little or no depth is perceptible. The convergence puller will change the rig settings while the cameraman follows the ball's trajectory. When the scene encompasses some depth again, the combined interaxial and zoom change is over and everything goes smoothly as far as the viewer's depth perception in concerned.

Rig Setup and Calibration

3D rigs, especially mirror rigs, are complex mechanical assemblies and, once installed on the shooting set, they require careful calibration. Motorized rigs are supplied with more or less complex control electronics that stores settings for a large choice of lenses, including zooms. It automatically pairs interaxial distance and convergence. All of these automatic functions save valuable time. Electronically controlled rigs also offer the advantage of storing all the settings for every frame and transmitting them to postproduction, which will greatly facilitate the special effects setups. But with or without automation, calibration and checking the calibration frequently are essential tasks.

After installing the lenses, mirror, control monitor, and all the wiring, you should check all possible rig movements, especially their mechanical stops, if there are any. Indeed, if an actuator starts in the wrong direction, it may well force two expensive lenses to collide and cause irreparable (and expensive) damage. With these precautions taken, you move on to calibration. The calibration can be checked using a computer, but it is done visually anyway, based on visual comparison of the left and right images. We therefore use the control monitor in semi-transparent mode to show both images one on top of the other.

Two Calibration Techniques

Two families of targets can be used to calibrate a rig: at infinity or close. Ideally the first method is favored because of its precision, but it cannot always be used because it requires very distant background.

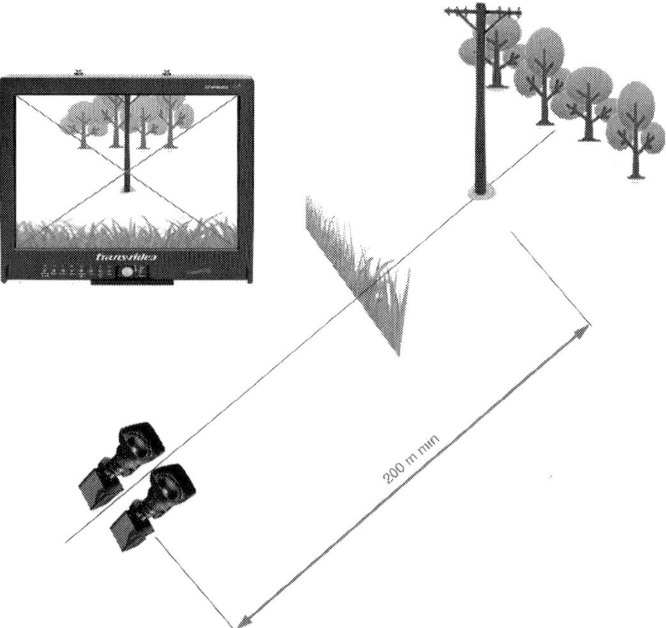

Rig calibration with target at infinity

Target at Infinity

The first method can be very accurate but it is mandatory to do it outdoors and with a clear view. You select a vertical, preferably narrow, guide located at least 200 meters away (distance needed to ensure a maximum error of about 1/100 of a degree with a 65 mm interaxial). A telegraph pole against the sky provides a perfect target. In the case of a mirror rig, you first set the interaxial to zero, then you align the two cameras until both images overlap perfectly in the center of the control monitor. For a parallel rig, the procedure is the same: convergence is zero but the interaxial is set to a minimum to reduce the error. If the interaxial cannot drop below 65 mm, the distance to the target should be increased proportionately.

Close Target

For a mirror rig, you place a white target with a vertical black line in the middle - a piece of gaffer tape, for example - at the farthest end of the set. Next, you place a vertical accessory stand about halfway. Then, with interaxial and convergence set to zero, you align the two vertical lines so that they overlap exactly at the center of the control monitor.

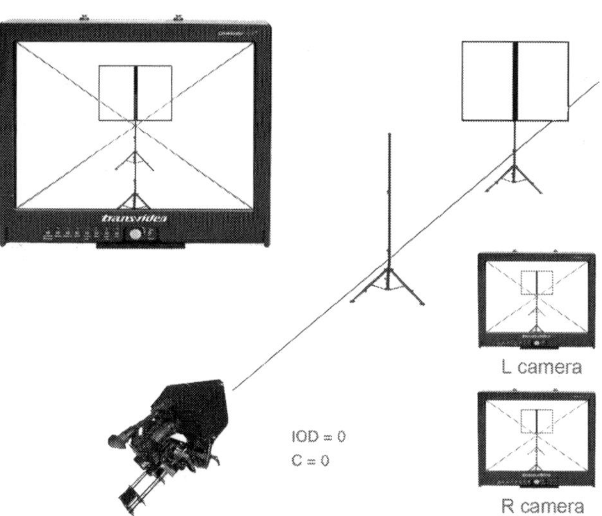

Rig calibration with two close targets

For a parallel rig, you make a rectangular white sign with two vertical black stripes, separated by exactly the interaxial distance (e.g., 100 mm). Masking the right camera, you align the left image in the monitor so that the left stripe on the sign is in the middle of the screen, at the center of the crosshairs. Then, without touching either the interaxial or the left camera (which must be masked in turn), you align the right image so that the right stripe is also in the middle of the screen. When the two images are displayed in semi-transparent mode, the monitor will display three black stripes; the central one will appear darker, since it consists of two superimposed stripes.

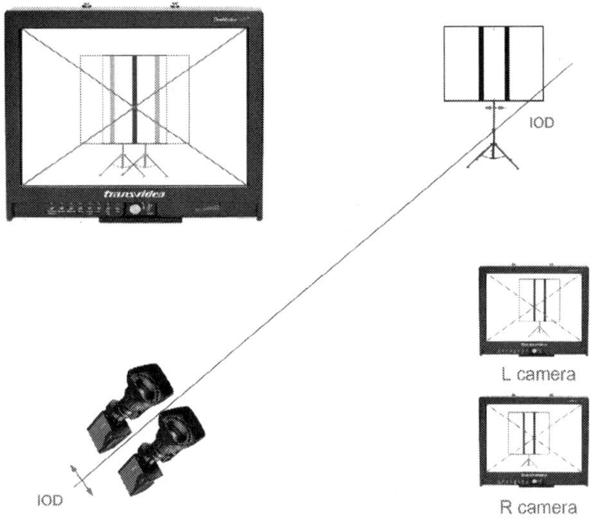

Parallel rig calibration (without convergence) with a dual band target. The gap between the two black stripes is equal to the interaxial distance.

Rig Calibration Software

Stereo3D CAT

Setting a rig requires great ability to analyze the produced images, scrupulous attention to details, and precision. These features are more usual in computers than in human beings. That is why Dashwood Cinema Solutions (*www.dashwood3d.com*) offers software dedicated to rig settings. Called "Stereo3D CAT" (Calibration and Analysis Tool), this program is coupled with a specific calibration pattern. In the first phase, sliders on the computer screen react to real-time setting adjustments. You adjust rotation, zoom, vertical offset, etc., successively, to bring the cursor in the acceptable area, which is colored in green. The accuracy of alignment adjustments can be as fine as a hundredth of a pixel. The second assisted calibration step is a semi-automatic parallax inspector. The calibration pattern board has rectangular holes to see the background. When convergence is set on the calibration target and the interaxial set to an appropriate value, the software measures the positive parallax in the hollow areas and compares them with the preset maximum.

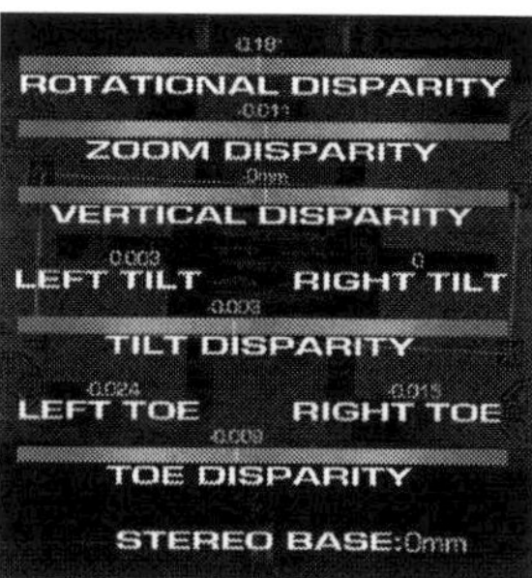

© Dashwood Cinema Solutions

The Stereo3D CAT disparity markers are displayed on top of the (here very dark) view of the calibration pattern board – with holes – viewed by the rig during calibration.

© Dashwood Cinema Solutions

The Stereo3D CAT depth analyzer shows positive parallaxes in green and negative parallaxes in red.

The rig settings are saved as metadata, but can also be sent to an iPad tablet connected to Stereo3D CAT and positioned in front of the test pattern board. Those metadata are embedded in the image and are guaranteed to be transmitted to postproduction. Stereo3D CAT is a Mac OS X application.

Disparity Tagger and Disparity Killer

Binocle (France, *www.binocle.com*) manufactures motorized 3D rigs, but also sells and rents the Disparity Tagger and DisparityKiller software. Disparity Tagger offers many visualization modes and, in addition to displaying 3D images from the rig, it performs real-time analysis of the stereoscopic pair and their differences. It also computes and displays the depth histogram of the stage and above all the more useful depth histogram in the projection room.

Disparity Tagger does not require user supervision; it diagnoses the quality of the rig alignment in real-time and comfort zone overruns according to the target screen size. Its control screen shows the depths of the various image elements with an intuitive color code: warm colors behind the screen plane, cool colors in front. Thus, the user is informed not only of depth overruns, but also of their exact locations.

Disparity Killer is an enhanced version of Disparity Tagger, offering a real-time correction function for vertical disparities and keystone in addition to diagnostics and calibration. It controls the motorized rig directly in order to cancel out unwanted disparities.

Disparities display and alerts on the Disparity Tagger screen. The dialog box in the middle of the screen is there to inform the software of the target projection screen size and depth budget limits.

Rushes Check

The tenth commandment of stereoscopy, "Always check your 3D shots on a 3D screen," should be implemented as early as on the shooting location. We will see in Chapter 6 various existing solutions to check the image at the camera level. But 3D is a difficult art requiring rigorous checking on a big screen. Installing a screening room near the shooting stage is not a luxury but a necessity. If the image is done with theater projection in mind, the control screen should be large enough to avoid accommodation/convergence conflicts and a 3-meter-wide screen seems to be the minimum compromise.

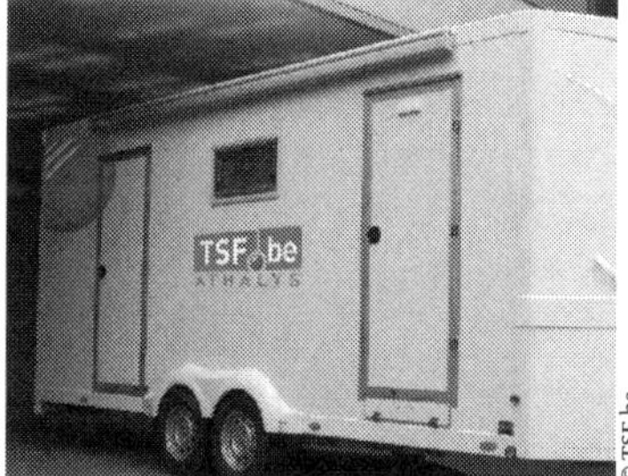

The TSF.BE VisioTruck where the movie "Asterix and Obelix: God save Brittania" was pre-checked on location.

© TSF.be

A comfortable solution is provided by a mobile viewing room equipped with a derushing, copy, and backup workstation that can be installed on site even during outdoor shoots.

Real-Time 3D Workflow

Shooting live events bears comparison with top level track & field performance: To reach a professional level, quick reflexes, endurance, and precision are required. Live production is, as we know, a stressful job for everyone involved. The many technical constraints of quality 3D only amplify the phenomenon. Fortunately, the equipment needed to relieve the production team of the most complex tasks is now available on the market. I shall describe a typical 3D shooting workflow for sports events, but of course dozens of variations are possible; they will depend on the type of event covered, available equipment, financial resources, and size of the production team.

Let's use a soccer game broadcast live for television by a major broadcasting network as an example. The event will probably already be followed by a HD production team and a mobile studio – referred to as an "OB Van" – specially dedicated to 3D will be present on location with its specific crew and equipment.

A typical installation of 2D cameras consists of up to twenty cameras to cover such an event. With 3D, the scarcity and price of equipment are combined with the desire to broadcast lengthier long takes to reduce the number of cameras to nine. Stereoscopy favors close-ups and does not like zooms, so most 3D cameras will therefore

OB43, a mobile 3D live production studio from Alfacam (Belgium)

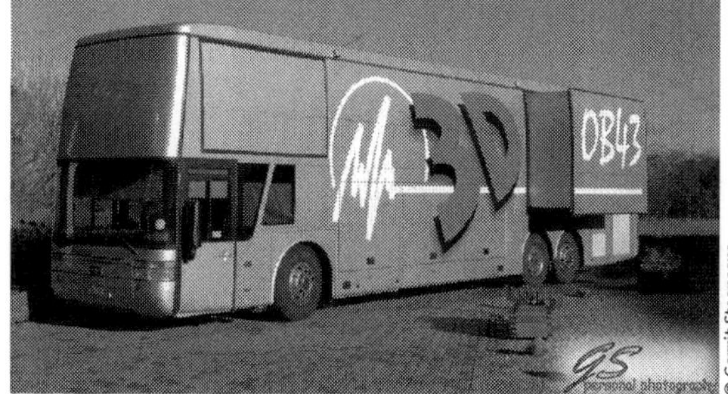

© Gerrit Steensens

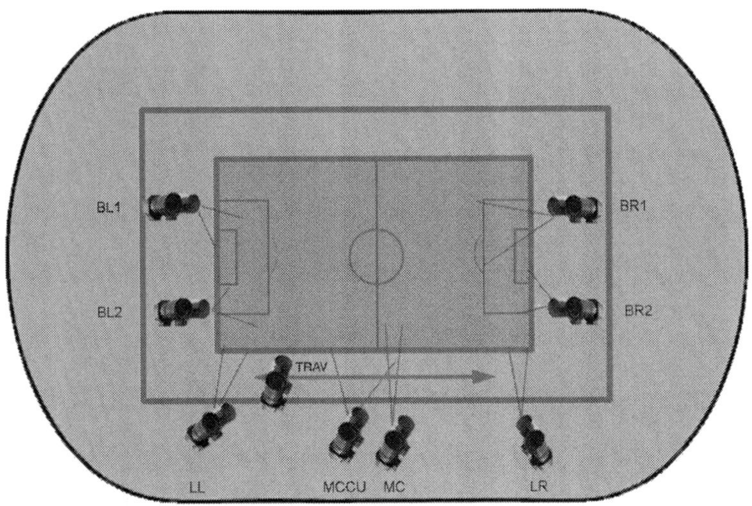

Example of a nine-camera setup around the playing field: The TRAV camera is mounted on a traveling dolly. The main camera is doubled: MC (large IOD, distant shots) and MCCU (mirror rig, shorter IOD, able to do the close-ups).

keep their focal lengths fixed for the duration of a take or even for the entire event. The camera labeled TRAV, which is closest to the action, will be mounted on a traveling dolly to keep the distance to the action area as short as possible (see diagram below). The cameras that do not need to shoot close-ups will not require a mirror rig and will be mounted in parallel, sometimes with an adjustable interaxial distance. An interaxial distance of up to 300 mm will give an acceptable depth range, even when shooting the opposite side of the field with long focals.

For live shooting, reliability is the main criterion. That is why every time it is possible to avoid mirror rigs, fixed or adjustable parallel rigs will be used, as they are always more robust and less prone to lose their settings, whereas the mirror is fragile, expensive, and relatively quickly soiled. Of course, during a league match, you cannot ask the players to do a second take! In avoiding mirror rigs, we reduce costs and unnecessary complexities and gain brightness, and thus depth of field: yet another advantage when shooting under conditions where the production cannot always control all the parameters.

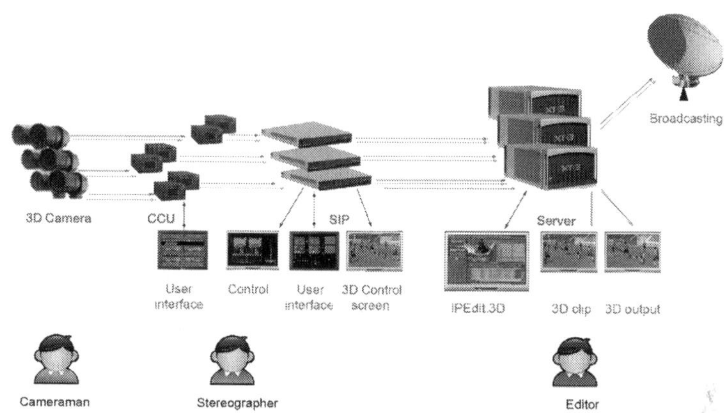

Example of a sports event live broadcasting workflow. The three production positions are manned by one or more people according to the number of cameras, editing complexity, and number of output feeds: On a big match, you will find ten to twenty camera rigs, six slow motion/replay servers, and a crew of twenty. Sound is generally handled independently (not shown here).

In the somewhat simplified workflow described here, we can see the camera control unit (CCU), which is usually located in the OB van and connected to the 3D rigs by two HD-SDI cables, a 3G-SDI cable, or an optical fiber. As often as possible, the CCUs are coupled together and therefore always receive synchronized commands, automatically eliminating parameter disparities between the two cameras. The image on the CCU control screen is generally reproduced on the camera by a return path in the optical fiber to facilitate the cameraman's work. This return path also includes the intercom signal, allowing the director and stereographer to communicate with the cameramen.

With real-time production such as in sports broadcasting, it is clear that no manual corrections can be made to the content in postproduction. Any corrections not made to the stream at once, on the spot, will cause errors and lower the whole production quality. Therefore, calibrating the cameras, 3D rigs, SIP (Stereo Image Processor), and entire 3D chain up to the OB van or studio's exit stream is vital. 3D control screens are thus critically important in all the production stages.

As a rule, a SIP processor is used to adjust images in real time. Examples are the Sony MPE-200, the Stereographer's Wizard from Advanced 3D Systems (*www.advanced3dsystems.com*),

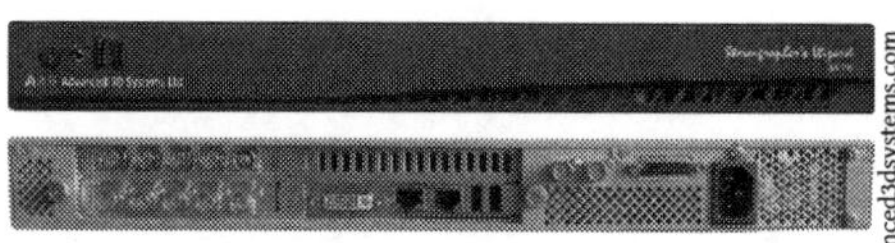

Front and back views of the Stereographer's Wizard SIP from Advanced 3D Systems

and the Miranda Density 3DX-3901. These processors' capabilities are evolving very quickly and they can deliver almost perfectly matching 3D streams with minimal human intervention.

The most common mistakes that a SIP can correct are vertical, rotation, and zoom misalignment of the cameras, as well as differences in color matching. The SIP can be used to reverse the image of one of the cameras if the camera cannot do that internally. It is usually equipped with a DVI, HDMI 1.4, or HD-SDI output with a side-by-side image to enable the stereographer to check directly the 3D image inside the mobile studio installed on the shooting location.

The stereographer has access to all useful information thanks to three screens, which are sometimes combined, namely:

- a user interface with keyboard and mouse or touch-screen to control and set up the SIP s/he is monitoring;
- a control screen with technical parameters and at least one "vectorscope" window to fine-tune horizontal disparities and the color balance of both left and right images; and
- a large full HD 3D control screen (50-65 inches).

Other errors are not fixed automatically; hence the stereographer's vital role. Incorrect calibration of a mirror rig or SIP or drift of one of the initially good parameters can sometimes be corrected directly by tweaking the SIP setting. If the problem cannot be solved in the OB van or studio, the stereographer will instruct the cameraman to change the convergence or interaxial distance, position of an anti-glare sun visor, or one of the many settings of a 3D rig.

3D video streams then pass through one (or more, depending on the number of cameras) EVS XT[3] streaming servers, the 3D version of the EVS (*www.evs.tv*) slow-motion servers used for replay in all OB vans. These servers store all incoming streams passing through for archiving but also for direct reuse on the production site as replays, that is, to show important action again in slow motion or from another standpoint. As all camera views are stored on the server, it is very easy to prepare summaries of the important phases of a game for replay at halftime or the end of the event. The XT[3]s are configured to handle two HD channels as a single 3D channel, so that the operator or editor can edit a 3D video stream as easily as s/he used to edit a single HD feed. On output, the server generates a frame-compatible video stream or two synchronized full HD streams for the left and right views, as one prefers.

© Gerrit Steensens

Inside Alfacam's OB43 3D mobile studio

Editing in the mobile studio is no longer limited to some simple viewpoint switches. The director continue to choose the source that will be broadcast, but, more and more often, the slow-motion server operator is taking on the functions of a true editor, using editing software that is far more sophisticated than a simple playlist manager. In the above diagram, the editor uses the [IP] Edit real-time editing software from EVS for on-the-fly editing. [IP] Edit looks just like any off-the-shelf video editor such as Final Cut, but it is able to select video clips from the incoming stream from the cameras and not just from files stored in a computer. The editor can prepare an action replay playlist including different angles, with slow motion if necessary, and add transitions and pre-recorded clips. Then, at the touch of a button, s/he launches the playlist, which is then sent directly on the air. Like the stereographer, the editor works with three types of screen:

- a user interface with keyboard and mouse or touch-screen to edit the video, with a standard timeline and a clip repository;
- a control screen to check the clip under editing; and
- a larger full HD 3D control screen to check the final edit.

It is worth noting that the real-time editing software does not work on locally stored clips, but directly on the incoming video streams and incoming and outgoing streams, and clips from – or those stored on – other video servers connected by the local Ethernet network. The two 3D control screens are connected directly to the production servers' outputs.

3D CGI and Stereoscopic 3D

CGI (Computer Generated Imagery) is the easiest way to produce perfect stereoscopic image pairs. Virtual cameras are always perfect by definition: no parasitic reflections, no mechanical alignment defects, no risk of collision between the lenses coming too close to each other, etc. The only drawback is that the world that they are filming is also virtual.

Video game and animated movie directors have mastered this technique and the complexity introduced by the double rendering requested by 3D production is easier to overcome than when shooting real scenes with real 3D rigs. Movies such as *Toy Story* and *Up!* are typical examples of 3D CGI made entirely by computer.

Virtual Cameras and Rigs

All CGI software uses a virtual camera placed in three-dimensional space and observing it. Nothing is easier than to pair two of them to create a virtual 3D rig that simulates an existing rig: same interaxial distance, same convergence, and same aperture. Only the mechanical flaws are missing! The preferred convergence method in a virtual rig is lateral offset of the image sensors. Thus, there is no keystone (trapezoidal distortion) as in real converged rigs, which simplifies postproduction.

In addition to their small size, which is reduced to a point, there is a difference between real and virtual cameras: Virtual cameras have two "magic" settings that real cameras do not have, namely, "'near plane" (distance from the camera to the nearest visible object) and "far plane" (distance from the camera to the farthest visible object). These two planes, one close and one distant, define the visible area along the vision axis. A real camera sees any object in its field, while a virtual camera may decide to ignore objects closer than a certain preset distance. You can therefore give in to the temptation of creating a near plane manually or automatically that will be located 30 times the virtual rig's interaxial distance from the camera. The famous 1/30 rule will be respected: No object too close from the camera will be in the way and disturb the viewer with outscaled out-of-screen effects. Conversely, you may also create a script that will calculate automatically the interaxial distance according to the near and far planes measured in the scene, as a stereoscopic calculator would do on a real shot. You can also add a Z0 plane to the set to represent the screen plane and adapt the script to set up the stereo camera to converge to the center of that Z0 plane. Be careful not to animate the interaxial value unnecessarily, for both a fixed camera and a variable interaxial give an impression of a size change of the whole scene that is very likely to be undesirable.

Websites, blogs, and newsgroups on 3D virtual cameras are legion. You will find them easily by looking for key words such as "CGI," "3D stereo rig," and "script" combined with the name of your favorite CGI software package. With Autodesk 3ds Max, I use the 3DHippie Stereocam script by David Shelton (*davidshelton.de/blog*).

Virtual Image Rendering

Virtual images are not so perfect that they don't require any postproduction. In fact, just the opposite tends to be true. This must be taken into account in the rendering phase:

- Image cropping (HIT) will be used more often than not; you will therefore take that into account by framing the scene a little wider than needed, *e.g.*, 2,200 instead of 2,048 pixels. The added width will be around twice the expected maximum positive parallax in size.
- Since depth map calculation is always done at some time during rendering (under the usual "Z-Buffer" name), you must never forget to record it in parallel with the left and right images; "Z compositing" operations done later in postproduction will need them and this will not increase the rendering time.

Z-Compositing

A virtual scene generated with the parameters collected during a real shoot can be embedded perfectly into the original shot. In this way, extremely realistic special effects can be produced. Knowledge of each pixel depth is used to insert virtual objects at their exact distance, so that real and virtual foreground and background objects will merge perfectly. Jim Cameron's *Avatar* is a perfect example of this: Real people were filmed against a green background, outlined, and then integrated into a virtual universe with perfect consistency.

Partial Transparencies and Reflections

Semi-transparent objects such as window panes, glass doors, and mirrors must be treated very carefully in 3D. To anchor the glass or mirror in space, it must be partially visible. As our brain needs fine detail in the image to locate such objects' depth, you must use a texture (*i.e.*, drops of water instead of steam) rather than uniform transparency, or else a few accessories that will give the glass or mirror a precise location in the 3D space, for example, a Post-It note, a sticker, or a curtain. Careful! Some portions of the scene reflected by the glass or mirror can be farther from the calculated scene than the far plane. You will thus make sure their positive parallaxes are not too large.

Almost all big-budget movies are packed with special effects, and thus a mix of real images and CGI elements. The most striking example is the simple backdrop: In 2D you could settle for a simple matte painting to replace the whole scene background, but in 3D you must now put in a full CGI model with its own detailed volumes, at least if it is partly within a few tens of meters of the camera or the camera is moving. You must therefore put great care into matching real and virtual cameras and compositing the scene using specialized software such as Nuke or Ocula (Chapter 5 will discuss this issue in more details).

Stereoscopy will require working ten times more accurately than multilayer compositing of a 2D movie, at least in the horizontal direction. Indeed, it is the horizontal offset between the two images that dictates the depth of an object. But with a depth budget of about 3% of the image width, the depth of the scene is represented by a few tens of pixels at most. It is not uncommon to have to adjust shots to 1/20 of a pixel to anchor a virtual character's legs onto the ground from a real shot or vice versa.

Green Screen and Overlay

Green screen shooting – shooting real scenes against a green screen for keying them in a virtual set – is becoming more and more common. The green background makes rotoscoping (or outlining) with automatic or semiautomatic tools very easy. Whether the rotoscoping is automatic or not, the method is the same: Each character's outline is detected by picking out the pixels that contain a certain proportion of green. The fine-tuning of this detection is influenced by the position and orientation of

Green screen shot for keying with the Panasonic AG3D-1A camera

light sources, including reflections of those sources on another character or an element of the set. With partially manual rotoscoping, the operator will tend to reuse the outline detected on one image when working on the other, by laterally shifting and slightly adjusting the contour if required. All these operations are very critical because in 3D any shift – even a fraction of a pixel – in the silhouette between the two views will be interpreted as a depth change. Consequently, each rotoscoping operation must be carefully verified by the operator on a 3D screen.

Rotoscoping and keying in a 3D scene can be done in real time, in which case the entire operation is of course done automatically. The real 3D rig's parameters are transmitted to the virtual camera in real time so that the rendering engine can render the set with the corresponding viewpoint. The overlay operation occurs in the last step, just before broadcasting, which thus takes place with a few pictures' delay.

Checking 3D in 3D

As in a real shot, a virtual shot involves constant monitoring of the results in 3D on a large enough screen. Some programs, such as Maya, natively offer a stereoscopic viewing window, but this is not the case for 3ds MAX. Fortunately for 3ds MAX users, there is a free plug-in, called "3D Vision Viewer" from Kostasoft (*www.kostasoft.com*) that manages a secondary monitoring window in various 3D modes: anaglyph, side by side, 3D Vision, OpenGL quad-buffer, etc.

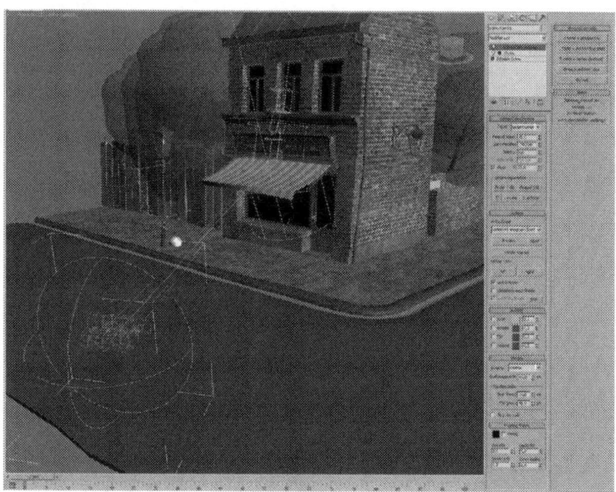

© L. Warnant

A 3ds MAX screen with a virtual 3D rig

© L. Warnant

The same scene displayed in red/cyan anaglyph in the rendering window

3D Presentation Software

Computer slideshow presentations are commonplace these days and Microsoft PowerPoint has the lion's share of this market. But when one wishes to prepare a stereoscopic 3D presentation, things are not so simple.

True3DPT, from True3Di (*www.true3di.com*), is one of the very few software packages to provide a complete solution. After importing a regular PowerPoint presentation, True3DPT lets you insert 2D objects into a multimedia presentation, each at a different depth level, in an arbitrarily defined range such as from -10 to +10. Titles, text, images, and videos can thus be made to stand out in front of a background that is likewise positioned in depth. 3D pictures and 3D videos, for their part, can be inserted as stereo pairs or in side-by-side format.

Various transition effects, including three-dimensional movements, are possible. So instead of highlighting a text by changing its size or color, you can make it jump out of the screen. And of course, in 2D-3D mixed content presentations, you no longer have to shuttle back and forth between the presentation software and Stereoscopic Player to switch from text to 3D videos. The available 3D visualization modes are interlaced, side by side (for 3DTV or dual screen plus passive glasses), and OpenGL Quad (for active glasses).

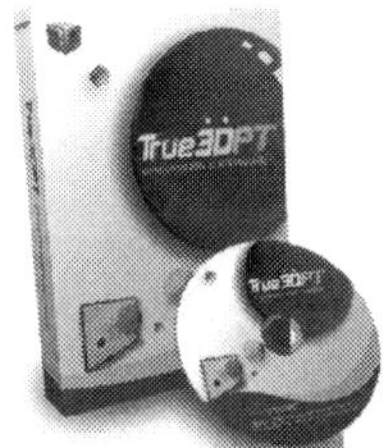

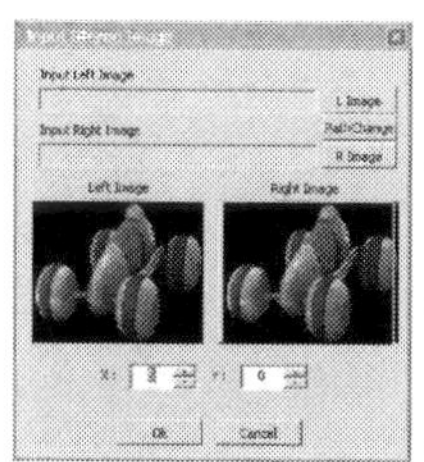

© True3Di

True3DPT software is a substitute for MS PowerPoint that adds the third dimension to your presentations.

An alternative to True3DPT is the French Taodyne software, a kind of presentation page generator based on writing simple scripts, and the images of which are compatible with most stereoscopic displays on the market (*www.taodyne.com*).

Yet another solution is provided by XpanD, the manufacturer of the active glasses of the same name. It is a plug-in for MS PowerPoint 2010 that adds stereoscopic functionality. It consists of two parts: a 2D-to-3D presentation converter (converter) and a 3D display module (player). However, this software is restricted to displays using the superimposed (over-under) display mode. It will therefore be usable on 3DTV or projectors working with active glasses. A free trial version, limited to three pages, is available on XpanD's website (*www.xpand.me*).

3D Photography

Still pictures do not compare easily with video. These are two different arts and when we consider the stereoscopic variant of both, the differences remain the same. Photography's huge advantage over video is its higher image quality. The perfect image can only be photographic, not video, and that is even truer with 3D. Unfortunately, for once, the transition to digital has somewhat leveled the quality. Stereophotographers project and use their images with exactly the same projectors as those used for movies. Of course, unique photographic applications, such as printing on anaglyph paper and on lenticular sheets continue to exist. The latter is the only static 3D reproduction method that preserves both the high resolution and the colorimetry of the original picture.

Nevertheless, taking still pictures is excellent training for the future stereographer. All the principles are the same, and once 3D technique has been mastered on stills, it will be possible to move on to 3D video with the best cards in hand. The thorough understanding of the limitations and constraints of 3D that is acquired with still pictures will accelerate the 3D video learning curve and reduce the - always high - costs involved.

3D Camera vs. Standard Camera

There are still very few digital 3D still cameras. Fuji has been a pioneer with its W1 and W3, offering 10-megapixel CMOS sensors, but they use rather small lenses, as on most compacts, with the accompanying compromises. The advantage of these devices is their ease of use. Being compact, they produce an acceptable result by simply pressing the shutter button. The only constraint is to avoid too close foregrounds that may cause excessive disparities between the two views.

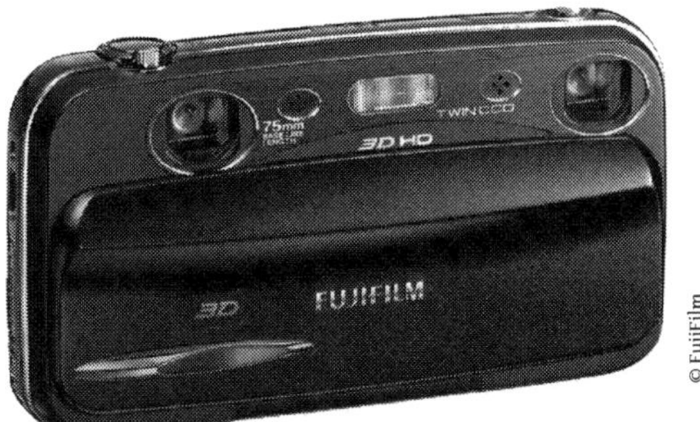

© FujiFilm

The Fuji Real 3D W3 compact camera

The 3D compact stores its images in a specific "MPO" format that is also used by Nintendo's 3DS stereoscopic game console.

Integrated devices offer perfect synchronization between both shots. However, this is not enough to guarantee that the two views are perfectly matched. For example, with the Fuji Real 3D W3, both zoom values are close enough to give acceptable, but not perfect, images. A zoom correction on the order of one percent is necessary to complete the pairing. This can be done by software afterwards.

The "Cha-cha" Method

If perfect synchronization is not required, for example, when shooting stationary objects or landscapes, you can use a single camera that you move laterally between the two takes. Of course, all aperture, focus, speed, and zoom settings will be locked in manual mode. As for the lateral displacement, it must be done carefully to preserve the orientation and alignment of views.

A simple solution is to use a large metal ruler against which the camera housing can slide. The distance between the two shots will depend on the distance to the subject. The standard 65 mm interocular distance is a good basis for subjects 2 to 10 meters away, but the following rule of thumb is most often used: move the camera laterally a distance equal to 30 times the distance from the nearest object. Mountain landscapes without foreground are sometimes photographed with a lateral shift as large as several meters. Keeping strict parallelism becomes particularly difficult under such conditions, in which case you can use visual cues on the horizon to align the two frames, while the remaining alignment errors will be corrected in postproduction.

The Fuji W1 (or W3) has a specific cha-cha mode. After the first take, the control screen shows an overlay of the viewfinder's view and the first picture, which allows a decent enough alignment. Once the alignment is achieved, the second take is triggered.

3D Shooting

Small objects. To photograph small stationary objects in 3D, the 1/30 rule will impose a small interaxial distance that is far too small to put two cameras side by side. Lateral displacement becomes the only solution, but the scenes' parallelism must be very carefully preserved.

The most accurate technique is to use a professional guide rail, such as those in the Novoflex Castel series. These rails are very rigid and provided with verniers and locking screws for reproducible precision work. The XL model, with a 380 mm total displacement, is appropriate for all scenes from macrophotography to architectural views with foregrounds as far off as 10 meters.

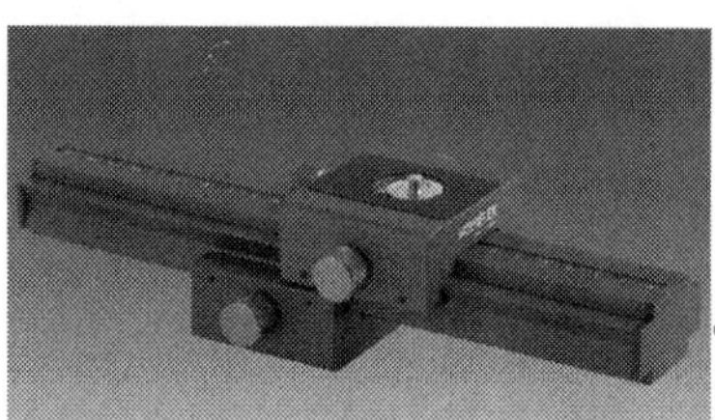

The Novoflex Castel XL guide rail with its 380 mm displacement capability

Big structures. To photograph large structures such as buildings or monuments, the 1/30 rule will impose an interaxial distance much greater than what is reasonably possible with a guide rail.

You can replace the rail by the motion of a vehicle. Preferably select a very stable vehicle that is solid and well suspended. With a little practice, you will learn to estimate the time between shots to set the correct distance between two successive positions. To keep a constant distance between two takes in a series of views, set the camera to burst mode. It will then acquire images at a perfectly constant rate, such as three frames per second. All you have to do then is to choose either two successive images or every third or fourth image to adjust the interaxial distance. Obviously, keeping the vehicle at a constant speed is essential. Photographer Michael Raymond shot a 3D stills series of Rome's monuments from a bus. The series, called "Rome (Stereos)," is visible on the Flickr website (*http://goo.gl/5IEIW*). This solution, taken to extremes, is used by NASA with a satellite orbiting around Earth, the Moon, or Mars as the support vehicle to take 3D pictures of those planets from hundreds or even thousands of miles away. Aircraft and boats can also serve as vectors for 3D photographs of landscapes.

Choosing the Right Focal. While the spacing between the two views can modulate the depth effect, zoom (or focal length) is used to select which portion of the scene will be seen. It is generally recommended to use wide-angle views for 3D, because it can encompass close foreground objects. But this may become a disadvantage, for if you follow the 1/30 rule, nearby objects will prohibit large interaxial values and far backgrounds will lose depth and appear too flat. To choose the optimal focal length, prefer to start from a wide-angle view and frame the subject while moving towards it, then once the viewpoint is selected, zoom in slightly so as to cast all foreground objects that are too close out of the frame.

Aperture, Sensitivity, and Lighting. Digital cameras, like their analog ancestors, offer a depth of field proportional to the sensor's sensitivity. As 3D requires a lot of light, high sensitivities and small apertures will always be preferred. As a consequence, the lighting should also be powerful. Do not hesitate to use a flash to lighten the shadows, even in full daylight.

Horizon. Most setup errors are recoverable in postproduction, but not all! The horizon must be perfectly horizontal in a 3D photo. In traditional photography, nothing is simpler than to rotate a misaligned image. But in 3D, while we set out to create horizontal disparities, we do not want any vertical disparities and any image rotation will create vertical disparities for all points that are not in the screen plane. You therefore will have to check the horizontality of your tripod meticulously before shooting.

Roundness. We have seen many methods to determine the ideal interaxial distance. But as our myriad experiences have shown, viewing conditions are as important as shooting conditions. And in still photography, unlike motion pictures, the viewer observes the same image for a long period of time. S/he thus has a chance to scan the entire image, its foreground and its background, and to appreciate the richness of every minute detail. The correct roundness and volume of objects can be assessed only at a precise distance from the observed image, a distance that depends on its size. Test this by looking at the image below: When seen from too far away, this image is distorted along the viewing line; when seen from too close, it offers flattened depths and the scene looks stretched laterally.

© B. Michel

A still-life photograph reproduced here in anaglyph. It lets you assess the roundness of objects and how its perception is influenced by the viewing distance.

3D Photo Rig

As soon as quality above that of a compact is sought, we are forced to fall back on traditional DSLR (Digital Single-Lens Reflex) cameras. We must then combine them in pairs on a suitable rig. A lateral guide rail may well serve as a rig, provided that it is fitted with two sliding heads. Countless more or less reliable variations exist in both professional and home-made equipment, and in numerous amateur photo clubs dedicated to 3D, such as the French Stereo Club, which is one of the oldest in existence (*www.stereo-club.fr*). A complete rig will allow correct positioning in X, Y, and Z; rotation; interaxial distance; and convergence. The rig must be rigid, fitted with lockable setting and preferably have graduated rulers.

A photo rig with manual adjustment from Redrock Micro used by Andrew Parke from DimensionWerks3D during filming in Korea

An example of a solution meeting these criteria is the SxS Micro3D rig from Redrock Micro (*store.redrockmicro.com*). One of the cheapest mirror rigs, ideal for schools because of its flexibility, is the Fastrig (*www.cine3d.ch*).

Z-Bars. A simple and light solution for coupling two cameras is the Z-Bar. Machined out of an aluminum profile, the Z-Bar lets you mount two cameras upside down to minimize the interaxial distance. A wide variety of Z-Bars can be found on the market, with some intended for a specific model, others being more generic. Some Z-Bar providers: Digi-Dat (*www.digi-dat.de*), APN3D (*apn3d.stereoscopie.eu*).

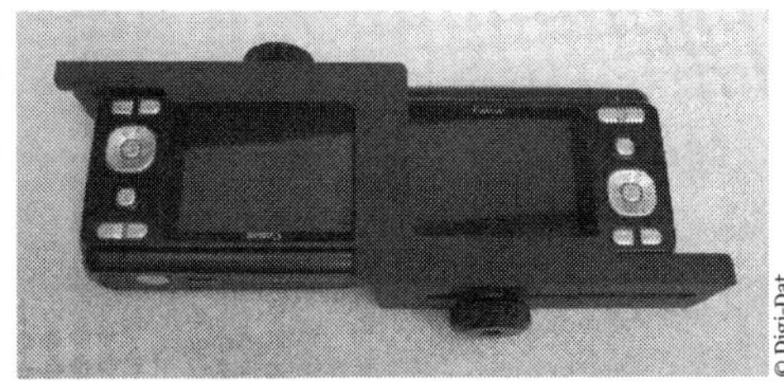

A Z-Bar for the Canon PowerShot A480. It lets you adjust the interaxial distance from 59 to 180 mm.

Photography mirror rigs. For macro or other very close scenes, as in video, the use of a mirror is essential. Stiffness and strength are the key words.

A mirror rig well suited to still photography is the Genus Hurricane, distributed by Manfrotto. It can be found in many photo equipment stores (*www.genustech.tv*).

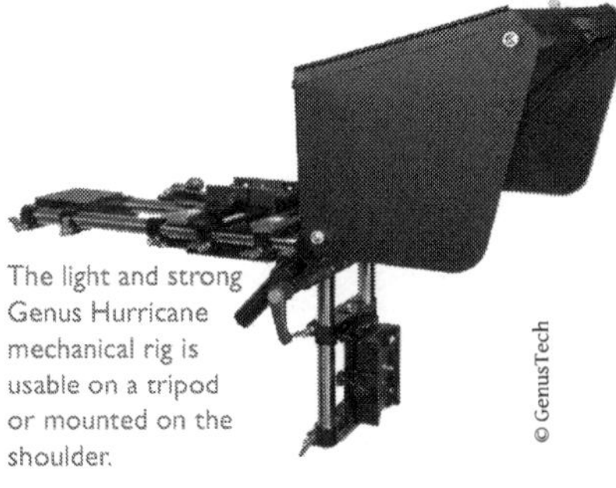

The light and strong Genus Hurricane mechanical rig is usable on a tripod or mounted on the shoulder.

For macro shots, the interaxial distance must be extremely small and the mirror is unavoidable. Some fans of 3D macrophotography manufacture their own mirror rigs, but many affordable ones are available on the market (under $700), especially if the cameras are small.

Some amateurs spend more time and effort than others on achieving their goals. The best example you can find is the Belgian stereoscopic macrophotography specialist Fotoopa. He manufactures his own mirror rig, including optics, precision mechanics, electronic controls, laser positioning sensors, firmware, and so on. His thousands of photos of insects caught in flight are well known to specialists. They can be seen on Flickr (*www.flickr.com/photos/fotoopa_hs*).

© www.ekeren3d.com

Anaglyph view of the 3D Macrobox, a mirror rig specially developed for 3D macrophotography manufactured by Co Van Ekeren.

© fotoopa

A dual mirror 3D rig intended for high-speed macrophotography of insects in flight. Flashes and shooting are triggered by laser motion sensors; focus is controlled independently by another set of laser sensors.

Calibration. Compared with video rigs, calibrating a still image camera rig is more difficult to do on site because the 3D image is not directly viewable. We therefore generally use very rigid rigs that can be calibrated in advance before moving to the shooting location. Fortunately, tools to overcome this problem exist, namely, multiplexers that can convert two HDMI signals – many recent cameras have a mini-HDMI connector – into a single DVI signal with an anaglyph image that can be displayed on a conventional computer monitor. If you are shooting outdoors, you will need battery-operated equipment if you want to avoid dragging a heavy, and rarely energy efficient, AC current inverter around with you.

© Delvcam

The DELVCAM 3DUX stereo multiplexer offers two HDMI inputs and one DVI output to display the resulting image in red/cyan anaglyph mode. Its 12-volt power unit makes it suitable for outdoor use.

Without a calibration check at shooting time, you must trust your setup and be prepared to correct any remaining defects in postproduction to remove all unintended disparities. This can be done using StereoPhoto Maker or similar software.

Synchronization of Two Still Cameras

As said above, the two cameras' synchronization is essential for all living or moving subjects, news reporting, and, in general, for all the most interesting topics. Aside from the few integrated 3D cameras on the market, we thus have to use some sort of electronic synchronization. The solutions used in video and described earlier in this chapter are valid here, too (see the "Synchronization without genlock" section).

LANC

The LANC protocol introduced by Sony is used for communication between one or more still or video cameras and remote controls. This is a serial communication protocol sending 8-byte-long messages. These messages are repeated continuously every 20 ms. A microcontroller is required to use this protocol. In our case, LANC will be used to synchronize the cameras, even though the remote-control functions are also useful for 3D shots. The LANC Shepherd is by far the most popular 3D photo synchronizer. This electronic unit is designed and marketed by Berezin Stereo, a California company, and retails for under $500. It controls two Sony photographic or video cameras and synchronizes then accurately. It has two cables connected to the two cameras by their ACC or LANC connectors. It synchronizes the start, stop, focal length, and zoom settings and the shutter. Its small LCD display indicates the level of synchronization accuracy obtained. It also includes a delay timer and a time-lapse timer able to trigger takes at intervals ranging from 2 seconds to 24 hours.

Compatible Sony cameras are Mavica MVC-CD500, Cybershot DSC-S75, DSC-S85, DSC-V1, DSC-V3, DSC-F717, DSC-F828 and DSC-R1 (the latter two only with the manual zoom), as well as the Alpha DSLR-A100 and Alpha DSLR-A700. But we can also use the LANC with the Nikon D70s, the Canon 10D, 20D, 30D, 40D, 60D, 1D, 1Ds and 5D, the Fuji S3, and the Konica/Minolta 5D and 7D (not exhaustive list).

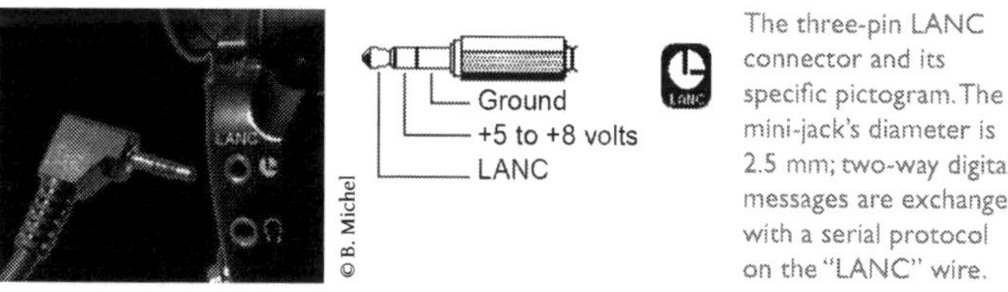

The three-pin LANC connector and its specific pictogram. The mini-jack's diameter is 2.5 mm; two-way digital messages are exchanged with a serial protocol on the "LANC" wire.

The LANC connector is a 2.5 mm stereo mini-jack labeled «Start,» «Control-L,» or «Remote» (on Canon cameras). The connector is labeled on the cameras with a specific symbol. LANC protocol is also available on some Sony camcorders through a ten-pin A/V connector.

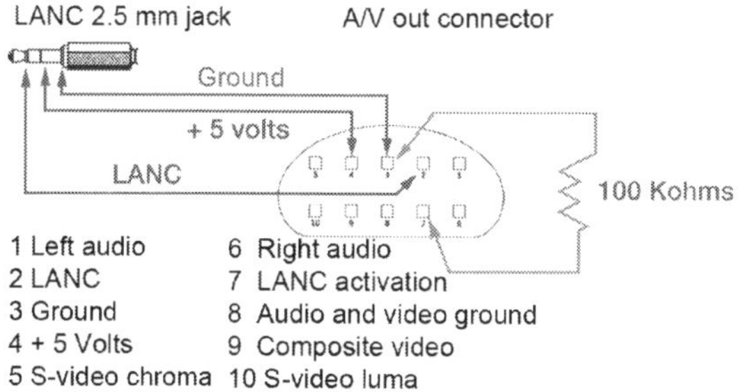

Correspondence between the A/V out connector and LANC jack. A resistor is required on the A/V output side to tell the camera that the interface is connected.

One advantage of LANC is the ability to trigger the shot automatically from an external event. So nature lovers can connect a photoelectric barrier or an infrared motion detector to catch animals in their natural setting. The LANC is also used by various accessories, such as the PowerPod-LANC from Applied Logic (*www.appliedlogiceng.com*), which is a pan-tilt base able to guide your 3D camera or 3D rig remotely. But the most flexible solution is to use a specific controller connected via USB to a laptop computer, which will automate the shots at will: every n seconds, when a specific event occurs, by remote controller, etc., The LANC Dual Camera Controller from Applied Logic costs around $250, including the Windows control software. Even when not connected to a PC, this controller can synchronize the two cameras and control their zoom and focus in sync. We regret only that the synchronization accuracy is displayed only by a few LED lamps instead of a digital display similar to the LANC Shepherd's; on the other hand, it costs only half as much.

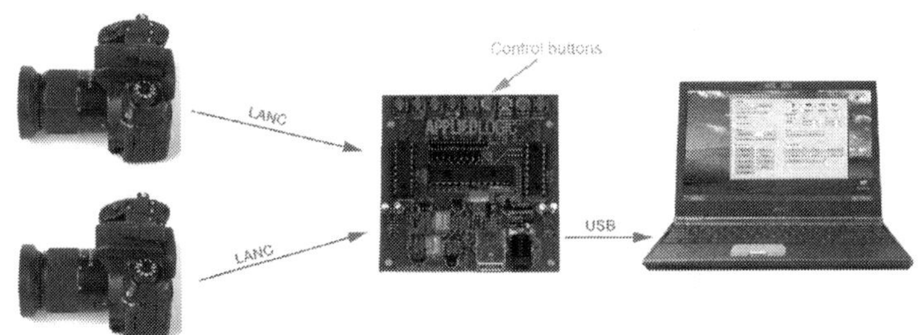

An Applied Logic controller connected to two Sony cameras and a PC

Canon and StereoDataMaker

3D shooting with Canon cameras is possible thanks to StereoDataMaker software (SDM for short). Developed by the Japanese Masuji Suto (***stereo.jpn.org***), SDM synchronizes almost all Canon cameras with an accuracy of 1/16,000 of a second. It has even been used to synchronize more than 50 devices for shooting Matrix-like scenes, where a frozen scene is seen from every angle. The only addition to the original unit is a switch and a cable to connect the cameras to sync. SDM is a modification of the famous (at least for Canon owners) CHDK (Canon Hack Development Kit), a temporary firmware that loads into RAM memory when the camera is powered on.

Supported Canon cameras are IXUS70/SD1000, IXUS800/SD700, IXUS950/SD800, A460, A550, A570, A610, A620, A630, A640, A700, A710, A720, and G7. In case of doubt, check the full list on the SDM's author's web page.

Installing SDM is simple: First find the version matching your camera model on the author's site, and then copy it onto a blank SD memory card. Then simply insert the card into the camera and start it. From that moment, the camera is temporarily under the SDM software's control and will remain so until powered down. Simply removing the battery will remove all traces of the software from the camera's memory, so there is no risk of damaging the unit.

With SDM installed on both cameras of a rig, you connect the two cables to the trigger button and check that both cameras are identically set (same sensitivity, same mode, etc.).

The only hardware addition required by SDM is a trigger with one button and two mini-USB cables that is available, *inter alia* from Digi-Dat (*www.digi-dat.de*).

© B. Michel

The Canon-SDM-Mini trigger from Digi-Dat

Photographic Stereoscopic Formats

JPEG

The JPEG format is universally used. The mere side-by-side juxtaposition of two photos in the same image is a guarantee of reliability, since the two images of the same pair will always be stored together. Some packages, such as Stereoscopic Player, usually displays side-by-side images correctly, but given the variety of resolutions and aspect ratios used in still photography, correct image display cannot always be taken for granted.

If the two images of the stereoscopic pair are crossed (the left picture is on the right and vice versa), it is possible to distinguish depth by squinting strongly. This is not for everyone, but people who regularly work with 3D photographs get used to this method to perform rapid checks without using any equipment.

© B. Michel

Honeybee in flight. Picture saved in side-by-side JPEG (the left image is on the right and vice versa).

By compressing images laterally by 50% in a 1920 × 1080-pixel format, we make sure they can be displayed on most 3D TV sets, as well as most dual projector setups. If 3DTV display is the main goal, images should not be crossed.

© B. Michel

Stereo pair laterally squeezed in a 1920 × 1080 JPEG image: an easy solution for display on 3DTV and other HD screens.
Disadvantages: Resolution is lower than the original, aspect ratio is mandatorily 16:9.

MPO

The MPO (Multi Picture Object, .mpo extension) format was standardized by the Japanese association CIPA (Camera & Imaging Products Association). The normative document and its English translation are available for free under the "CIPA DC-007-2009" and "CIPA DC-007-2009-Translation" names. MPO format is an extension of the JPEG (Joint Picture Expert Group, .jpg extension) format well known to photographers. It encodes N images in a single file, where N is a positive integer. With N = 2, it is a stereo pair, but with a higher value of N, the .mpo file may contain many images, making it a convenient format for photos targeting multiview autostereoscopic displays. Like JPEG images, MPO images contain EXIF metadata characterizing the image. If the camera has the information on hand, you will find one thumbnail image for preview, shooting time and date, orientation, GPS positioning, etc., in each image MPO file. Even more interesting for stereophotographers, the metadata for each image contain –

or at least are likely to contain – its position relative to the first frame inside the MPO file and the convergence angle, allowing more effective handling in postproduction.

The MPO format is composed of little more than two JPEG (.jpg) files glued one behind the other. If you have an MPO file available, do this test: Rename the file by changing its .mpo extension to .jpg. It should then be readable by any program. Obviously only the first image, the left one, will be displayed, as image editing programs do not imagine that a second image is hidden behind the first one! Various software packages, most of them free, convert an .mpo file into a pair of ordinary .jpg files or a single .jpg anaglyph one. For example, MPO Toolbox is very easy to use: a simple drag-and-drop is enough to convert images (*www.stereoscopynews.com*).

Remark
.mpo file manipulation programs are legion. MPO2JPEGS is worth mentioning as it converts an .mpo file into a pair of JPEG images on a Mac or Linux. So is StereoPhoto Maker, which manipulates .mpo images in various ways on PCs.

JPS

The JPS format is a standard .jpg image containing a pair of stereoscopic images put side by side. The images must be of identical size and crossed, that is to say that the left image is placed on the right and the right image is placed on the left. This provision allows trained observers to perceive 3D directly, without glasses, by squinting strongly. For those (almost everyone) who find this method difficult or uncomfortable, if not completely impossible to use, you can just use Stereoscopic Player to view them. You can also use the freeware Free JPS Viewer, which displays 3D .jps images in anaglyph mode (several variants).

Post-Processing of 3D Images

Integrated 3D cameras and two-camera rigs alike are never perfect. You must therefore do your utmost to remove unwanted disparities such as colorimetry differences, vertical shifts, rotations, etc. Various software packages manage this with various degrees of success, but StereoPhoto Maker, which is free from Japanese Masuji Suto (*stereo.jpn.org*), is very suitable for this purpose. Its automatic adjustment capabilities are excellent; its error detection functions discover and correct alignment, perspective, and rotation errors of 0.1°, zoom mismatches of 0.1%, and 1 pixel vertical alignment errors. In automatic mode, StereoPhoto Maker brings the points that jut out most in the image back into the screen plane, but it is very easy to adjust the parallax using the left and right arrows keys to get any element of the scene at the zero parallax point.

StereoPhoto Maker accepts as input the .mpo format, which is used increasingly by integrated 3D cameras, and by the Nintendo 3DS game console. Of course, it also accepts side-by-side images and separated pairs of left and right images.

StereoPhoto Maker after alignment of a 3D picture taken by a Fuji W3 camera. Corrections applied: rotation -0.1°, perspective 0.3°, vertical shift 4 pixels. Total parallax is 3.54%.

Narrative Grammar

4

What is immersion?

How do we write a 3D movie storyboard?

Is it possible to previsualize a 3D movie?

How should a 3D shot be framed?

What are the Ten Commandments of stereoscopy?

Does a 3D grammar already exist?

Designing for 3D

Movie design, editing, lighting, and technical setup become completely different when a third dimension is added. The feeling of being immersed in a scene is greater in 3D. The viewer is transformed from a simple spectator to an actor who quickly identifies with the action's protagonists. Suddenly, the distance between the camera and the actors no longer makes sense: Instead of watching the scene through a window, the viewer becomes part of it and feels surrounded by the decor. From a technical point of view, we quickly realize that some methods suitable for a conventional movie become unusable when stereoscopy is used: Filming dialog in reverse angle with numerous fast cuts no longer works. The viewer is unable to identify with either character; instead, s/he becomes the third man, the quiet observer on the set, always present but not interacting.

Another example: The good old matte painting technique, where a distant landscape is replaced by a painting or photograph picture, is simply out of the question in 3D. If it were used, viewers would spot the fake scenery's flatness instantly.

Invent New Rules

The aim of this volume is to present the technology associated with 3D, not give a film course. Nevertheless, we will go through a series of concepts involved in shooting a movie to highlight the influence of technology on stereoscopy's visual grammar.

There is no good movie without a good story, but the way the story is told has evolved in step with technology. The transition from silent movies to talkies revolutionized film writing. The same is occurring with the transition to 3D: Instead of describing the action taking place in the frame, we describe it in terms of volume, the volume behind the screen, but also sometimes the volume in front of the screen. In this last case, the action is constrained inside the invisible pyramid that links the viewer's

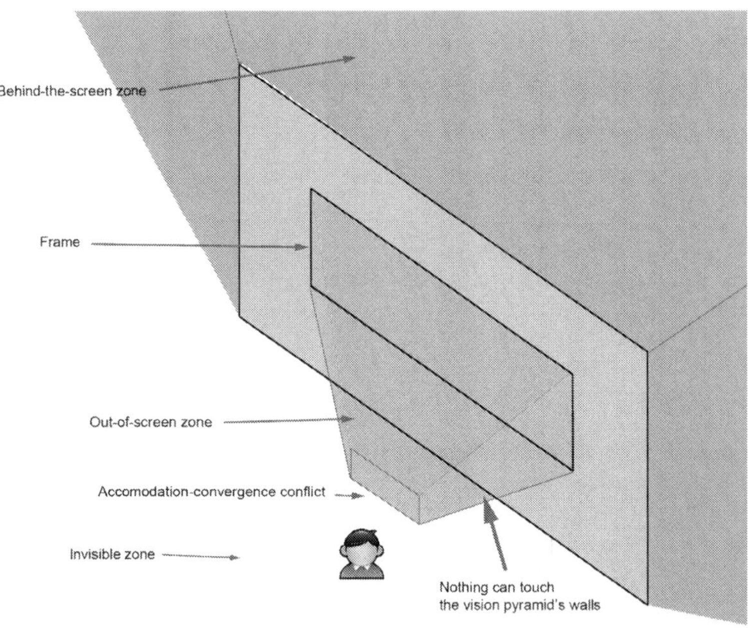

To maintain scene credibility, foreground objets may not be truncated by the pyramid's sides. The tip of the pyramid is also a forbidden area in order to avoid accommodation-convergence conflict.

eye and the screen frame. This strange constraint is the source of a completely new grammar, with its own rules and exceptions. Sadly, the situation is new and the experience of our elders is of little help. A large part of this new grammar is still unwritten.

We have just framed the 3D cinema scene's limits in space, but stereoscopy also imposes constraints on the fourth dimension, that is, time. The pace of editing and transitions between shots must allow for the way our brains perceive depth. It takes an ordinary human between half a second and two seconds to recreate a three-dimensional representation of the scene after a cut. A faster pace may induce fatigue and even headaches. Michael Bay's ultra-dynamic editing is therefore no longer appropriate. What is more, long takes will give the viewer plenty of time to explore all of the scene's depth and feel immersed in the action.

Forget the Rules of 2D Cinema

Today, every person with access to visual media understands the grammar of cinematography easily. Our kids spend so much time in front of animated images on a broad spectrum of screen sizes that they learn the rules of moving pictures at the same time as they learn to speak! Zooms, flashbacks, picture-in-picture inserts, slow-motion takes, and other effects all look natural to children and adults in today's hi-tech societies.

Yet such effects never occur in actual reality: Our eyes positions never change instantly when following a dialog, and we need binoculars to watch distant events. Going back in time and other timeline manipulations, for their part, are simply beyond reach.

The immersive 3D cinema grammar – stereoscopic today, maybe holographic in the future – needs to be reinvented from scratch by observing the real world. It cannot be created by adapting the existing 2D movie grammar.

Immersion

With a 2D movie, the viewer becomes a bodiless observer watching the scene. With a 3D movie, s/he immerses her/himself in the scene as either an actor inside the scene or an invisible observer on location. Her/His eyes scan the set, ground, and close and remote objects frequently in an inconscious process that helps to recreate an internal representation of the scene with all its depth and volume. Consequently, her/his gaze is fixed less often on the main character or staged action. The shooting stage is no longer merely a set; it becomes the reality in which the viewer is immersed.

Suspension of Disbelief

Rule number one of immersive cinema is to keep the viewer plunged in the presented artificial reality without falling back into the surrounding real world. All the rules given below exist only to keep the viewer in this suspension-of-disbelief state. Any breach of this state should be considered a failure for the film director.

The expression "*willing suspension of disbelief*" is attributed to the poet Samuel Coleridge, who contended that a story could be unbelievable without the reader's (or the viewer's, in our case) noticing as long as the story was interesting enough. The audience suspends judgment on the story's implausibility as long as it continues to be captivated by the story; the viewers' subconscious minds continue to postpone the decision because they want to know first how the story ends. If keeping everyone enthralled throughout a 2-hour-and-40-minute feature film is possible for James Cameron, it can be for others, too, but talent and extensive experience are essential ingredients.

Cuts

Maintaining the suspension of disbelief entails having the viewer become an observer or even one of the actors in the scene. Changing the point of view in the same scene should thus be done for a reason. Switching from camera 1 to camera 2 because it is better located is not the best solution; it is better to move the camera from point 1 to point 2. This will facilitate identification by an observer moving inside the scene. A cut required by spatial or temporal stage changes will shatter the illusion less, for the viewer will understand that the place or instant has changed, especially if the scene is completely different. As cuts disturb the suspension of disbelief, they must be done sparingly and justified by the needs of the script.

Framing

Framing is more difficult for a 3D movie because of the depth, which adds constraints that are absent from 2D. While the vision pyramid – the portion of space seen by the camera – remains the same, introducing the third dimension adds the notion of distance with respect to the physical window frame. The window, which is embodied by the screen's contours, is located at a fixed distance from the viewer. The brain likens it to a real window frame that reveals only part of the world behind it and leaves all foreground objects in plain view. Therefore, if the action takes place in front of that window, it can conceivably leave the frame. As this is technically impossible, the director has to force the action to remain inside the pyramidal volume between the camera and the window, never crossing its sides. In other words, the edges of the vision pyramid can never be touched by any scene element, on pain of a "window violation," which is a sure way to break the suspension of disbelief.

Floating Window

If the above rule is broken, the remaining solution is to use a floating window. This method, which is applied in postproduction, creates a virtual frame in front of the physical screen frame and enlarges the volume in which action is perceived as occurring behind the frame. Overusing this method may lead to discomfort, however, because the virtual window can introduce disparities that become nuisances on screens that are subject to ghosting, such as polarized screens.

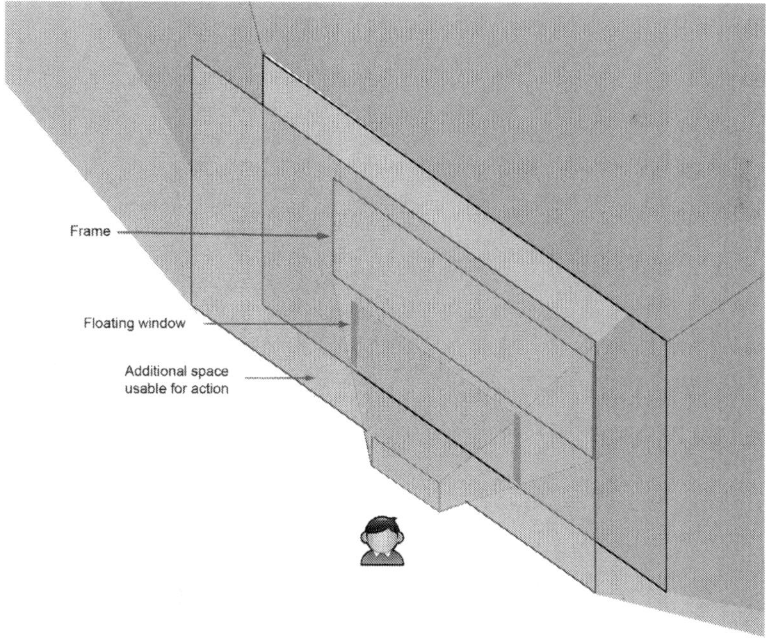

A floating window broadens the action zone during shooting. Overusing this technique will generate discomfort because of strong contrasts and the ghosting that often follows.

Caution. A floating window can often rescue a shot where an unwanted foreground object is cut by the left or right edge of the screen. However, if the top of a character's head is cut by the top of the screen when he pops out of the screen, the error is going to be tough to correct. The only thing to do is to move the whole scene backwards to the point where the actor is in the plane of the screen. Unfortunately, this is not always possible. For example, if the horizon is already using the maximum allowed positive parallax, no leeway remains.

Reference
Consult Chapter 2 for the practicalities of creating a floating window (page 64).

Window violations at the bottom of the screen are less problematic: Indeed, the bottom of the image is often hidden by the heads of the people sitting in the preceding row. Viewers blocking the view of the bottom edge of the screen create a closer physical frame for the screen, one that is usually far closer than the scene's foreground, even with strong in-your-face shots. Visual discomfort will therefore be felt only by the first-row audience, for whom the image depth is shallowest.

Center the Point of Interest

Disbelief is not suspended if the action or main characters leave the vision pyramid at the back of the frame; that merely makes the viewer believe the characters are hiding out of sight behind the edge of a real window. Despite this, actors coming into and going out of sight too often will make viewers aware of observing the scene through a window. That will increase the viewers' distance from the action and reduce the feeling of immersion in the scene. You should therefore try, as soon as possible, to encourage entering and exiting the scene behind an item on the set and not directly behind the edges of the frame. If we make the actor disappear behind a tree to the right of the screen, the viewer's gaze will stop at the tree rather than at the edge of the frame, so that s/he is no reminded that s/he is watching from the other side of the window, and her/his feeling of immersion will be undisturbed.

Depth (Dis)continuity

3D strength can vary greatly during a movie. Some scenes will offer strong depths, with distant horizons and very close foregrounds; others will be very shallow, with only a few depth layers that are fairly close to each other. Still, each time, scene cuts should bring elements in sight at depths similar to those in the previous scene to enable the viewer gradually to adapt her/his personal representation of the 3D world. For example, when cutting from an outdoor shot with a distant horizon to an indoor one, if you keep the center of attention – usually the main character – at the same distance in both shots, the transition will appear smooth. If the script requires a complete depth change between the two shots, you can try to ease the transition by bringing an item into the field at an average distance at the end of the first shot and another one at approximately the same distance at the beginning of the next shot. The basic rule is to ensure continuity of distance for one or more elements, preferably the ones that are most likely to attract attention. Under these conditions, it won't matter if the transition is an abrupt cut, a slide, or a slow cross-dissolve, and everything will be fine.

In the time/depth diagram shown below on the left, an indoor shot follows an outdoor shot. At cut time the first shot's main character is slightly behind the screen and a foreground object is slightly in front of the screen plane. In the second shot another character – on which the viewer will probably focus – appears on the same side of the screen and more or less at the same depth. The foreground is not prominent and the camera engages in a slow forward tracking movement that brings the character and set closer to the camera. In this context, the transition is easily accepted as there is true continuity between the focal points' depths before and after the transition.

In the right time/depth diagram, a shot fades into another one in order to shorten the duration of the forward traveling motion in an indoor shot. Depth continuity is provided by the cross-dissolve and facilitated by the fact that the point of interest's movements are smooth and parallel, and consequently very predictable.

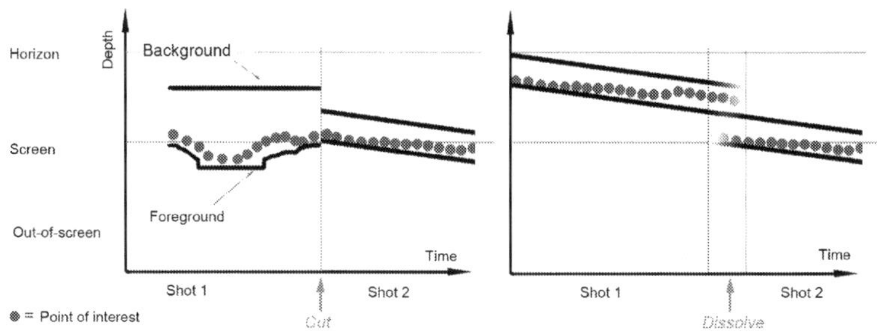

On the left: Change in the depth budget during an easy-to-take cut transition.
On the right: Change during an acceptable cross-dissolve.

Horizontal Horizon

The feeling of immersion in a virtual world is possible because our brains reconstruct a three-dimensional reality consistent with what our senses perceive. The stereoscopic image accounts for about 80% of our sensory stimuli, but the visual stimuli are not alone: Our ears perceive sounds that must stay consistent with the perceived images, as well as with our senses of touch and balance.

Theme parks add mechanical special effects to their stereoscopic presentations, *e.g.*, moving seats, jets of air and water, vibrations, etc. These effects need to be generated carefully, because any inconsistency between them and the image will break the suspension of disbelief. The slightest disturbance in the consistency between the image we see and our inner ears' perceptions of the vertical will break the illusion. Alternatively, and perhaps even worse, if the suspension of disbelief is not completely broken, the brain is forced to choose continously between two conflicting realities. Almost instant nausea is guaranteed!

An unsteady or simply tilted horizon is enough to shatter the virtual world's consistency. When shooting in 3D, always double-check the horizontality of your camera mount. A horizontal horizon is even more important in 3D than in 2D!

High-angle View

Horizontal camera, centered horizon, immersion as an actor in the scene.

High angle shot, high horizon, immersion as a scene observer.

© B. Michel

The viewer/set relationship is different when the camera has a high-angle view with the horizon at the top of the frame.

In outdoor scenes, the position of the visible horizon is important, as it gives us a more or less conscious reference of the camera's horizontality. A horizon that rises gently and then faster and faster suggests a feeling of falling. If the immersion is really effective the viewers will grip their armrests to keep from falling. James Cameron plays on this effect successfully in *Avatar*: The highly visible horizon at the start of the dive situates the action. Then, once the fall starts, the horizon is no longer visible, but the camera remains above the whole stage, giving the viewer a sense of being present in the scene while suggesting unconsciously that her/his observer position is safer than the hero's, which is lower and thus closer to the danger.

Some directors are in the habit of shooting many scenes with a slight high-angle camera. The higher horizon in the frame and more widely visible soil reduce the overall scene depth: The many parts of the set located in the area where the stereo effects are most pronounced reinforce the sense of presence as an observer.

Zooms and Camera Movements

The depth cues that are present in a stereoscopic shot combine to help the brain to create a 3D mental image. Scene depth will be appreciated better if many different but consistent cues are present. For example, motion parallax induced by a tracking shot will complement binocular vision and make understanding the scene easier. The progressive viewpoint changes in a long take constantly reveal new scene elements that our brain adds one by one to its mental representation. This will be less tiring than restarting the whole process from scratch at each cut in a fast-paced edit.

Let's not forget, either, that even though we are accustomed to zooms in television coverage and even in movies, they are not natural to the human eye: The human eye never zooms! Immersion in action will be better if the camera moves within the scene using a fixed focal length instead of staying put and zooming in or out. Zooming also adds severe technical constraints to the shooting. First, the two zooms of a 3D rig will never be perfectly matched and automatic or manual correction will be required. Second, increasing the focal length will reduce the feeling of depth, which will have to be offset by increasing the interaxial distance. If the frame encompasses foreground objects at the beginning or end of the zoom, the cameras' convergence will also require gradual adjustment.

We may conclude from the above that zooming should be reserved for special cases where it is technically or artistically impossible to replace zooming with real camera moves. However, if used with care, zooms also remain useful creative tools.

Storyboard

If they are to be more than a passing fad, 3D cinema and TV must serve the needs of storytelling. Beyond the spectacular effects that attracted the first audiences, 3D moviegoers are seeking the feeling of immersion and sense of having a piece of the action.

The media industry is engaged in a race for ever better stories and storytelling. Strengthening the immersion that 3D provides can only contribute to this goal, but many obstacles must be over-

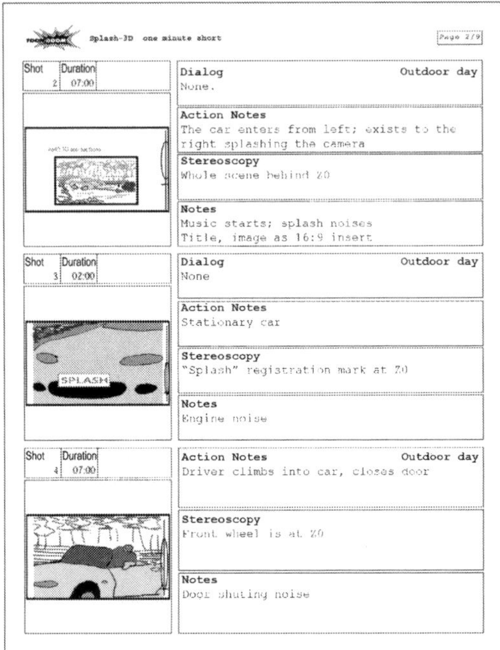

A storyboard page crafted with Toon Boom Storyboard. Information about 3D is added as a vertical graph on the right side of the picture in the left sketch square.

come, some technological – as explained above – and some semantic. How is as important as why in storytelling. With an added dimension, improvisation is no longer possible: The clever use of depth has to be integrated into the movie as early as the script-writing phase.

Fortunately, there are previzualisation tools to support the director in this creative task from the very outset. They will also facilitate collaboration among the director, author, director of photography, and stereographer. But let's remember the first rule of cinema, whether 3D or not: A movie will be good only if the story is good!

3D's success in movies obviously requires technological mastery, but, even more important, it calls for creativity and the refinement of innovative storytelling methods using all available assets. The transitions from silent movies to talkies and from black and white to color both successfully overcame this challenge. The coming years will tell us if the transition to 3D will succeed as well.

3D's use must be modulated throughout the movie. Everything is possible: moments of intense depth, more intimate moments with gentle 3D effects and close backgrounds, breathtaking moments involving dizzying dives into bottomless abysses, ... However, the first step will be always to include the third dimension as of the writing stage.

3D and Storyboard

A script does not render the nuances of a dialog or the visual feelings the author had in mind easily. Translating it into a storyboard is thus readily chosen as a way to communicate among all the parties involved. Of course, using the appropriate software facilitates the first creative steps, even if the starting point is often the simple sheet of paper. Toon Boom Storyboard is a very simple storyboarding software package (about $120, *www.toonboom.com*). It is not specific to 3D movie production, but as it is based on sketches and free text, including the additional 3D information is easy. Everyone will use their own conventions but the most important thing is unequivocal communication among the various production members.

A storyboard is a series of annotated illustrations describing a whole movie, whether fiction, animation, documentary, or live reporting. Descriptive details, dialog texts, and indications about framing, staging and 3D can be very detailed, very vague, or even completely absent, but, even in this last case, the storyboard is one of the most useful tools we have to familiarize ourselves and the other team members with the movie's subject. Illustrating a scene, even with a simple rough sketch, shows the structure of the visual message you want to put across and highlights the director's priorities. It is also a tool for assessing the durations of the different scenes and their balance in the final cut quickly. Obviously, if you do some location scouting, you can use photographs as illustrations or as the backgrounds on which to sketch the scenario.

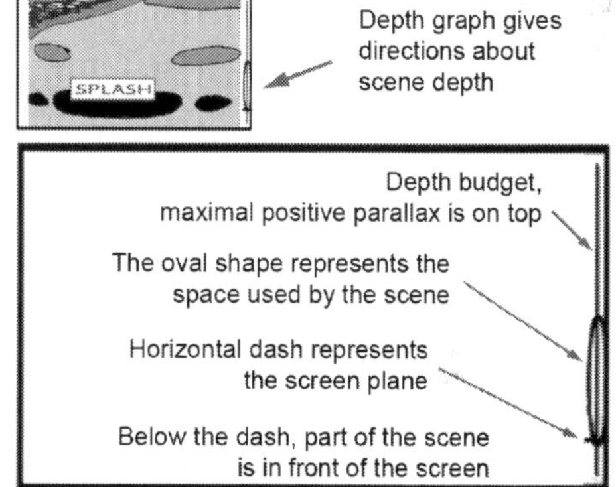

Example of a conventional graph for representing scene depth in the depth budget.

Specific indications about 3D are included in a frame we called "Stereoscopy" and are supported visually by a simple graph along the vertical edge of each sketch of the storyboard. This graph shows the depth range used in the shot over the full depth budget, with the foreground, background and screen plane (also called Z0 level because there is no parallax for objets in the screen plane) clearly marked.

Previsualization

Once a storyboard that includes some 3D indications has been created, the next level of sophistication in preparing the film is to use previsualization software. This kind of tool models the set

Previsualization software FrameForge Pre-viz Studio 3 handles 3D for several cameras.

in three dimensions more or less faithfully and then generates a storyboard with much more accurate technical information than a hand-drawn storyboard.

The time spent making this CGI layout of the whole movie take by take is often a profitable investment. Indeed, it offers a precise framing vision and confirms or refines the technical choices. The best known previsualization software is FrameForge Pre-viz Stereographic 3D Studio, which takes all the specifics of 3D shooting into account (*www.frameforge3d.com*).

For each scene, you set up the characters or simplified or purpose-made CGI models of the set and a mirror or parallel stereoscopic rig for each planned shooting position, and then you set the 3D parameters, lenses, and lighting. The screen display shows the location layout on one side and the 3D scene as viewed by the camera (in anaglyph, on a polarized screen, or with another 3D display method) on the other side. Scene elements can be created with Maya, 3D Studio Max, or any other modeling tool or merely extracted from the extensive library provided with FrameForge. Over a thousand objects and characters are available, as well as the most common models of cameras, 3D rigs, lighting sources (faceted spotlights, follow spotlights, the sun, etc.), plus dollies, booms, cranes, and other accessories.

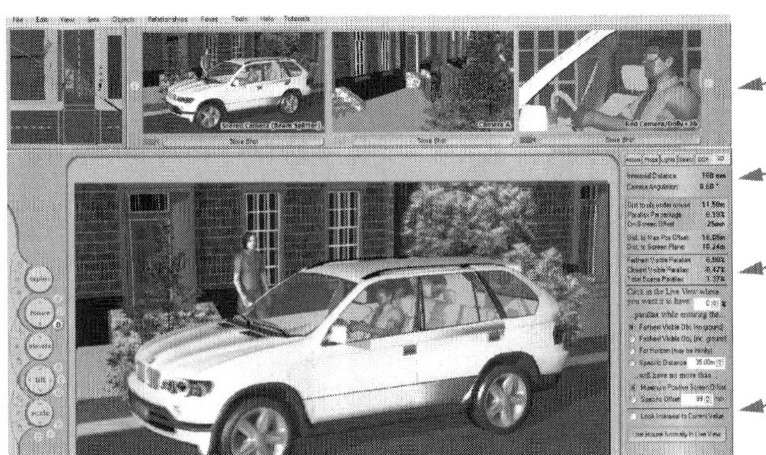

FrameForge previsualization software at work

Reports and Previsualization Documents

The previsualization program can export an illustrated storyboard with 2D or anaglyph pictures ready for printing or viewing on a PC screen, but it can also export an animatic, that is, a kind of draft video that complies with the timing and planned camera moves. Various formats are possible, including dual full HD.

Besides its ability to model the set accurately, the previsualization software's main advantage is that it allows real validation of the technical setup. Having chosen a mirror rig with Red cameras and prime lenses, you check that the frame is correct in the simulator, the tracking movement does not reveal unwanted elements, and, as a bonus, you can even validate the stereoscopic settings, such as convergence and interaxial distance, directly. FrameForge also generates a list of all equipment used, including cameras, rigs, dolly tracks, and lights.

The stereo verifier is another report generated by FrameForge: It scans the scene for all visible objects and reports the minimum and maximum distances, their parallaxes, and on-screen image offsets. Results from the integrated 3D calculator, such as

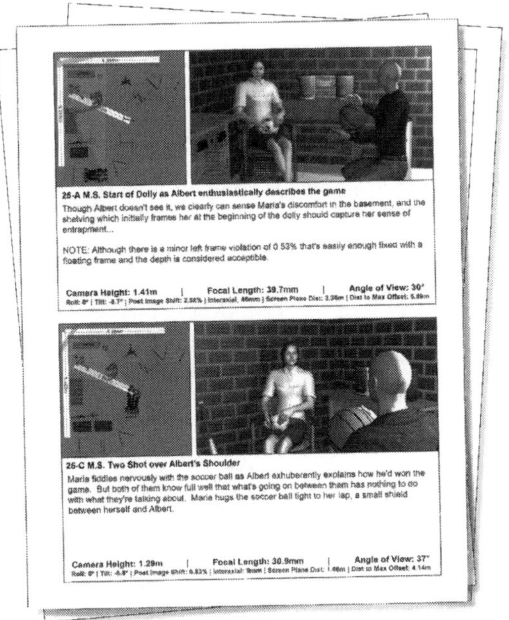

Example of a FrameForge storyboard with the 3D rig parameters as annotations. Thanks to anaglyph rendering, 3D effects are visible even in the paper version.

convergence and interaxial values and window violations, and technical data, such as minimum and maximum parallax values in pixels or percentages, can be exported to an Excel® spreadsheet. The most important technical data can also be included at the bottom of each storyboard box.

3D Video Shooting

With the introduction of 3D, new technical constraints have been added to the usual rules of traditional photography. For example, the final screen size will affect the director's choices and the way shots are framed can change the suspension-of-disbelief feeling. Below you will find some tips to avoid the most common traps in 3D shooting.

Film in Parallel

All stereographers warn cameramen against the temptation of zooming the same way as in 2D. Stereoscopy gives a better immersive feeling with a wide-angle lens than with a telephoto lens. Of course, wide-angle shots mean that the camera is positioned very close to the subject, which is not always possible.

To shoot most outdoor sport events, a side-by-side rig with an adjustable interaxial distance is preferred to a more complex mirror rig with adjustable convergence, which is not really needed when the scene is more than ten meters away (mirror rigs remain useful for tennis and basketball, where close-ups are more common). If you are forced to increase the focal length during a take, increase the interaxial distance at the same time as you zoom in and keep the cameras parallel. Objects and characters will keep their volumes and shapes and the parallel setup will reduce the risk of bad disparities introduced by convergence. However, care must be taken to move the camera (forward or oblique tracking) during the rig parameter changes to mask the changes in the scene's apparent size through the introduction of additional depth cues. Otherwise, the scene scale change could be too disruptive if the camera stayed put.

Convergence can bring the main subject into the screen plane and increase its roundness. While this may be appropriate for movies, it is rarely a good idea for sports or other outdoor events. If the action is far away (in the case of football, for example), it must remain behind the screen plane. Some foreground elements may remain visible to increase the feeling of depth without unduly attracting attention. Of course, you should not stress the viewer's visual system unnecessarily by giving too much parallax to the main subject. Always keeping the main subject in the screen plane reduces the usable part of the depth budget to a fraction of what is possible and will look unnatural.

Manage Your Parallax Budget According to the Screen's Size

The maximum parallax values must be set according to the projection room with the largest screen. As discussed in Chapter 3, the British broadcaster Sky 3D stipulates that the parallax budget must not exceed 3% of the screen's width, *i.e.*, 2% behind the screen and 1% in front. In short graphic sequences, such as rotating logos or other brief visual effects, the depth budget may rise briefly up to 4% positive parallax (behind the screen) and 2.5% negative parallax (objects in front of the screen).

The acceptable values for movie theaters are far smaller, especially for the part of the depth budget that is behind the screen. On a ten-meter-wide screen, the maximum parallax of 65 mm is reached with only a 0.65% shift. With the exception of IMAX theaters, most movie theaters have screen widths under 20 meters; it is thus advisable to limit the parallax to 0.32%. If a small amount of divergence is tolerated, 0.5% may be taken as an extreme limit.

If the content is intended for movie theaters only, without any distribution on TV or other small screens, out-of-screen effects can be increased and the entire scene pulled forward by means of a simple horizontal shift that will reduce the positive parallax to an acceptably small value. This is possible because the accommodation/convergence correlation problems that are seen on 3D TV sets do not exist in movie theaters, where the viewer is far from the screen.

To simplify the depth budget management issue, we can recommend limiting the depth budget to a 3% spread between the minimum and maximum parallaxes when filming and then, during the postproduction step, to limit the positive parallax to the 65 mm stop value measured on the largest screen on which the movie will be projected. As shown in the table below, the larger the screen, the smaller the allowed parallax budget.

From the above table we see that an excellent feeling of depth is given on an IMAX screen with a parallax budget as small as 1%. At the other end of the scale we notice that presenting truly distant

Parallax Values as a Function of Screen Size

Perceived Distance in % of the Screen-Viewer Distance	Parallax (mm)	Parallax on a 1-meter TV Screen		Parallax on a 5-Meter Screen (Home Cinema)		Parallax on a 10-Meter Screen (Cinema)		Parallax on a 20-Meter Screen (Cinema)		Parallax on a 30-Meter Screen (IMAX)	
		PIXELS	%	PIXELS	%	PIXELS	%	PIXELS	%	PIXELS	%
< 20% (avoid!)	< -260										
20%	-260	-532	-26.0	-114	-5.6	-57	-2.7	-28	-1.4	-19	-0.93
30%	-152	-311	-15.2	-65	-3.2	-32	-1.6	-16	-0.8	-11	-0.54
40%	-98	-200	-9.8	-43	-2.1	-21	-1.0	-11	-0.5	-7	-0.34
50%	-65	-133	-6.5	-28	-1.4	-14	-0.7	-7	-0.3	-4.7	-0.23
75%	-49	-100	-4.9	-12	-0.6	-5	-0.2	-2	-0.1	-1.3	-0.06
87%	-10	-20	-1.0	-4	-0.2	-2	-0.1	-1	-0.05	-0.6	-0.03
Screen plane	0	0	0	0	0.0	0	0.0	0	0.0	0.0	0.00
144%	20	41	2.0	8	0.4	4	0.2	2.0	0.10	1.2	0.06
200%	32.5	65	3.2	14	0.7	7	0.3	3.5	0.17	2.4	0.12
500%	52	104	5.1	23	1.1	11	0.5	5.6	0.27	3.8	0.19
1,000%	59	118	5.8	25	1.2	13	0.6	6.4	0.31	4.3	0.21
Infinite horizon	65	133	6.5	28	1.4	14	0.7	7	0.34	4.8	0.24
Extreme envelope of the parallax budget			32.5%		7%		3.4%		1.7%		1.1%

Note: Values not highlighted are usually not recommended.

horizons and strong out-of-screen effects at the same time on a one-meter TV screen (a 50" model) will require a parallax budget as large as one-third of the screen. For many reasons discussed earlier, such wide shifts will be extremely difficult for the viewer's brain to merge. Tthe depth budget will have to be strongly limited, as Sky3 recommends. This will give viewers a less pronounced

3D feeling but avoid unpleasant viewing experiences. A movie designed with a 3% parallax budget will allow for true infinite horizons and good out-of-screen effects on large screens, barring IMAX theaters and some very large venues.

To avoid annoying divergences in the largest theaters, you need only to shift the two images toward each other enough to bring the positive parallax under the 65 mm bar, which will pull the whole scene forward and thus bring it closer to the audience.

Be Gentle with Your Camera

Rapid lateral movements (panning) are not appreciated in 3D movies, as the brain needs time to integrate distances from quickly appearing objects. It is wise to leave enough time for the viewer to take in each new appearance one at a time before returning her/his attention to the main character. To take this into account, fast pans and lateral tracking shots should be avoided, especially if a lot of foreground objects keep coming into view.

In order to keep the viewer's focus on the main subject, you have to find a way to bring foreground objects into view without disturbing the action. For example, a lateral tracking shot that gradually brings a playing field and its white lines into view will not be surprising and will not disturb the viewer's focus. However, if it reveals unexpected characters or objects, the viewer will instantly leave the main subject to focus on these new elements.

The rule is to program moves two times slower than we are accustomed to doing in a classical sport event shot. This rule should be followed even more strictly when numerous close obstacles spring up in the frame.

In the coming years, when shooting at high frame rates, let's say at 48 or 60 frames per second, becomes more common, motion will become more fluid and kinder to the eye. In any case, fast pans over a complex scene force the brain to process a large amount of depth and volume information to reconstruct its mental image of the set. So, excesses will quickly tire viewers and be likely to cause headaches.

Prefer Zoom Out

Editing done inside the on-location OB van should favor cuts after the camera finishes zooming on the subject instead of before or during the forward zoom. After settling on the subject, the camera will start to zoom out slowly, revealing surrounding elements one by one and giving perspective and adding depth to the action.

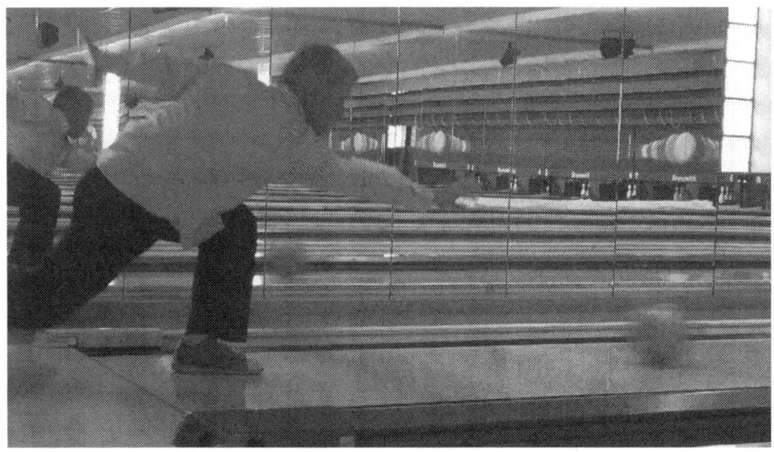

© Crig.be

A low camera and some foreground increase the feeling of depth and help to give a good 3D effect.

This will strengthen the 3D effect, whereas a forward zoom would reduce the feeling of depth by bringing part of the foreground and background elements out of the frame and squashing depth at the same time.

Use a Low Camera for Wide Shots

Placing the camera a bit closer to the ground than usual will bring more foreground elements into view. Soil, game lines, etc., will provide receding lines that increase perspective and draw the viewer's gaze toward the action. Once the shot is set, you may zoom slowly forward and then cut to another camera located closer to the action. However, don't constrain your composition to follow this rule at all costs: Artistic considerations and the message that you want to put across should prevail!

Be Creative

3D rigs are expensive. Why not add a few pairs of fixed rigs fitted with a pair of fixed-focus wide-angle cameras for short insert shots giving an audience perspective? Such cameras are cheap, portable solutions for unusual points of view favoring a more personal view of the action, the same way as some beauty shots are used to raise the emotional level at key moments during big matches.

The Ten Commandments of Stereoscopy

Each stereographer has her/his own list of rules to follow to avoid the pitfalls of 3D shooting.

Despite unavoidable differences of opinion, the ten basic rules of stereoscopy are commonly accepted and most stereographers will agree with the ten commandments below (and especially with the eleventh!)

1. "Thou shalt match your cameras to geometric perfection."
2. "Thou shalt limit your parallax according to thy screen's size."
3. "Thou shalt not covet shiny objects and strong contrast."
4. "Let light flood thy entire stage."
5. "Thou shalt prefer the wide angle to the zoom."
6. "Thou shalt never let your camera remain motionless."
7. "Honor slow-take sequence shots above all others."
8. "Flee frame edges like the plague."
9. "Use out-of-the-screen effects wisely."
10. "Always check thy 3D shots most verily in 3D."

    but

11. "Thou mayest yet transgress these rules one day with discernment."

These rules are only guidelines that you should follow – and transgress – with full knowledge of the consequences. All producers and stereographers will confirm that the most important thing is to tell a good story and to tell it well. If a scene is going well while violating several of the rules, then it is OK! And there are numerous examples of this. In *Avatar*, for example, James Cameron decided to underlight several out-of-screen foregrounds, such as in the military headquarters, just to keep the focus on the main characters rather than the surrounding decor.

"The Ten Commandments of Stereoscopy" as published by the author many years ago, but more relevant today than ever.

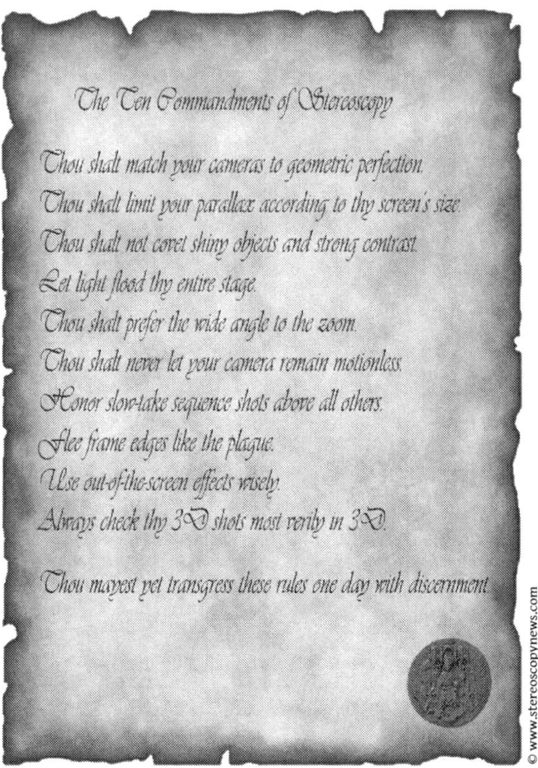

The Ten Commandments of Stereoscopy

Thou shalt match your cameras to geometric perfection.
Thou shalt limit your parallax according to thy screen's size.
Thou shalt not covet shiny objects and strong contrast.
Let light flood thy entire stage.
Thou shalt prefer the wide angle to the zoom.
Thou shalt never let your camera remain motionless.
Honor slow-take sequence shots above all others.
Flee frame edges like the plague.
Use out-of-the-screen effects wisely.
Always check thy 3D shots most verily in 3D.

Thou mayest yet transgress these rules one day with discernment.

The 3D Grammar is not Here Yet

Cinema is an art and all we know is that art cannot be reduced to rules. New directors are now experimenting with digital stereoscopy and just starting to invent its new grammar. In fine arts and music, beginners are advised to follow the example of the masters before creating their own styles. Tomorrow's 3D filmmakers will follow suit. Fortunately, some big names have already cleared the first hurdles. James Cameron and his *Avatar* immediately come to mind. Cameron understands that viewer satisfaction is strongly linked to the ease of immersion in the action and sustained suspension of disbelief. To achieve that, 3D use must be controlled and not allowed to run wild, as several past movies were famous for. Cameron's stereoscopy is not aggressive and most of the time it remains behind the screen plane. However, it is modulated quite cleverly, with strong 3D at times and almost too little depth at others.

In fact, the art of 3D lies in the ability to play with the viewer's ability to understand and perceive depth, almost as if this capacity were not constant but depended on the effort demanded of the visual and cognitive systems in the previous moments. We clearly lack perspective to determine the scientific basis of this understanding of 3D. For the time being, we can only see that it works: Audiences of all ages loved *Avatar!*

Concept of 3D Capital

Let us start with the idea that at the beginning of a movie the viewer has a rather limited ability to appreciate depth. The opening credits and first scenes setting the stage for the plot will ramp up 3D effects in a matter of minutes. Once the brain is accustomed to these visual gymnastics, the director may jump into the action and start to adjust the 3D effects, especially 3D transitions, high and low angles, traveling shots, and zooms, by playing with the depth values. The 3D effects will be strong at times to strengthen the actions and characters' movements with respect to each other, and weak at others, in order to give the viewer some time to rebuild her/his depth capital. At the end of the movie, the viewer must leave the room with her/his depth capital fully recharged. Therefore one should avoid throwing swords and axes right over the audience's heads during the closing credits!

Physiological studies on 3D effects are neither numerous enough nor old enough to draw firm rules, but any avid viewer of 3D movies will confirm that specific elements tire the brain and reduce one's 3D capital drastically.

Many of those elements are projection-linked, *e.g.*,

- a screen that's too close and crushes depth;
- a screen that's too far away and exaggerates depth;
- a viewing position that's too low;
- a viewing position that's too far to the side;
- projection brightness that's too dim; and
- ghosting.

Other tiresome elements come from the movie itself, to wit,

- window violations;
- fast-paced editing with numerous cuts per minute;

- cuts with strong depth budget jumps;
- duration of strong 3D;
- use of the whole range of the depth budget;
- fast depth budget changes;
- (overly) large parallax values on heavily contrasted objects;
- large negative parallaxes lasting one second or even longer;
- horizon level and slope changes; and
- character entries and exits through the field's edges.

Some parameters will increase or fully rebuild the depth capital:

- a good, well exploited depth budget that the vewer may explore on several levels without out-of-screen elements;
- stage-setting shots with slow camera moves;
- sets rich in near and far detail;
- a depth budget that remains well within its limits;
- slow, natural changes in the depth budget; and
- actions remaining centered in the middle of the screen for a long time.

The director will thus have to juggle the items on these lists to help viewers to create their depth capital during the first minutes of the film, and then to use it and restore it without exhausting it completely, and finally let them go at the end of the show with a depth capital strong enough to be able to return to the real world! S/he must take into account the fact that the public is made up of very different audiences: moviegoers who are more interested in the scenario than in visual effects, hard-core stereophiles, teenagers who adapt easily to the fastest paces, children with low interocular distances, and so on.

Modulate the 3D

As mentioned above, 3D cinema is an art, and controlling all its aspects is no trivial task. It is never an equation with a single solution, but a blank canvas on which each director finds a way to adjust her/his use of stereoscopy.

James Cameron (*Avatar*) takes great pains never to exhaust the audience's 3D capital. He usually meets most of the Ten Commandments. His cuts are very respectful of depth continuity. He avoids long, strong, positive parallaxes. Most *Avatar* scenes take place in a forest with rarely visible distant backgrounds. He builds a consensual 3D feeling that will tire neither the cinema audience nor home viewers in front of a 50-inch TV, while providing highly immersive entertainment on very large screens.

Henry Selick (*Coraline*) and his stereographer Brian Gardner, who is the inventor of the animated floating window, play with depth to switch from the real world (low depth budget) to the much deeper imaginary world. They also enjoy playing with shallow depth of field, leaving the background slightly blurred. The viewer's gaze is channeled to the main character as much by the depth of field as by convergence. Moreover, the two are usually synchronized.

Ben Stassen (*Sammy's Adventures*) is in favor of long takes, a restrained depth budget, and well-centered action - but very often presented in front of the screen. He does not hesitate to leave some depth jumps between shots. These features encourage viewing on the big screen rather than a TV set.

Tim Burton (*Alice in Wonderland*) causes dizziness in the scene where the characters change size by manipulating zooming and tracking in the opposite direction simultaneously while changing the interaxial distance. Not only is the immersive feeling not broken, but the viewer physically feels the set and characters' resizing, a very disturbing feeling if there ever was one! In this scene Tim Burton works deliberately against established rules, but succeeds in making his dream world credible.

Wim Wenders (*Pina*) uses a remote-controlled crane a lot as a shooting tool in order to keep the action centered on the screen without ever cutting his characters with the screen edges. With his stereographer Alain Derobe Wenders does not attempt to coordinate focus and convergence the U.S. way, but lets the viewers' gazes explore the whole set volume. On the contrary, he reinforces his subject with carefully studied lighting. Thanks to its success, *Pina* will certainly be among the reference movies for the future 3D grammar.

Joseph Kosinski (*Tron: Legacy*) and his director of photography Claudio Miranda used James Cameron's shooting equipment (Cameron-Pace rigs, Sony F35 cameras), but in radically different ways. They often used wide-angle lenses (14 mm) but - as as fequent followers of the eleventh commandment of stereoscopy - promoted often shallow depths of field, except in the stage-setting shots, where the viewer's gaze is free to explore the whole set volume.

Coupling between convergence and shallow depth of field therefore forced the director to underlight the background or even blur it strongly to mitigate divergence. He did not hesitate to shoot in front of an 18 kW spotlight and get great lens flares, which are completely acceptable in 3D. They were also the first to reinvent over-the-shoulder shooting for 3D, keeping the camera just far enough from the character behind to avoid encroaching on his personal space. Both stress the absolute necessity to validate the 3D footage on a big screen directly on the shooting site.

There are as many creative directors as there are creative ways to interpret or ignore the Ten Commandments, but they are all, each in her/his own way, writing the first pages of the future grammar of stereoscopy.

2D to 3D Conversion 5

How are 2D movies converted to 3D?

Are these methods efficient?

How does one design a movie with 3D conversion in mind?

What are the advantages of 3D conversion?

What software is used to convert movies to 3D?

Converting movies shot with a single camera to new stereoscopic versions is a potentially extremely profitable business and a number of companies are engaged in it. Indeed, the range of available 3D stereoscopic content is limited, experienced stereographers are few and highly sought after, and the market seems vast. In addition, 3D films are difficult to plan financially, while a subsequent conversion has a known fixed cost per minute. No wonder that the conversion market is valued at an estimated 35 billion dollars for the next five years! Prime Focus, which recently started working in this area in India, now has 1,200 people on the company payroll!

Unfortunately for gullible investors, 2D to 3D conversion is a difficult art, or science. This explains why the price of movie conversions is negotiated between $50,000 and $150,000 per minute, according to scene complexity and the quality required. The sector's pioneers are all close to Hollywood, with In-Three Inc., Legend 3D, and Sony Imageworks being the best known.

2D to 3D Conversion Principles

Converting a two-dimensional image to 3D actually means turning it into two images that differ only by the horizontal displacements of their pixels, i.e., the parallaxes. Since the parallaxes are generally small, at around 1 to 3 percent of the image's width, we can consider one image of a stereo pair to be the original one and then simply create the other image. Generally, when we stick to this method, the original 2D image is taken to be the left one and we must create the right image from scratch. Of course, the reverse is also possible.

Depth Map

Excluding simplistic methods with dubious results, all conversion methods are based on the creation of a depth map, usually followed by segmentation of the scene into its various components, and then creating a second image by introducing offsets deduced from the depth map. Various automatic and manual methods can be used at each stage. Good-quality conversion involves a huge amount of manual work, which explains the high cost charged per minute of conversion in the filmmaking industry.

An image and its depth map (black = close; white = infinity)

Photo and Video Conversion

We shall now review the main steps of the 2D to 3D conversion of a video sequence. The principles remain the same in the case of converting a single photograph, but the quantity of available information is smaller. The task is thus often more difficult. However, as there is usually more time available to do such jobs, particularly careful work and the operator's talent can usually compensate for this disadvantage.

Well-Kept Secret

Efficient conversion methods are almost all jealously protected by their developers, who tend to provide conversion services instead of selling their methods. This is easily explained by the fact that quality conversions are necessarily semi-automatic and require thorough knowledge of the tools and undeniable know-how. Under these conditions, selling or licensing one's secrets as "turnkey" software can lead to serious disappointment. Nevertheless, there are various programs on the market, and even some released as open source, that provide partial answers to the conversion problem.

Several vendors offer real-time conversion solutions based on the best algorithms known at the moment and implemented in powerful graphics processors. In some well-known or relatively predictable environments, such as a football stadium, these methods may give satisfactory outcomes, but none can yet compete with the best semi-automatic procedures described below.

Inefficient Methods

Over the years, various simple but rather unconvincing methods have been applied to the 3D conversion problem. The simplest one is to take the next or previous frame in the same video sequence as the right picture of the pair. If the camera pans from left to right, the disparity between two consecutive frames will give a pseudo-depth feeling. This approach works really well only in rare cases, and if the camera moves up or down, vertical disparities will give very uncomfortable results. Of course, if the camera reverses its direction, depth will also be reversed and the background and foreground will trade places! This method does not have a bright future, unless it is to be used to cause headaches and nausea deliberately.

A slightly more sophisticated variant retains only the horizontal component of the image shift and thus lessens adverse effects without preventing depth from disappearing completely when the camera stops. The most salient take-home message from these harrowing experiences of the past is that 2D to 3D conversion is anything but easy to do and sophisticated means are required to achieve a satisfactory level of quality. Vestiges of commercial attempts based on these ineffectual principles can still be found on the Web, among other places, usually in the form of a set-top box for connection to an analog TV. Inefficiency is guaranteed!

Fortunately, modern computers and, even more importantly, their graphics boards now have enough processing power to tackle the task more intelligently.

2D to 3D Conversion Steps

Adding depth to an image in fact means creating a second view from the first one. The work is broken down into six steps, as follows:

1. Selecting keyframes in the original sequence (the left images).
2. Segmenting the keyframes into separate elements.
3. Determining the depths of the keyframes' main elements.
4. Propagating the depth information to all the points in all the images.
5. Generating all the new images (the right image of each pair).
6. Patching the "holes" created by the generation step.

Time Slicing

The video is divided into segments in which the movements of the various objects and characters in the scene – a car moving away, a person walking from left to right, etc. – are simple and easy to find by interpolation. We select a small number of keyframes at the beginning of each of these simple movements.

Segmentation

Then we segment each keyframe into its main elements, e.g., sky, ground, objects, and characters, and determine the distance from each to the camera. Segmenting the image into separate components is easier in a video sequence than in a single image, as objects moving in front of each other make it very easy to find their contours. At the same time, the segmentation software determines the relative distances of objects relative to each other: Character A is in front of the table, which is in front of the sky, etc.

Many segmentation methods, based on various visual criteria such as color, shape, and other characteristics of objects and characters, exist. Following the silhouette of a moving object in a video sequence is easy for the human eye to do. The most promising automatic segmentation methods are based on optical flow techniques. Optical flow methods compute the trajectory of each pixel of an image from its movement history. This enables us to determine where and when an element appears and disappears when it is hidden by another.

A specific "optical flow" calculation applied to all pixels of an image was long considered impossible, as huge computational power is required. The appearance of massively parallel GPU graphics processors has changed the situation and this type of computing is now possible in real time. Knowing all the pixels' precise paths makes it possible to interpolate their positions and thus detect the contours of objects moving in front of a static or slightly mobile background.

Remark
Note also that optical flow methods are also useful in other areas besides stereoscopy, such as creating intermediate frames to slow down the video or generating intermediate pixels to increase image resolution when zooming.

In this optical flow example computed on a sequence from the film *Elephants Dream*, the direction of movement is color-coded. Even on this black-and-white copy, it is easy to pick out the background contours moving to the left (grey zones at left and right sides), while the character holds his upper body position (white) and lowers his head (dark grey).

Segmentation is not an easy task. In many cases, the edges of objects or people are not sharp; you get hair flying in the wind, lace veils, clouds of smoke, and other partially transparent elements that are difficult to separate from the background.

Depth Determination

Once the elements' relative distances are known, we must assign absolute distances from the camera. Of course, this step can be done manually, but intelligent software can already be very effective, as it will use a lot of depth cues that the human brain also uses for this purpose, *e.g.*, prior knowledge of objects, relative sizes of several identical objects, etc. For example, a human head in close-up will obviously be near and a human figure with a quarter of the screen height is around ten meters away, a car a few pixels in length will be in the distant background, and the sky will be considered to be at infinity.

An automatic algorithm first looks for a horizon, as outdoor scenes are frequent. It then applies a first estimate: the sky is at infinity, while the ground approaches linearly from the horizon to the foot of the camera. All possible depth cues are explored and used in turn, either automatically or manually.

- If the camera has a limited depth of field, blur can be used to determine which elements are the center of attention and therefore located near the screen plane.
- Perspective, vanishing lines, and vanishing points are very useful for locating the relative depth of buildings, roads, edges of sports field, the interior of a house, etc.
- For objects at an angle to the screen plane, we determine a depth gradient rather than a depth value. This is typically the case of the ground or walls.
- The movements in very dynamic scenes are used to determine the distance of an object of known size such as a vehicle, balloon, or character. For example, if the position of the ground is already determined, moving characters will be located at the distance where their feet touch the ground.

At this point, we already have a depth map made of several planes that are recognizable by their different brightness values.

Spatial Propagation

Each element of the image must now be refined. The elements are generally copied to several layers, similar to Photoshop layers, and the depth of each layer is refined by imposing a brightness variation in the depth map to give it some roundness. Again, monocular depth cues, such as variations of light and shadow to determine the roundness of a face, help refine the results.

On the left, a depth map created in two independent layers: forest and house. On the right, the same depth map after a shift in the house's gray values in order to match the gray shades of the house's foundations and the forest floor. Once this is done, the house appears to be anchored to the ground.

Prior knowledge of objects and characters is also very important. It will give roundness to faces or balls just by drawing them on the depth map by hand or by creating computer models of these elements, from which it will be easy to extract the depth map. As a rule, this operation is supervised and corrected manually. Thus, the operator ensures that the characters are firmly on the ground by giving their feet the same depth as the ground on which they are placed (and thus the same shade of gray in the depth map).

After this stage, the keyframe has its perfect depth map, and a first result can be viewed on screen without waiting for the next step.

Temporal Propagation

From validated depth maps of keyframes, a (more or less) simple interpolation creates the depth maps that correspond to all the intermediate images. Human validation is always welcome here, as a movement that seemed linear at first glance may not be, for some reason.

Right Image Generation

All the original images are now accompanied by their depth maps. The only thing left to do is to shift each pixel of the original horizontally by a distance proportionate to its depth.

If the depth map is coded in 8 bits, it has 256 depth levels. This does not necessarily mean that level 128 corresponds to the screen plane or level 255 corresponds to infinity. Each sequence of a movie will have a depth budget to follow. This budget is determined by a stereographer who takes the size of the screen onto which the image will be projected and the amount of depth desired from an artistic point of view into account. The stereographer also sets the screen plane's value. For example, he may decide that the main actor's face is at the zero parallax. Then the gray level at this point on the depth map will be selected as the reference. As a depth budget often totals just a few pixels or tens of pixels at most behind the screen plane, the offsets will often be small and fractional. It is not uncommon to have to locate dozens of characters or objects at different distances using only a few pixels of difference in positive parallax.

Once the gray values of the depth map corresponding to the minimum, maximum, and zero parallaxes are determined, an automatic procedure applies the desired offset very easily to all the pixels of all the frames in the sequence. A 3D check of the whole sequence on a 3D screen is immediately possible.

On the left, a depth map using the full depth budget (gray levels from 0 to 255); on the right, a depth map pushed behind the screen (gray levels from 128 to 255); and the corresponding depth histogram under each map.

Hole Patching

The sometimes major shifts made in images' components are not without some problems. When two neighboring pixels move different distances, one of two things can occur: either an overlay or a hole. In the first case, the shifting algorithm, having started automatically with the most distant pixels, draws the closest pixel in front of it, which is normal, for near objects always hide more distant objects.

A hole is more disturbing, for in this case a part of the scene that was not present in the original view is revealed. Automatic methods simply copy down a neighboring pixel to fill the gap. They may also generate a pixel by averaging the colors of neighboring pixels. The most intelligent algorithms will search the next or previous frames in the sequence for the missing pixel and fall back on the previous methods only if this search fails. The most important thing when one creates part of an image from scratch is to avoid retinal rivalry: You must always check that the added pixel and its counterpart in the original image are not too different in brightness and color.

Remark
Some processes apply half the shift to each of the two images instead of just modifying the right image and keeping the left one undisturbed. With two weaker and more symmetrical changes, artifacts will be less extensive and corrections less visible.

Methods that create the depth map layer by layer simplify the problem, as each layer can be adjusted independently from the others.

None of these cases give perfect results, but a visual inspection – on a 3D display, of course – can detect the most embarrassing mistakes. Typically, this step is semi-manual. Careful correction by an experienced operator remains the best solution.

Practice of 2D to 3D Conversion

2D to 3D conversion comes in several variants depending on its end use and the type of content provided. Thus, feature films will require maximal quality, while television will require real-time conversion. Some movies may not have been designed for 3D (*Titanic*), whereas others were shot for the sole purpose of being converted (*Conan the Barbarian*). Conversion also has a role to play as a correction tool to restore a wrong depth budget in rushes shot in 3D with incorrect parameters.

Real-Time 2D to 3D Conversion

Despite the compromises necessary to get a 100% automatic process, it is possible to achieve what are considered acceptable results in some cases. Therefore, some algorithms based on the depth map method have been implemented in hardware as electronic components. The process is based on the aforementioned six steps, but all the manual adjustments are replaced by automated decisions based on heuristics, which merely select a solution from several possibilities according to the context.

One example is the 2D-to-3D conversion features included in some Samsung TV sets under license from Dynamic Digital Depth (*www.ddd.com*), which also markets the Tridef Photo Transformer software, and the JVC IF-2D3D1 device for the television broadcasting market.

Not all subjects are converted acceptably by these automatic methods. For example, vegetation usually offers such complex shapes and volumes that large errors are common. Subjects with simple geometry, such as a sports field, are more suitable.

The fact that the camera is in continuous motion, as during a tracking shot, improves the conversion because motion-based algorithms allow for better depth discrimination than the purely monoscopic clues that are available when shooting with a fixed camera.

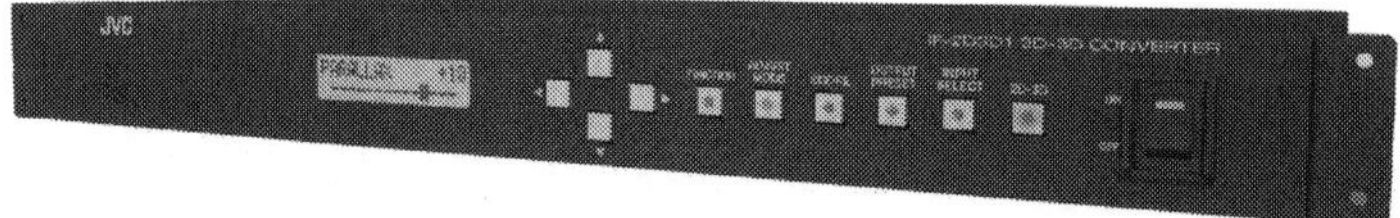

The JVC IF-2D3D1 rack-mounted processor does automatic 2D to 3D conversion.

Optimal Content for 2D to 3D Conversion

It is obvious that some video sequences are better suited for conversion than others. If you are forced to shoot a sequence with a monoscopic camera in order to convert it afterwards, here are some rules to follow to increase the chances of a successful conversion:

- Use long sequence shots. Shots under three seconds will not contain enough information to extract depth information effectively. Moreover, it takes several seconds for viewers to absorb the various levels of depth of a scene, so feel free to let your video proceed at a gentle pace.
- Light the whole scene properly, both foreground and background. Dark areas will lack details and be more difficult to convert.
- Ensure that all objects in the field are well textured and contrasted. Flat, uniform areas do not contain much depth information.
- Leave a maximum number of depth cues in the field: receding perspectives, geometric objects, and so on.
- Favor wide-angle lenses rather than telephoto lenses, which greatly reduce most depth cues.
- Keep the camera in constant, steady motion. Tracking and panoramic shots add a lot of information on perspective.
- Use depth of field to blur the background slightly. This will improve depth perception for foreground objects.

When a movie is shot with the idea of converting it into 3D in post, it is obvious that everything is done to simplify the conversion. Thus, the effects shots composited from several layers or integrating CGI elements are already separated into layers and, of course, CGI elements automatically have depth maps provided by the modeling software.

G-Force, from Walt Disney Pictures, is an example of partial conversion performed by the company In-Three (*www.in-three.com*), a pioneer in 2D to 3D conversion in Hollywood, using a process that it calls "Dimensionalization" and for which it has filed several patents. In *G-Force*, 2D shots were converted and then integrated into a computer-generated virtual world.

Advantages of 2D to 3D Conversion

2D to 3D conversion is often criticized for not being "real 3D." In fact, we must admit that purely automatic methods are unable to reach a high level of quality. However, with time and skilled operators, conversion can yield excellent results, up to the point where James Cameron has scheduled the conversion of Titanic, and nobody will question his commitment to quality. We must also recognize that some subjects are better suited to conversion than others, either because converting is easier or because circumstances render 3D shooting too difficult or impossible.

Some scenes that are difficult to shoot in 3D will benefit greatly from a semi-manual conversion, which, through its synthetic side, will prevent a series of unpleasant effects and problems. This is, for example, the case of close-ups and over-the-shoulder takes where distant backgrounds are visible. In contrast, opening sequences filmed in 3D with a long lens do not pose technical problems and if they were filmed in 2D would be very complicated to convert at an affordable cost, especially if they contain a lot of vegetation, snow, smoke, water droplets, or other numerous and complex elements.

2D to 3D Conversion to Rescue 3D Shoots

Conversion can also be an advantage in highly technical takes in which image elements must be removed from the scene in postproduction. In many action films, nets and safety ropes have to be removed from the image by rotoscoping, a well-known task done by postproduction studios. However, rotoscoping is extremely difficult to do in 3D shots because the two images must be corrected in perfect synchrony. It is therefore tempting to shoot the scene with one camera and then convert it to 3D after removing undesirable elements.

Depth map methods can also be used to adjust and balance the depth budget of a movie shot in or already converted to 3D. A stereographer can adjust the amount of 3D used in a scene or improve transitions between scenes to comply with the director's wishes and avoid any sources of audience discomfort.

Finally, when we discover important technical error in one of two views of a stereoscopic sequence late into postproduction, it may be more economical to discard the bad image and convert the good one than to reshoot the whole scene. James Cameron, for example, used conversion to correct problematic issues in some *Avatar* scenes.

3D Grading

3D grading – or depth grading – is the adjustment of the amount of depth, and thus parallax, available in a scene. After shooting, we grade colors so that takes offer the color shades that the director wants. The same goes for depth.

Hit (Horizontal Image Translation)

As we have seen before, the commonest 3D grading operation is horizontal translation, which is also called "HIT" (Horizontal Image Translation). By shifting the image horizontally from right to left, we are able to move the depth of the scene toward and away from the viewer.

Handling the Depth Budget

Further 3D grading operations are more complex, the trickiest one being to revise the depth budget by compressing or expanding it in absolute values. So, to increase or decrease the depth of all or part of a scene, we find ourselves facing the same problems as with 2D to 3D conversion, albeit with the invaluable advantage of having two images instead of one! The disparities between the two images are used to generate a depth map, which we will then be able to modify afterwards in various ways. The steps to follow are more or less identical to those of 2D to 3D conversion, but they are facilitated by having two source images: For example, holes may be filled by retrieving the missing pixels from one or the other image, according to which part of the image is missing.

Role of 3D Grading

3D grading is indispensable in a number of cases, such as,

- when the director wants to adjust a scene's depth for artistic reasons;
- when the transition between two successive scenes requires it;
- to adapt the parallax budget to the size of a projection screen not originally planned, whether this means reducing it to avoid divergences on a large screen or amplifying it for TV or Blu-ray versions; and
- to reduce the parallax budget in some headache-inducing cases that have not been solved during filming. This may be the case during an adaptation for young audiences, since children have a smaller mean interocular distance.

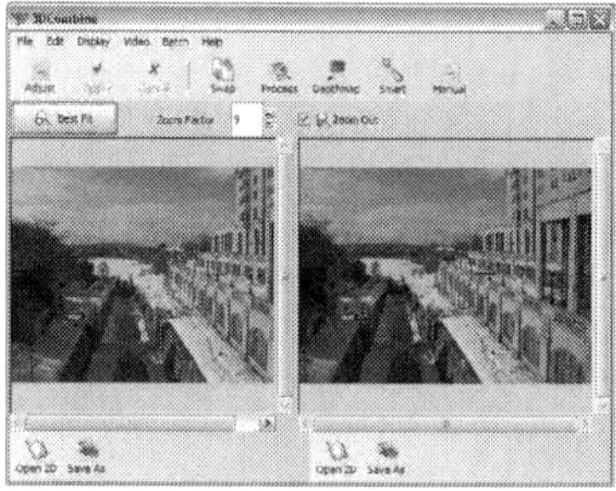

3DCombine software will be efficient only if you craft a very precise depth map manually.

2D-to-3D Conversion Software

Several 2D-to-3D conversion software packages, all based on the depth map method, are available to the general public. They usually leave the user the essential task of creating the depth map, *i.e.*, the hardest part of the operation. This kind of software, which is pretty cheap or even free in some cases, lets you try conversion on a handful of pictures. The hole-filling step is automatic and carried out with varying degrees of success. We can mention, as examples,

- 3DCombine (semi-automatic 2D to 3D photo conversion, $50): *www.3dcombine.com*; and
- DDD TriDef Photo Transformer (2D to 3D photo conversion, $50): *www.tridef.com/photo/transformer/overview.html.*

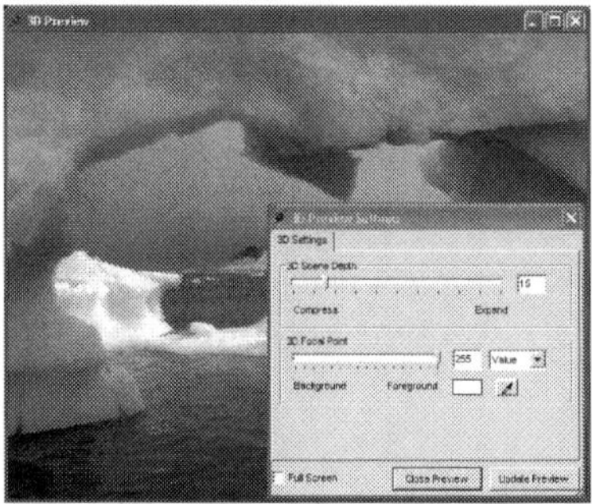

TriDef Photo Transformer is also based on the semi-automatic creation of a depth map.

Stand-alone conversion software is not the only alternative: There are also online services, such as the one from YouTube, that let amateurs convert their uploaded 2D videos to 3D automatically. Once it is processed – this can take several minutes or even hours – the video will then be available in 3D. The quality of the conversion will depend on many factors, such as the source's resolution and the number and kinds of available depth cues. Scenes with many camera movements are converted better, as the automatic algorithms receive more depth information.

6 Postproduction

How do we edit 3D movies?

How do we correct stereoscopic errors in postproduction?

Are rotoscoping and 3D image editing really impossible?

What are the best 3D postproduction tools?

Is it possible to add closed captions to a 3D movie?

How can we validate the depth of a 3D movie and be sure it won't cause headaches?

Movie processing in a studio includes all the steps after shooting up to and in-cluding preparing the final images for distribution to their users. This sequence of operations is called the *workflow.* The workflow's complexity will depend on the goal: feature film, TV series, webcast, amateur short film, and so on. Despite big differences in complexity and quality, several important constants emerge in 3D content production. First, producing in 3D means producing twice as many images, resulting in higher hardware and software resource use. A simple 2DHD content can be handled with a 50 Mbits/s stream. In 3D, the required flow rate of 100 Mbits/s will be handled by the I/O interfaces, disk drives, monitors, etc. And for cinema, where one is working with uncompressed images of at least 2K resolution, the data rate going through the hard drives is around 1,000 megabytes/second when editing two 2K DPX streams. Major productions require even higher rates and more often working on 4K images

The data rate, which is already doubled for 3D, is quadrupled when one goes from 2K to 4K, requiring the whole chain to accept a rate of about 3.6 GBytes/s. As for storage capacity, the 3,000 Gigabytes required for a 2D 2K production will increase to 24,000 Gigabytes in 3D 4K! We then need multiple disks in RAID configuration, which will provide greater speed and greater capacity. For large productions, RAID-dedicated servers provide the required performance; they will be connected to the workstations through a high-speed network (SAN, 10-Gigabit Ethernet, SAS, or Fibre Channel).

Systems must obviously use the most powerful multi-core processors and memory as large as possible. In the Windows world this means opting for a 64-bit system, the only one to use memory larger than 4 GB. The graphics card will also be powerful, especially if active glasses calling for an image refresh rate of 120 times per second are used. Of course, the monitors will also have to accept the 120 Hz rate required by active glasses. And if the work requires previewing on a professional 3D projector, the graphics card will be equipped with a dual HD-SDI output, reducing the range of possibilities to some of the most expensive ones.

Usual Image Sizes

Image formats are numerous and not yet fully standardized. Often the sizes of the images captured by the camera are not the same as the format of the film distributed. There are many reasons for this, *e.g.*, artistic choice, technical constraints, distribution and financial constraints, and so on.

The format of an image – or a pair of images for 3D – is defined by two variables, namely, the aspect ratio and size in pixels. The aspect ratio describes the extension of the image and is often expressed as a ratio of two numbers, such as 4:3 or 16:9. The image size is given in number of horizontal and vertical pixels, for example, 1920 × 1080. If the image aspect ratios measured in pixels and in meters are the same, the pixels are square. But this is not always the case and sometimes the image pixels are vertical or horizontal rectangles. To add to the confusion, some movies' formats are specified in number of pixels with an indication of the pixels' aspect ratio. So there is enough information provided to infer the aspect ratio of the projected image indirectly:

$$\text{Image aspect ratio} = (\text{width}/\text{height}) \times \text{pixel aspect ratio}$$

$$\text{Example (PAL D1): } (720/576) \times 1.07 = 1.33 \text{ (or 4:3)}$$

Most Common Image Formats

Usual name	Description	Width (in ixels)	Height (in pixels)	Image Aspect Ratio		Pixel Aspect Ratio
IMAX						
	IMAX 11K	11753	8772	1.33	4:3	1.00
	IMAX 5.5K	5868	4386	1.33	4:3	1.00
High Resolution						
	RED 4K RAW 16:9	4096	2304	1.77	16:9	1.00
	RED 4K RAW 2:1	4096	2048	2	2.0:1	1.00
4K	4K DCI Container	4096	2160	1.89	1.89:1	1.00
	2K Film Full Aperture (Apple ProRes 4444)	2048	1556	1.31	1.31:1	1.00
	RED 3K RAW 16:9	3072	1728	1.77	16:9	1.00
	RED 3K RAW 2:1	3072	1536	2	2.0:1	1.00
	RED 2K RAW 16:9	2048	1152	1.77	16:9	1.00
	RED 2K RAW 2:1	2048	1024	2	2.0:1	1.00
2K and HD Formats						
2K	2K DCI Container	2048	1080	1.89	1.89:1	1.00
2K Academic	2K DCI Flat	1998	1080	1.85	1.85:1	1.00
2K Flat	2K DCI Flat	2048	930	2.2	2.20:1	1.00
2K Scope	2K DCI CinemaScope	2048	858	2.39	2.39:1	1.00
HD	Full HD	1920	1080	1.77	16:9	1.00
HDV 1080i and HDV 1080p	HDV 1080	1440	1080	1.77	16:9	1.33
HDV 720p	HDV 720	1280	720	1.77	16:9	1.00
DVCPRO HD 1080i60	DVCPRO HD 1080i60	1280	1080	1.77	16:9	1.50
DVCPRO HD 1080i50	DVCPRO HD 1080i50	1440	1080	1.77	16:9	1.33
DVCPRO HD 720p	DVCPRO HD 720p	960	720	1.77	16:9	1.33
Low Resolution						
	NTSC	640	480	1.33	4:3	1.00
	NTSC DV Widescreen	720	480	1.8	1.80:1	1.20
DV	PAL DV/IMX	720	576	1.33	4:3	1.07
DV large	PAL DV Widescreen	768	576	1.86	1.86:1	1.42

Aspect Ratio and Image Size

The aspect ratio is the ratio between the height and width of the image seen by the viewer. In television, this ratio is expressed as the width divided by height, separated by a colon (L:H). The most widely used aspect ratios are 4:3 and 16:9, which correspond to PAL or NTSC TV and HDTV. In cinemas, many formats coexist and are defined usually by their aspect ratio relative to a unitary width. The aspect ratios of 1.85 and 2.39 correspond to the academic and Cinemascope formats, respectively.

With 3D, the 1.89 format, which corresponds to 2048 × 1080 pixels, is the one that uses the entire surface of the active element of the projector and therefore produces the best brightness.

The aspect ratio of the image on its support, whether physical or digital, is not always the same as the displayed image, for the pixels themselves can be rectangles or of varying widths. Fortunately, in today's full-digital world, square pixels have become the norm almost everywhere. The only notable exceptions are HDV and DVCPRO formats.

Often, for artistic reasons, the director decides to use the academic format 1.85. If the camera records a squarer format than the desired image, for example 1.37, horizontal bands above and below the image will be removed in postproduction.

The best final quality will always be obtained if one keeps the full camera resolution up to the last postproduction step. However, for various reasons, this is not always done. We must then aim to do the most destructive operations before any resizing. Cropping and horizontal offsets (HIT) will degrade the image much less if they are carried out on the full-resolution images.

Professional Grading Workstations

These high-end computers use multiple CPUs, multiple powerful GPUs and large RAID disk arrays to work on original content in 2K, 4K, or even exceptionally, in 8K. These include, for example, (in alphabetical order):

- Baselight, from FilmLight (www.filmlight.ltd.uk);
- Da Vinci Resolve, from BlackMagic (www.blackmagic-design.com);
- Mistika, from SGO (www.sgo.es);
- Pablo, from Quantel (www.quantel.com);
- Scratch, from Assimilate (www.assimilateinc.com); and
- Speedgrade XR, from Iridas (www.iridas.com).

These workstations are optimized for working on an interconnecting network. They usually share files through very high throughput connections such as Fibre Channel SAN (Storage Area Networks).

Professional 4K workstations are used only for finishing and color-grading tasks, when images of original size and resolution are required. Shot selection and editing are usually performed on proxies, *i.e.*, smaller versions of the original images. The lists of operations to be performed (*edit lists*) are established with Final Cut Pro or other editing software. Only at the end of the chain are operations in the edit list, often including color grading, applied to the original images. What is more, this is done in a single step in order to minimize handling of the huge and valuable original files.

Postproduction and 3D

Working on a 3D movie means not only handling twice as much information, but also adding a number of specific steps, such as clip matching and 3D validation, to the workflow. It also involves creative editing in stereoscopic depth budget, depth transitions, and overlays, as well as positioning subtitles carefully at the correct depth.

Each studio organizes its own workflow according to deadlines and production requirements, while coping with a budget that is almost always fixed beforehand. We can divide the postproduction workflows into two main categories: direct image handling and processing with proxies. Both methods are presented in the two simplified diagrams below.

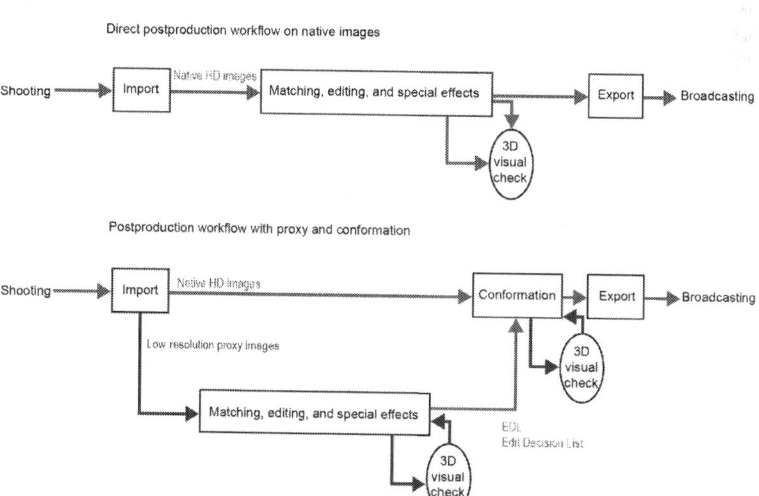

Two types of workflow coexist: with proxies for large productions and with direct processing (closer to live editing) for short-cycle productions.

At the end of postproduction, images and sound are combined in the so-called "DI" (Digital Intermediate), which contains the finished film in its best quality. The distribution chain will be responsible for reducing the amount of DI data to the minimum necessary required by each involved channel (DCP 2K or 4K for theater distribution, AVCHD for TV broadcasting, Blu-ray Disc or DVD for sales and rental stores, etc.).

Direct Postproduction

Once the rushes are imported and checked, the *workflow* is dedicated to special effects and editing. Workstations manipulate the images in their native formats directly. Editing software, also known as *edit suites*, fortunately comes increasingly with more and more 3D-specific features. When this is not the case, plug-ins and other third-party software add-ons improve the range of available tools. The most common editing suites are Apple Final Cut Pro and Avid DS. The stereoscopic add-on of choice for Final Cut is Neo3D from GoPro-CineForm. The 3D add-on tool for Avid is free and called MetaFuze.

Postproduction with Proxy Files

In the case of big productions such as blockbuster feature films, data size is such that native file editing is very detrimental. So, low resolution proxy copies of the original are used for editing. All editing operations are recorded and then applied to the original images in an offline rendering step.

The list of recorded operations is often called the *edit list* or EDL (edit decision list). It is quite common to edit – or at least to do a first draft of the edit – in 2D, as a non-stereoscopic film. The 3D check and adjustment are done in a second round, and only on the takes selected during the first edit. Color grading and 3D corrections are performed simultaneously on a professional editing station (see box above) powerful enough to handle the huge data flow of native-size stereoscopic images.

Importing and Matching

The most expensive sentence used in the cinema industry, "We will correct it in post," takes effect as of the first step, when images are received from the shooting stage. Strictness is always mandatory during this importing step, and even more so in 3D.

File Names

The left and right files of each imported stereoscopic pair must be named identically with "-L" and "-R" suffixes to distinguish the left one from the right one. If the files are already named correctly, it is mandatory to check on a 3D display that no switching has occurred. Indeed, inverting two files afterwards will become difficult, if not impossible, after a series of destructive handling steps such as overlays, fades, cropping, etc.

Metadata

Ideally, the various tools used in a workflow should keep all the ancillary data accompanying the images, *i.e.*, the metadata, intact. Unfortunately this is not always the case. So, to avoid many problems, we create a database in which we save all metadata as soon as they become available. Thus, as the footage is being imported, the metadata from the shooting location (cameras' convergence angle and interaxial distance, aperture, zoom, focus, position, and orientation of the camera) are saved. A number of devices insert this type of data in the images but, alas, many programs ignore them at the very next step. After a half-dozen steps, the workflow has lost track of most of the information painfully inserted at the beginning; hence the advantages of storing them in a separate database that stays accessible in all workflow steps. Such data are precious, as all shooting conditions, including convergence and interaxial distance of the cameras, are very important parameters when you want to mix images from different sources in postproduction, *e.g.*, an overlay of two takes, addition of CGI elements, etc.

Matching

Depending on the matching quality of the image pair at shooting time, the amount of corrections required will be large or small. The base rule is that between the left and right images of a pair, the only permitted differences are horizontal parallax shifts. All other discrepancies must be removed before any edit. Therefore we will correct:

- geometric disparities: zoom, rotation, and vertical position differences;
- colorimetry disparities: exposure, gamma, gain, and contrast;
- retinal rivalry: objects, reflections, light sparkles, etc., present in a single image or present in different places in the two images; and
- synchronization.

Most rivalries are corrected using purpose-built tools such as the CineForm FirstLight add-on for Final Cut Pro or just the free utility StereoMovie Maker.

In some cases, especially when significant airbrushing is anticipated, we had better check whether another take may be usable, as touching up a single image is tricky and may introduce other unforeseen rivalries. Painting a pair of stereoscopic images is almost impossible to get right in a reasonable amount of time without using specific software. Fortunately, such tools are beginning to appear!

Color Grading

Grading – or color adjustment – can be done independently from color disparity corrections. However, since these operations are in the hands of the same operators and performed on the same machines, they are often coupled.

Synchronization

Left and right shot synchronization is theoretically perfect when the cameras are time-locked by a control signal called "genlock." When cameras are not synchronized at shooting time, time alignment is required. If both shots are accompanied by a soundtrack, you can superimpose the waveforms of the audio signals to check for any synchronization error. Otherwise, locate in the left shot a fast motion, such as a clapboard closing at the beginning of the take, then look for it in the corresponding right shot and align the two shots. The accuracy will be of course less precise, with the risk of a half-image shift in the worst case. Here again, various software packages can do the job automatically.

Storage

Once both shots are matched, we can save them in the trim bin. With a specific 3D codec, such as CineForm Active Metadata, which is used by the combined Final Cut and Neo3D, the pair of shots is mixed in a single file. With less 3D-savvy editing software, use the least destructive codec allowed by the limitations of the available hardware; then mark the left and right files clearly. In addition to names ending in "-L" and "-R," we can also categorize the clips in various ways. For example, with Edius Grass Valley, the red color is attributed to the left shots' icons and cyan is associated with the right ones.

Import of 3D Sources for Traditional Software

If you use standard editing software that doesn't have any specific 3D features, such as Adobe Premiere or Grass Valley Edius, file import, matching and correction will not be easy. Fortunately a free tool, *StereoMovie Maker*, lets you make basic adjustments at the beginning of the postproduction step without investing a single dollar! This software (*stereo.jpn.org*), written by Masuji Suto from Japan, is a must for amateurs. StereoMovie Maker will perform, shot by shot, most of the timing, geometry, and color corrections required before editing. StereoMovie Maker accepts as input separate left and right clips from any pair of cameras, or avi files coded with time interleaving from some 3D camcorders and cameras, such as Fuji W1 and W3. It supports many output formats, including all avi formats readable by your PC, as it uses the same codecs as Window's MediaPlayer. It saves the corrected files as two separate left and right clips or as a single side-by-side clip. Visualization during work can be done in anaglyph or with Nvidia's 3D Vision kit and a suitable120 Hz 3D monitor.

Remark
StereoMovie Maker is easy to download from the author's web site (stereo.jpn.org). The English and Japanese versions are the most complete ones. It is best to avoid the French version, which is updated less often.

Both the left and right files that are exported by StereoMovie Maker can then be edited on traditional 2D software in multitrack mode, one track being dedicated to the left eye and another one to the right eye. Care should be taken to perform all edit operations on both tracks simultaneously!

A screenshot from StereoMovie Maker

Check Your 3D on a 3D Screen

Checking content in all steps in a production workflow is essential. For stereoscopic production, the 3D check is as important as, if not more important than, the color check. In some cases, checking depth in anaglyph mode may be sufficient, but more often we will favor the use of a workstation fitted with a good full-color stereoscopic display. One of the easiest methods is to connect a 3DTV or a 3D projector through the HDMI output of the PC's graphics card. You can also use a 120 Hz screen connected by DVI or DisplayPort and Nvidia's 3D Vision system. The choice will depend mostly on the display options supported by the software you are using.

As 3D quality is affected by screen size, you must also check the images under real conditions far more often than you would for a conventional 2D edit. Conditions planned for the end user must be reproduced as well as possible, using a large projection screen for a movie and a large 3DTV set for broadcast content. If the same 3D content is distributed through several channels, it must always be tested on the largest targeted screen.

The control monitor should be calibrated carefully, taking the type of glasses used into account. In most cases, colorimetry will be adjusted specifically for this screen to provide a look as similar to the final render as possible. This adjustment is done nondestructively using a specific3D LUT (Look-up Table).

On PC, Nvidia Quadro graphics boards offer the ability to display 3D in a window – and not only in full screen – with some software. On the Mac, boards such as the AJA Kona 3 and the BlackMagic Decklink HD Extreme 3D offer directly stereoscopic HD outputs.

Editing, Compositing, and Rotoscoping

Editing a stereoscopic image pair is fraught with danger. Indeed, any involuntary difference between the two images introduces either a retinal rivalry or a depth error. The smallest detail of an image must be changed in an almost identical way from one image to the next, but not completely, because the parallax, that is, the lateral offset of each pixel, depends on its depth, which also changes with time. In practice, manual painting operations are virtually impossible if quality is required. Very few software packages can do a decent job of painting or rotoscoping in 3D.

Silhouette, a professional program from the SFX company, and StereoPaint ESP, an inexpensive software tool written by Karl Lilje of Zenratai, are the best known among them.

Silhouette software at work

Silhouette is a professional program offering many features, such as semi-automatic clipping of hair, smoke, and reflections. In painting, it offers several special brushes with blur effects and color correction. The filters can be animated and configured using keyframes. The program is compatible with both types of workflow, *i.e.*, direct and with proxies. In both cases, the results are compatible with all major professional edit suites.

StereoPaint is a great program, especially given its cost ($30 for the full version), which allows for consistent touch-ups of image pairs or sequences of images pairs in various formats (.jpg, .png, .bmp, .tiff) and unlimited size. It offers several brushes, filters, and gradients; facilitates translation and rotation; and can use a graphics tablet as an input interface. Although it does not offer the same level of service as Silhouette, StereoPaint saves rushes degraded by an inappropriate reflection, a microphone in the field, and similar errors.

3D Painting and Rotoscoping Programs

Software Name	Operating System	Editor and Web Link	Use	Price
Silhouette	Windows, Mac, Linux	SFX, *www.silhouettefx.com*	Professional painting and rotoscoping	US$1,000
StereoPaint	Windows	Karl Lilje, South Africa *zenratai.com/software*	Painting	US$30

StereoPaint software at work

After cropping, inlaying and overlaying real or CGI images are essential steps to create special effects. If the above tools are able to remove a security cable in a stunt scene, even more powerful solutions will be required to overlay real or synthetic flames, explosions, and vehicles debris for an action scene. Here again, conventional 2D tools fall short of flawless results. A champion in this category is the NUKEX software from The Foundry (a UK firm: *www.thefoundry.co.uk*), along with various plug-ins from the same editor for the best known editing programs such as Adobe After Effects and Apple Final Cut Pro.

NUKE is well known for its compositing ability. In addition, its ability to slip seamlessly into existing workflows on Mac, Windows, and Linux makes it a tool of choice. Its variation, NUKEX, is dedicated to stereoscopy and includes automatic 3D camera tracking and a depth map generator. Importing and CGI object overlays are done very easily, with matching of perspectives and depths in a true 3D workspace where objects can be moved in the X, Y, and Z planes. Final rendering can be done with Pixar RenderMan Pro, a guarantee of quality. With its depth maps, NUKEX is able to work on more than two views and hence to achieve special effects for eight- and nine-view autostereoscopic screens, if necessary.

3D camera tracking by NUKEX is able to stabilize a shot and overlay volumetric objects at precise depths.

Depth Control

The purpose of this book is not to explain in great detail how to edit a movie, but to explain the steps specific to 3D movie editing. We shall thus take a closer look at the editing step that is absent from the 2D workflow, namely, depth adjusting, which is also known as depth grading.

The minimum and maximum horizontal parallaxes present in an image are more useful measurements of depth than depth itself. Indeed, it is easy to measure the gap between the left and right images of the same object in pixels on the screen.

Maximum Positive Parallax

As noted above, positive parallax (for objects behind the screen plane) should definitely be limited to a value that depends on the largest screen on which the images will be projected. To understand this better, consider the example of a 20-meter-wide theater screen. With 2K images (approximately 2000 pixels), one pixel is one centimeter wide, so positive parallaxes can never exceed 6.5 pixels if we do not want the offset between images on the screen to exceed 65 mm. If a shot exceeds this value, we will have to bring the two images closer to each other to narrow the gap. In so doing, we see that the pixels to the left and right of the image have no correspondents in the other view. We must therefore enlarge both images slightly and then crop them. This HIT (Horizontal Image Translation) operation degrades the image quality to a variable extent and will therefore be done only when necessary. It would not be required if the cameras were calibrated correctly from the start, but...

101%

In some cases, we may decide to shoot with slightly too wide a frame in order to allow for some HIT flexibility in postproduction. An often-used compromise is to frame 1% too wide, giving a margin of 20 pixels on an HD image for later cropping in post. In this case, the clips are imported with a zoom of 101%. A margin of about 10 pixels on each edge of the image will be lost, but a left or right shift of just a few pixels should not entail any consequences for the picture. The same margin will also be used during some readjustments in the matching step.

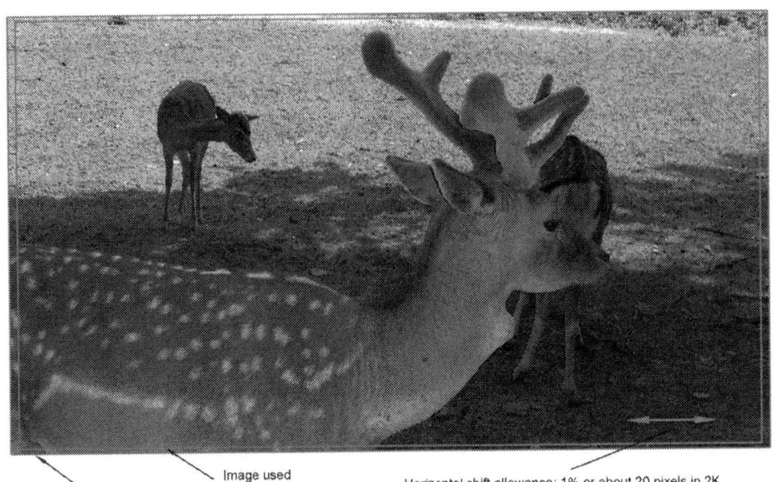

Principle of the 101% import, giving some horizontal reframing allowance

Fades and Depth

The edit's main goal is to concatenate a collection of takes. The transition between two shots will be a simple cut, a more or less quick fade, or some other transition effect. In all cases, the transition from one 3D shot to the next should not change the range of depths visible on the screen abruptly, especially in the area of the image on which the viewer is focusing at the transition time. The depth of that focus point is the most important one. For better visual comfort, care should be taken to maintain this value's continuity during transitions, and, as always, we will check that the transition is acceptable on the big screen.

Two possibilities to improve depth during a transition exist: You can either shift the entire depth budget of a shot or shift the depth only as you approach the transition. The latter is necessary when the depth of one of the shots is already imposed, for example by the transition from the preceding one. In the case of a depth shift during a shot, we must act very carefully, for if the depth shift effect is too slow, too fast, or too strong, it can be mistaken for a back or forth movement of the camera, which is not the desired effect.

Where is the focus point in a scene?

The point of interest at a givent instant is not unique and absolute. It may vary from one viewer to another according to age, cultural level, or just because the movie is watched for the first time or not. The center of attention is often the main character's face, but this may occasionally be an object that suddenly appeared in the foreground or an object moving very fast. Checking transitions is thus a rather subjective issue. When in doubt, ask a few people not yet aware of the sequence in question for their opinions.

Correcting depth jumps during transitions

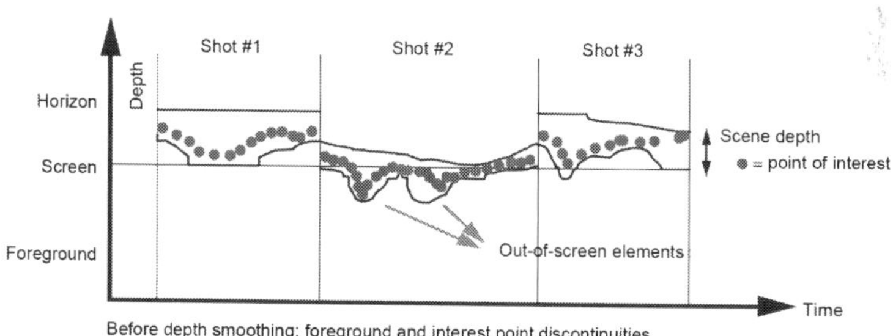

Before depth smoothing: foreground and interest point discontinuities

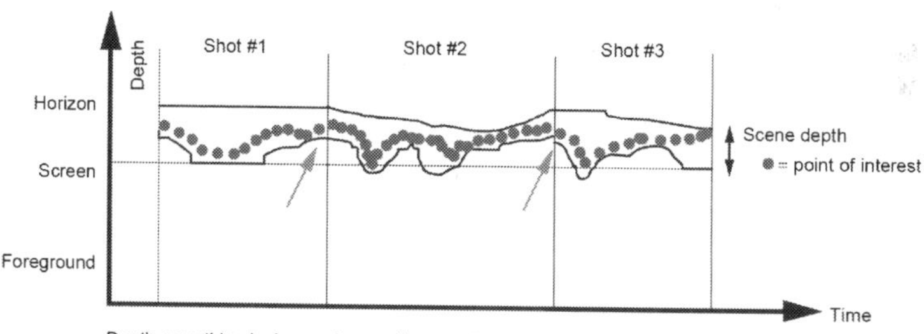

Depth smoothing by increasing positive parallax: interest point distance is now continuous.

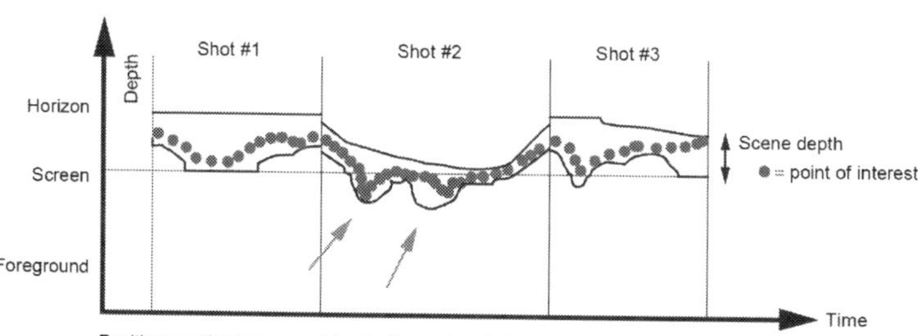

Positive parallax increase at beginning and end of shot 2 only: in-front-of-screen effects are less reduced than in the preceding case.

HIT and Projection Screen Size

At the end of editing, we will check the depth calibration according to the size of the screen on which the movie will be viewed. The larger the screen, the lower the allowed positive parallax. For each production, we set the maximum viewing screen size, from which we can then deduce the maximum positive parallax. For example, as noted above, a large movie screen will impose no more than 7 pixels (2K). HIT shifts added during editing to smooth depth transitions should heed this constraint. It is useful to review the whole movie at the end of the edit to make sure that no scene contains horizons that exceed the imposed threshold. If this is the case, we must correct the problematic shot with a finely tuned HIT that will reduce the gap between the two images.

Content intended for television or computer screens is much less likely to reach its maximum positive parallax limit, as the latter is much larger. In some cases, we can also separate the two images for the movie's entire duration to push the scene back and reduce out-of-screen effects. As always, we will check depth on a 3D display of the same size as the targeted one.

3D Editing

NUKE and Ocula

NUKE, from The Foundry, is a powerful compositing tool with a modular architecture that lets you write complex scripts assembling various operations on images serially or in parallel. It is used in large high-resolution productions and has adapted to the recent 3D cinema demand from the studios.

NUKE has a series of optional plug-ins called "Ocula" that can solve common stereoscopic problems effectively. James Cameron's *Avatar* is the first reference to come in mind, as it includes many shots that went through Ocula plug-ins to rectify color and geometry disparities. These plug-ins provide the following functions:

- vertical alignment and correction of keystone due to convergence;
- color matching, especially after shooting with a mirror rig, which can alter the colorimetry;
- display and correction of geometric disparities, including those due to cameras' convergence; this step is essential before compositing two images shot – or calculated – with different convergence values;

- creating a view interpolated between two existing ones, for example to reduce the interaxial distance or to create sequences for autostereoscopic multiview displays; this process is much more complex than a simple horizontal translation HIT, but it saves many shots that have been filmed with too large an interaxial distance;
- depth map creation from parallax; and
- creation of two images from an image and its depth map.

Neo3D, Stereo3D Toolbox, and Mac Editing Software

Apple's Mac computers are very popular in the major studios.

Running on Mac computers, the Final Cut Studio software suite from Apple is very complete. It accepts almost all existing formats (HDV, DVCPRO HD, AVC-Intra, XDCAM HD, XDCAM and XDCAM EX HD 422, HD Uncompressed 8-bit and 10-bit, etc.). The number of possible effects is endless and all provide very fine control. However, despite its power, it does not offer the 3D editing features that we are interested in. Fortunately, there are at least two programs that supplement it to fill this gap: Stereo3D Toolbox (*www.dashwood3d.com*) and Neo3D. Final Cut exports edit decision lists (EDLs) in all possible formats, so it is ideal for editing in workflows with proxies, with off-line conformation being performed on native images in high definition after lighter (from all points of view) editing done in Final Cut.

Stereo3D Toolbox

Tim Dashwood's Stereo3D Toolbox provides essential functions for editing and viewing in 3D using most of the best-known edit suites, including Final Cut Studio. Its price is on the same order of magnitude as that of Final Cut Studio (about $1,500 list price, although a Lite version is less than $99), but its unique features are such a time saver in post that the cost is quickly recouped.

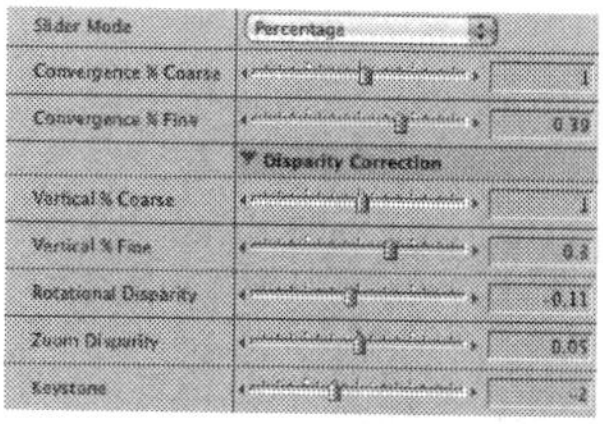

As seen in this figure, Stereo3D Toolbox handles disparities as fine as a 100th of a pixel!

Used by major Hollywood studios, Stereo3D Toolbox handles demultiplexing of 3D side-by-side, interlaced, and over-under formats, and exports in all the usual formats. Its shifts can be controlled up to the subpixel level; it generates rectangular or complex floating windows with very fine control on all parameters, and it even offers a deghosting option to reduce retinal rivalry. As it also generates depth maps, it is used to prepare 2D-to-3D conversions or effects and CGI object overlays.

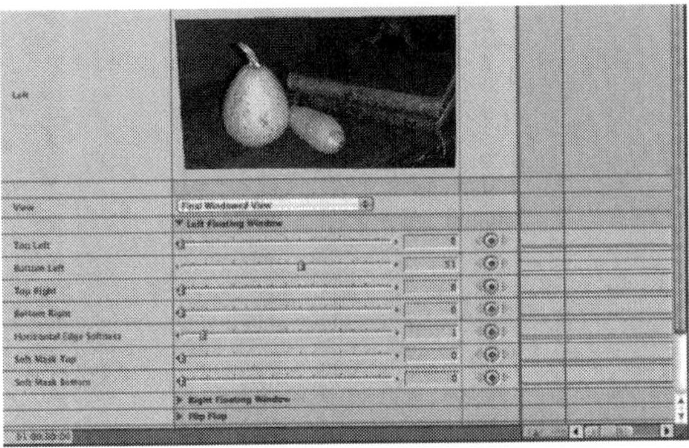

Nonrectangular floating window modification with Stereo3D Toolbox

Among the advantages of Stereo3D Toolbox, it should be noted that, as in Final Cut Studio, all parameters, such as the position of a floating window's corners or light levels, can be animated by keyframe. So setting some carefully chosen images correctly will suffice to adjust all the intermediate frames automatically.

Neo3D

Neo3D accepts all types of input files, but its primary advantage is its integration with the Silicon Imaging cameras, the SI 2K 3D, which record their images on a Mini Deck recorder that guarantees the left and right takes' synchronization and encodes with the Cineform Active Metadata codec. This is no coincidence, for Neo3D was also developed by GoPro-CineForm! It integrates seamlessly with Final Cut Studio and After Effects, Adobe Premiere, Avid Media Composer, and Sony Vegas Pro.

For visual checking, Neo3D provides an interlaced or side-by-side image that a suitable graphics card is able to transmit to a 3DTV through its HDMI interface for 3D previewing. A professional graphics card such as the AJA Kona 3 can even display the full HD resolution for each eye using the HD-SDI outputs. Neo3D is split into two applications: Remaster and FirstLight.

CineForm Active Metadata

This codec has the unique ability to encode the left and right images of a stereoscopic pair in a single frame that poses as a single image, although it contains two! Thus, applications using QuickTime or AVI files encoded with Active Metadata think they are handling a single clip and, in so doing, unwittingly keep them in sync throughout. This trick can be used when editing 3D movies with programs that are not 3D-aware. Files are exchanged simply by cutting and pasting between the edit application and Neo3D, which handles everything stereoscopic.

Remaster is the import tool that transcodes any video format into CineForm Active Metadata. The quality of this encoding is not open to question: It is almost lossless compression, retaining up to 12 bits of precision per component.

FirstLight is the adjustment program: It takes pairs of left and right clips and corrects exposure, gamma, contrast, etc. Its monitoring and checking tools are very comprehensive: waveform oscillograms and vectorscopes controling the colorimetry of both views may be superimposed to highlight any inconsistencies. Even self-correction is possible.

Neo3D displays RGB oscillograms for colorimetry fine tuning.

Sony Vegas Pro

Starting with version 10, Sony Vegas Pro offers stereoscopic features. You will find detailed information on the latest Vegas Pro release on the Sony website: *www.sonycreativesoftware.com/vegaspro.*

When importing, Vegas Pro lets you select two clips and lock them as a 3D Subclip. In the case of images from two separate cameras, the two clips can be synchronized easily by visual matching of the waveform soundtracks. Vegas Pro can also import already matching stereo pairs in a single interlaced, side-by-side, or over-under file. This is useful for sequences downloaded from YouTube or a 3DTV channel, which are almost always presented side by side. Vegas transforms these clips directly into two separate ones before making a 3D subclip. It also imports still stereoscopic

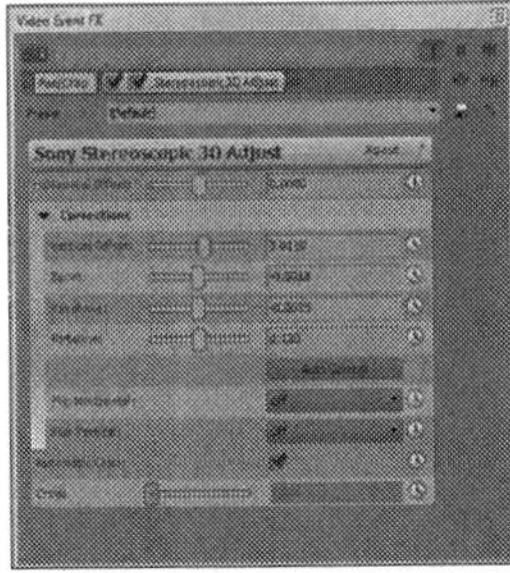

The Stereoscopic 3D Adjust dialog box from Sony Vegas Pro

3D images in jps, pns, and mpo format. Vegas Pro also supports natively 3D clips encoded with the CineForm Neo3D codec. Once a clip is imported, the matching is adjusted with the *Stereoscopic 3D Adjust* effect. The *Horizontal Offset* slider will be used for depth tuning. An AutoCorrect button can be used to kill vertical shifts, as well as trapeze, zoom, and rotation errors, automatically.

3D preview is possible in red/cyan anaglyph, on a 3D TV, or on a computer screen with Nvidia 3D Vision or 3D Vision Pro shutter glasses. In the latter case, as the application runs in windowed mode and not full screen, you must have a Nvidia Quadro graphics board. All editing operations can be done with 3D checking directly in the 3D window. Multilayer compositing operations are possible with easy depth positioning of each layer.

At the output, you can generate two separate files or views combined in side-by-side, interleaved, or over-under modes. A special option allows you to burn a Blu-ray disc with a side-by-side image or even a true 3D Blu-ray with MVC-3D coding (starting with release 10.0d) directly. If Neo3D is installed on your computer, CineForm formats will be available at both input and output.

Vegas Pro costs about $600, but a smaller version, Movie Studio HD Platinum, is available for $130 and retains almost all the 3D features.

Magix Video Deluxe

Magix Movie Edit Pro MX (*www.magix.com*) is a typical consumer editing software package priced at a around $70 with a very good price/quality ratio. All the usual import, video and sound editing, effects, and export features are present, including a menu authoring feature for DVDs and Blu-ray discs. A more complete but pricier version (Video Pro X4, *www.videopro-x.com*) adds hardware acceleration and support for the Blackmagic output breakout boxes and extended format support, including ProRes, MVC, and DnxHD, for around $400. On the stereoscopic side, the Pro X4 version adds improved perspective correction features and better colors in anaglyph mode, but the most important gains come from the speedier handling of MVC files.

The Magix Movie Edit Pro MX Plus interface with the 3D Stereo Align dialog box

3D Movie Subtitling

Audiovisual media and broadcasting in particular are increasingly calling for subtitling. 3D subtitle standardization and automation are, however, still in their infancy.

As we have seen, the depth of an object is given by an appropriate parallax shift. Subtitles are artificial elements added to the image, but they are not really part of the image. They should be read but not seen, at least not consciously, so as not to disturb the scene. For this, they must appear for the shortest possible duration, be very concise, be very readable, and not interfere visually with the main subject. For example, putting a subtitle in front of a main character's eyes is forbidden, even if it is popping out of the frame from below.

It is difficult to subtitle a film without disturbing the supension of disbelief for the viewer. We should therefore favor dubbing or place the subtitles in an area with few depth cues, such as on a black frame. Some even advocate writing them in the black bars underneath the film, when the aspect ratio allows this. But this is of course not always possible or desirable. For example, in James Cameron's *Avatar*, the Na'vi dialogues are subtitled, even in the original movie. The example below shows that the number of words stays under ten and the duration under five seconds.

Excerpt from the English subtitles file used on the Na'vi dialog in *Avatar* from James Cameron

```
43
01:19:32,490 --> 01:19:37,063
<i>You are now a son of the Omaticaya.</i>
44
01:19:38,185 --> 01:19:42,672
<i>You are member of the Na'vi people.</i>
45
01:27:03,577 --> 01:27:06,033
<i>Tsu'tey will lead our army.</i>
```

Display Time

A subtitle should appear just long enough for it to be read. However, in 3D, as the viewer focuses her/his gaze on different levels of depth, we must give her/him more time to accommodate her/his gaze on the subtitle, which is usually in front of the scene's focus point. This reconvergence takes between one and one and a half second, so the subtitles should be displayed longer than what is suggested by the good practice rules in 2D.

Positioning

Subtitles should appear in front of the scene. We will place them wisely to prevent them from crossing depth planes already occupied by foreground objects or characters.

In the case of subtitles superimposed on top of the scene, the depth positioning will by done by modifying the horizontal parallaxes of all the subtitle pixels simultaneously and symmetrically in both views, so that the subtitle stays centered on its original position. As a rule, if the whole scene is positioned behind the screen, the subtitle will remain in the screen plane. However, if elements of the scene in the subtitling area are out of the screen (which typically occurs 5% of the time), we will be forced to move the subtitle, giving it a negative parallax greater than that of the most protruding foreground object. Still, we will take care not to exceed a value after which the text becomes uncomfortable to read and, especially, to stay far enough away from the screen's edges. The limit values are strongly dependent on the end-user screen width, so TV captioning and movie subtitles can never be the same. A parallax set for a one-meter-wide TV set will be unacceptable on a twelve-meter-wide theater screen.

Foreground Positioning

The most obvious solution is to find the depth position of the farthest forward object in the subtitling area in the whole movie and set all the subtitles at a constant distance slightly lower than that. That way we will be sure that they are never occluded by any part of the scene.

Unfortunately, this is unacceptable in practice because, for example in a long sequence shot outdoors, the viewer might scan the landscape and simultaneously have to move her/his gaze back and forth between the far away scenery and the subtitle. The resulting eyestrain would be too great to be acceptable and, as the refocusing time is estimated to be about 1.5 second, nearly three seconds would be lost in total at each subtitle reading.

Dynamic Positioning

Avoiding obscuring significant parts of the scene and minimizing eyestrain are the two main reasons to adopt dynamic positioning of subtitles. The subtitle is each time positioned horizontally, vertically, and in depth. The default position, which should be used as often as possible, will be centered, at the bottom of the screen, and in the screen plane.

For perfect results, the adjustment will be made by an experienced captioner with good 3D experience and a stereographer will use a large enough 3D display to check duration and position, that there is no action or major depth cue occlusion, and that eyestrain stays acceptable.

Depth positioning cannot usually be evaluated on the pixel level. To stick as best as possible to the scene, parallax adjustments are often necessary. The software used must be able to manage subpixel lateral shifts (at least a quarter-pixel precision is required).

In 3D, even more than in 2D, we avoid subtitles during cuts because they are too much of an overload for the visual system.

Size

Dynamic positioning in depth leads to a logical but annoying side effect: When subtitles appear in the screen plane, they are interpreted as tangible objects with a precise size, as if they were present in the scene; but if we move them forward without changing their size in pixels, the brain will think the subtitles narrow gradually as they draw closer. If we are to avoid disturbing the viewer, we will have to vary their size, making them larger as they come closer and smaller again as we bring them back to, or even behind, the screen plane. Dynamic positioning is thus not as free in practice as it is in theory. Indeed, the size of the larger subtitles must remain limited and should never invade the screen.

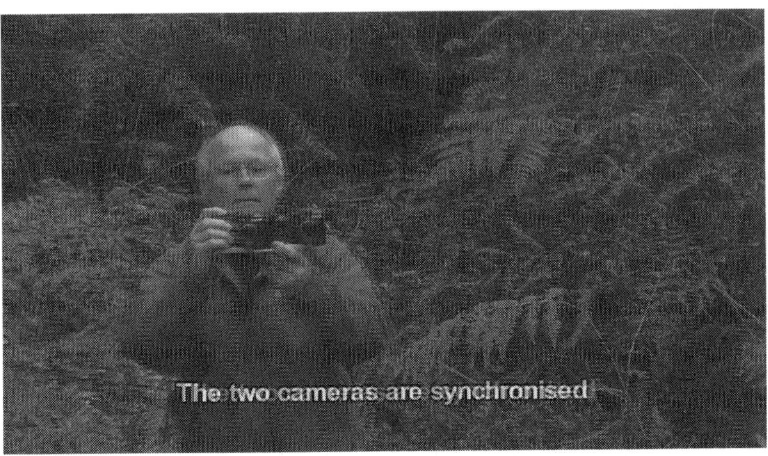

© C. Oury

Subtitle located just in front of the main character, itself a bit out-of-screen

Volume and Texture

Our interpretations of the positions and volumes of objects in a scene can be disrupted by the two-dimensional nature of subtitles. If the scene is rich in textured materials such as wood, cloth or stone, a subtitle that is in a solid color and strictly flat may break the suspension of disbelief.

We can try to give the subtitle a slight depth effect, such as emboss, and apply a light plaster-like texture. This will strengthen the depth cues and help the viewer to locate the subtitle in the three-dimensional space. This type of effect must obviously be applied very precisely and – as always – validated on a screen of appropriate size and in 3D.

This remark also applies to logo insertions, such as the channel name in the upper right corner of the image, a phenomenon that is used more and more in television.

Stereoscopy does not mix well with strong contrasts. A white subtitle on a very dark background may cause unwanted ghosts, especially if the film is slated for projection on the big screen with passive glasses. It is therefore necessary to manage the subtitles' contrast cleverly and to darken them if necessary when the background is particularly dark, as in night scenes.

3D Subtitling Software

Software with specific 3D features usually includes subtitle generation. When this is not the case, it is of course possible to make subtitles conventionally and to shift both instances slightly to the left and to the right to position them in depth. But a 3D subtitles-dedicated plug-in will save a lot of time and effort. The 3D ISP Title plug-in (*www.isp.co.jp*), for example, integrates seamlessly with Adobe Premiere CS5 or After Effects CS5. It allows for easy positioning adjustment, offers an anaglyph preview, and supports various 3DTV display modes, such as side by side and over-under.

The ISP 3D Title plug-in for After Effects allows easy insertion and adjustment of subtitles.

Subtitles' Distribution

When the subtitles are not printed directly in the image of the film, they are distributed through various channels, such as teletext, DVB, or in the form of raster images to be superimposed on the movie in the projector.

In the case of teletext for European television or US Closed Captions 608/708 for the United States, we are supposed to provide metadata such as the font name, position, size, and parallax, with the latter specified as a fraction of a pixel. This information can be encoded in the metadata stream, but this is currently neither a standard nor the subject of consensus among set-top box manufacturers. It is therefore unlikely that this pathway will be developed in the short term.

The DVB transmission standard for television provides subtitles for side-by-side content. The subtitles are then transmitted as bitmap images along with parallax information. There is not, to date, any agreement on how to implement the standard in detail and various interoperability issues have still to be resolved. Consequently, DVB subtitling is certainly promising but not yet mature.

Remark
In practice, standard recommendations are inadequate and the only method ensuring maximal quality is to burn subtitles in the image at postproduction time.

Moreover, while side-by-side formats appear to be manageable in the medium term, formats that interleave the left and right images in various other ways are prone to serious aliasing issues. Indeed, shifting subtitles by a tenth of a pixel requires image resampling on the display level, which is very difficult when the image is not stored in contiguous areas inside the device's memory. These formats are therefore not recommended if subtitles are being considered.

For cinema, things seem to be better but all is not perfect yet. Indeed, DCP, the distribution format used for digital movies, is defined in SMPTE DCI 428-7, a document that provides for the use of 3D movies subtitles, including a Z parameter giving the depth to which the subtitle should be overlaid on the image by the projector.

After television and cinema, the third medium using subtitles is Blu-ray. The latter is more likely the future medium of choice for a stereoscopic film transport channel to the home, and the variety of commonly offered languages inevitably requires the use of subtitles superimposed on the display by the player itself. The first release of the Blu-ray 3D standard exists, but the parts covering subtitles have not yet been published.

XML by Sony

Nevertheless, Sony Creative Software offers a standardized XML schema that provides the subtitle's depth in Z as a function of time, given as a timecode. This schema may be used in the future for digital cinema, DI (Digital Intermediate) files in postproduction, TV broadcasting, and Blu-ray discs. The professional application, which Sony calls Z Depth, generates metadata compatible with this proposal (*www.sonycreativesoftware.com/zdepth*).

Below is an XML file sample following Sony's proposed specification. Note that lines such as <Event TC="01:01:21.08" ZValue="-3.5"/> indicate how deep the subtitle should be at a given time (here at timecode 1 hour, 1 minute, 21 seconds and 8 frames); subtitles will have a negative parallax of 3.5 pixels. As the Blu-ray specification prohibits the use of shifts in fractions of a pixel, the value will be rounded to three before use. The fractional part will be used only for cinema applications.

```
<?xml version="1.0" encoding="UTF-8"?>
<Document xsi:noNamespaceSchemaLocation=»
zvalue.xsd» xmlns:xsi=»http://www.w3.org/2001/
XMLSchema-instance»>
  <Project>
          <Title>My Film Title</Title>
          <Content>feature</Content>
          <Version>Original</Version>
          <Resolution>2K</Resolution>
          <FrameRate>24</FrameRate>
          <DropFrame>NDF</DropFrame>
          <MediaNumber>123456</MediaNumber>
          <Revision>1</Revision>
          <Language Attribute=»inserts»>en-US</
Language>
          <FFOP>01:00:08.00</FFOP>
          <LFOP>01:11:17.22</LFOP>
          <Reel>1</Reel>
          <TotalReels>6</TotalReels>
          <ValueType>pixel</ValueType>
  </Project>
  <Events>
          <Event TC=»01:00:00.00» ZValue=»0»/>
          <Event TC=»01:01:21.08» ZValue=»-3.5»/>
          <Event TC=»01:01:35.13» ZValue=»3.8»/>
          <Event TC=»01:01:37.08» ZValue=»4»/>
          <Event TC=»01:01:59.12» ZValue=»8.4»/>
  </Events>
</Document>
```

Subtitling with Stereoscopic Player

Stereoscopic Player is able to overlay subtitles onto a stereoscopic movie at display time. However, it accepts subtitles in .sub, a stricly 2D format, only.

Currently, displaying subtitles is restricted to the over-under mode and to the "ffdshow MPEG-4 Video Decoder" codec. The properties dialog box of the ffdshow codec is located in the "File>Video>Filter>ffdshow Video Decoder>Details" menu. You can check the "Subtitles" and "Stereoscopic" boxes and specify at which parallax the subtitles should be displayed. A value of 2 pixels is usually the best choice. This value is unfortunately impossible to change dynamically; it thus applies to the entire movie.

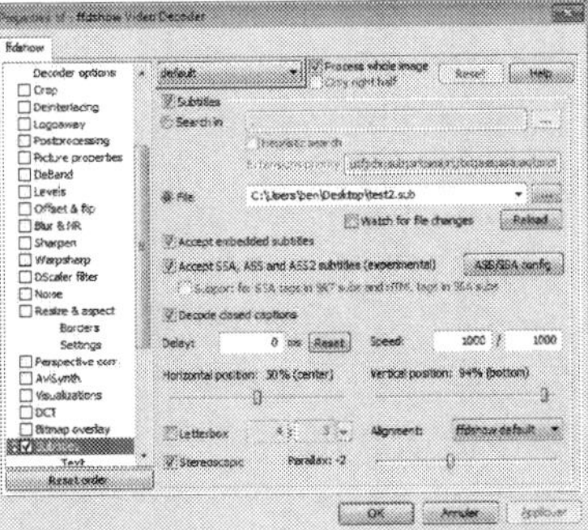

The stereoscopic Player dialog box gives the choice of displaying subtitles and select the parallax values for them.

Subtitle files using the "SubViewer 2.0" format (file extension .sub) are simple text files with no depth indication. This format was adopted by the DivX codec, widely used by Microsoft's Media-Player, and has become a must since its adoption by YouTube, thus giving foreign languages access to a large number of videos. This format is also extremely useful when retrieving subtitled video sequences on DVDs. In this case, the free SubRIP program can extract automatically all the subtitles from the disk in this format.

The SubViewer 2.0 syntax is simple. It includes an informative header, the keyline "[SUBTITLE]", a "[COLF]" line to declare the font, and then the subtitles, with two lines per subtitle: a first line with start and end timecodes separated by a comma and a second line with the subtitle text.

"Subtitle Workshop," which is free software from URUWorks (*www.urusoft.net*), is handy for subtitle creators. It supports over fifty different formats, including the .sub format.

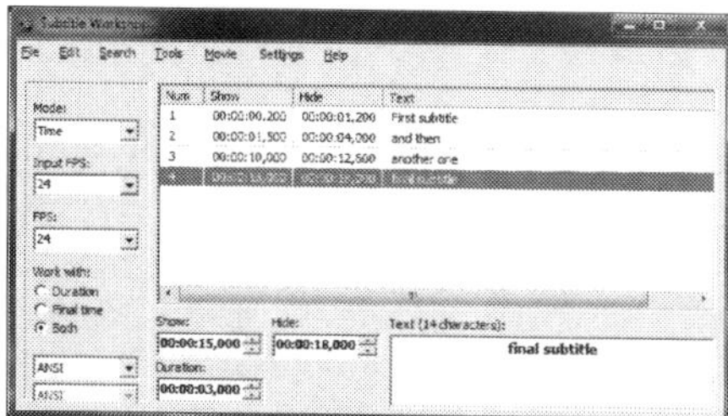

Subtitle Workshop in action: Data displayed on screen generated the file below.

Example of a .sub file:

```
[INFORMATION]
[TITLE]Test video with subtitles
[AUTHOR]Benoit Michel
[SOURCE]www.mywebsite.com
[PRG]
[FILEPATH]
[DELAY]0
[CD TRACK]0
[COMMENT]first subtitle
[END INFORMATION]
[SUBTITLE]
[COLF]&HFFFFFF,[STYLE]bd,[SIZE]18,[FONT]Arial Narrow
00:00:00.20,00:00:01.20
first subtitle
00:00:01.20,00:00:04.20
and then
00:00:10.20,00:00:11.20
another one
00:00:11.20,00:00:12.20
final subtitle
```

Sound Editing

At first glance, one might think that a film's sound editing was not influenced by the switch to 3D. Yet it is! The placement of a scene in depth, and especially that of the story's main characters, is of prime importance. This will require, wherever possible, a 5.1 surround mix giving depth to the soundstage. Stereo sound, which by definition is limited to the flat single plane of the screen, will struggle to restore a sound environment with sources located at various depths.

The Impossible Perfect Mix

For sequences located mainly behind the screen plane, changing the sound's location is not too important. However, when a character or any sound-emitting object jumps out of the screen, the sound source must be rendered in 3D space between the screen and the viewer. Imagine a speaking character centered and placed at 25% of the distance between screen and viewer. The corresponding sound source should be centered and located around the first 25% of the theater. In this case, the sound source will be correctly located for a viewer of the last row. But for a viewer sitting at the left end of one of the first rows, the sound will seem to come from the right and the effect will not seem at all natural.

In real situations, it is almost impossible to render proper surround sound sources in front of the plane of the screen to the entire audience. The ideal solution would be to send the sound mix to each individual viewer through headphones. To be quite complete, we should even add detection of the viewer's interocular distance, since, for a child with a 48 mm interocular distance, an object is located 25% closer than for an adult whose eyes are 65 mm apart!

Influence of Sound on Staging

The above limitations are important, and a poorly balanced soundtrack can ruin an entire production. The most obvious solution is to minimize the number of point sound sources in the volume between screen and viewer. A perfect example of creative staging is found in the animated feature *Monsters vs. Aliens* where, early in the film, a jokari ball goes back and forth in the out-of-screen space. The ball's impact on the racquet is loud and well localized, but, as it is placed exactly in the screen plane, the sound is correctly positioned for all spectators.

A sound source located in front of the screen coincides with the image only for viewers who are exactly centered in front of the

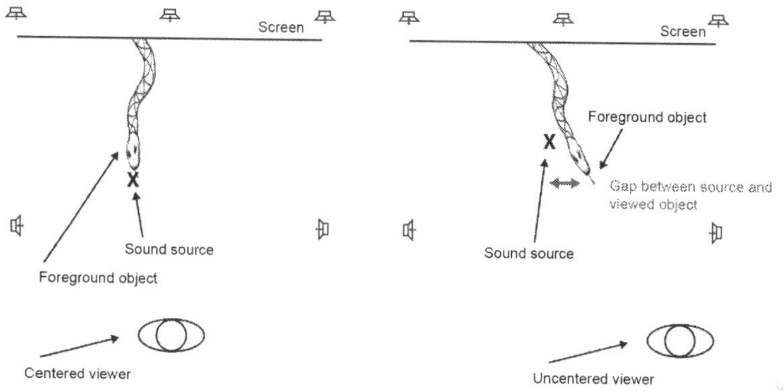

Sound sources and out-of-screen objets are not located in the same place for viewers not placed on the screen axis.

screen; the more eccentric the viewer's position, the greater the distance between the perceived sound source and out-of-screen object becomes. The stronger the pop-out effect, the more this defect is accentuated. Therefore, one should avoid placing very spatially-limited sound sources – such as punches, impacts, the shattering of glass, etc. – too far out of the screen plane.

In conclusion, it will be much easier to produce a soundtrack for a 3D movie if the main characters interact in a plane close to the screen and we avoid highly circumscribed sound sources in the volume in front of the screen. As usual, the balance between artistic creativity and technical constraints is a matter of taste.

Off-screen Sound Sources

Excellent spatialized sound editing can add a lot to a 3D movie. Once the pitfalls we have just studied are avoided, we should locate in space as many sound sources as possible in order to strengthen and complement visual depth cues. For example, a helicopter attack in an Avatar-like movie requires placing sound sources in the 5.1 surround mix well before the helicopter appears on the left edge of the screen and then keeping them in after it has flown off on the right side. We will strengthen the spatial location effect by passing the sound source from the rear left to front left just before it appears. Thus, based on hearing alone, viewers will expect the helicopter to appear in the correct place and their brains will already be unconsciously prepared to focus their eyes on this entry point. A good front-back mix is therefore not only useful, but may even help reduce visual discomfort by "setting the stage" before objects appear in the visual field.

Validation

Stereoscopic 3D needs validation at many control points in the production workflow: the shooting stage, postproduction, packaging, and distribution. A visual check is essential but automatic tools may find and locate accurately even the smallest issues.

Automatic Validation Check

Several stereoscopic analyzers are on the market and should be used on all productions, from the largest blockbuster to the smaller indie film, because if you don't double-check your 3D stereoscopic production, you are at risk of giving your audiences' eyes a really hard time. And that means only one thing: They may never watch a 3D movie again!

Software-assisted validation is never enough, but it can save a lot of time and money and protect you against big mistakes. Also, it is easy for a software tool to make sure that your parameters stay within imposed margins at all times, so that BSkyB and other distributors will not be able to argue about your depth budget overruns.

Stereoscopic analyzers such as STAN (STereoscopic ANalyzer), from the Fraunhofer Institute (*www.hhi.fraunhofer.de*), and Cel-Scope (*www.cel-soft.com*), in Germany, offer various histograms displaying the changes in 3D parameters, *i.e.*, depth budget, vertical disparities, and depth spectrograms, over time. Alarms can be set off when parameters reach some predefined level, such as maximum positive or negative disparity.

The Cel-Scope logging option adds comprehensive reporting and analysis functions to the basic analyzer's overall measurement and display facilities. A log report is a finished document showing the events, measurements, and system settings, along with time codes and optional thumbnail pictures from the media being evaluated.

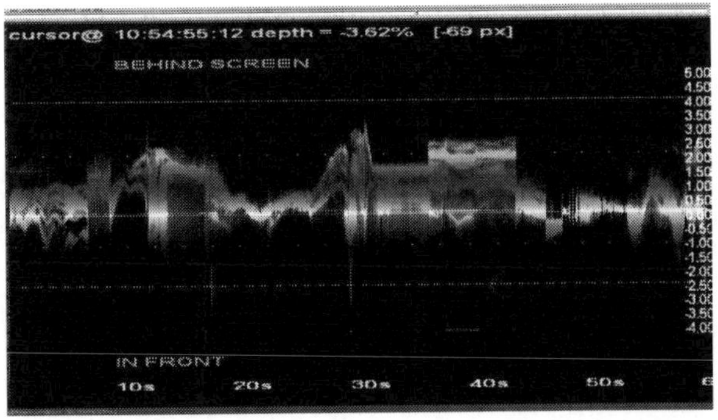

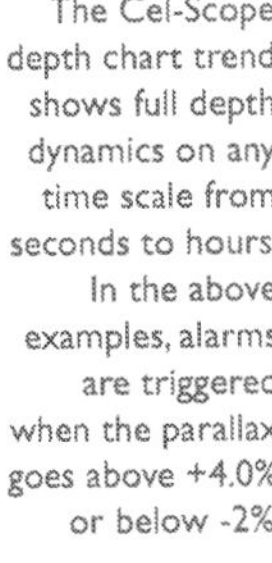
The Cel-Scope depth chart trend shows full depth dynamics on any time scale from seconds to hours. In the above examples, alarms are triggered when the parallax goes above +4.0% or below -2%.

STV (Stereographer Toolkit & Viewer, *www.touchedu.com/stv*) is a simpler but very cheap ($9.99) tool offering checking capabilities for a series of still images only. STV runs on iPad and checks stereo pairs up to 4K resolution.

Be careful!
Validating all the points outlined in the following table in a single vision pass is almost impossible for any human being, including experienced stereographers, and even automated software. For major productions, a double visual check supported by a dedicated validation program is required.

Visual Validation Check

A visual check in 3D on a large screen is essential. The viewing conditions must be cinema-like without parasitic lighting and sound equipment should be calibrated similarly to that in an end-user theater.

Technicolor provides a similar fifteen-point checklist and offers a certification scheme, called "Certifi3D," validating all shots of a production. Technicolor uses specific quality control software that determines the depth budget and creates a report containing all unwanted disparities.

Most Common Stereoscopic Errors and Corresponding Remedies

Stereoscopy Errors	Correction	Importance	Correction Difficulty
Synchronization error: whole number of frames mis-sync	Resynchronize	**	*
Synchronization error: fractional frame mis-sync	Redo the whole shot	**	***
Vertical disparities on the main subject	Realign	*	*
Vertical disparities in the background	Blur the background	*	*
Brightness and color inconsistencies	Redo color grading	*	*
Focus or depth of field disparities	Blur the sharper image	**	**
No 3D effect (the two images are identical)	Correct left and right image management	*	*
Pseudostereoscopy, inverted left and right images	Switch left and right images	***	*
Positive parallax greater than 65 mm on screen	Bring both left and right images closer through HIT, then reframe	***	*

Stereoscopy Errors	Correction	Importance	Correction Difficulty
Shallow 3D effect in some layers of a composition	Redo the compositing	*	**
Partial pseudostereoscopy: incorrect 3D in some layers of a composition, typically titles or subtitles	Redo the compositing	***	**
Large retinal rivalries: missing objects in a multi-layer composition.	Redo the compositing	***	**
Small retinal rivalries: incorrect pixels, flares, or optical aberration inconsistencies	Correct with StereoPaint	**	***
Unmatched lateral edges: incorrectly placed floating window	Move the floating window	***	*
Out-of-screen objects obscured by lateral screen edges	Add a floating window; reframe	***	*
Out-of-screen objects obscured at the top of the screen	Reduce brightness or contrast of the object; reframe	**	**
Out-of-screen objects obscured at the bottom of the screen	Reduce brightness or contrast of the object; reframe	*	**
Out-of-screen effects too strong	Move both left and right images farther apart through HIT, then reframe	**	*
Ghosting: A parasitic view of the left or right image is partly visible to the other eye	Reduce brightness or contrast in that area; for titles and subtitles, "decontrast" the text or move it elsewhere	**	**
Soundtrack: Stereophonic effects placed incorrectly left-to-right or sound sources too far on one side	Recenter and bring back the source towards the screen plane; redo the 5.1 mix	***	*
Soundtrack: Sound source placed incorrectly in depth (too far in front of or behind the screen)	Bring back the source towards the screen plane; redo the 5.1 mix	***	*

Stereoscopic Movie Transmission

7

How are live images recorded or broadcast by two cameras at the same time?

What formats are used for the storage and live transmission of 3D images?

How does one broadcast a match live in HD-3D?

Is a new decoder required to watch 3DTV feeds?

Is HDMI 1.4 required to display 3D content on a 3D TV set?

What is a 3D DCP?

How can I encode a 3D DCP to send my movie to a theater?

How can I burn a 3D Blu-ray disc?

Distribution Problems

Context

Barriers to the distribution of 3D movies from the production site to homes or concert halls are linked to the absence of recognized industry standards, the high cost of required changes in existing infrastructure, and technical incompatibilities between the various makes/types of equipment. Choosing the right formats for transmission depends on the quality required now and in the near future, but also on storage and transfer efficiency, compatibility with existing 2D equipment, and, finally, the costs of all the required changes. The main distribution channels are cable, satellite, DTT (Digital Terrestrial Television), and optical disks, a family in which only Blu-ray discs offer enough capacity.

Standards

International standards are therefore essential. Without them, there would be no mass distribution to movie theaters or individuals; with them, equipment interoperability will lead to the technology's rapid mainstreaming and lower costs. Ideally, standards should be global and established very early to keep one product from engulfing the market with a proprietary solution, leading to a standards war and, ultimately, delays in adoption by the greatest number. We remember the standards war for the videotape market in which VHS eventually prevailed over BetaMAX.

Fortunately, standards bodies began working in the digital stereoscopy field early on, and we can already count on the HDMI 1.4 standard for the connection between set-top boxes and televisions and the Blu-ray 3D disc for storing and transmitting feature films. Standards for real-time transmission, which is mandatory for television, are already on track.

Main Standards Organizations

Acronym	Organization	Target	3D Standards	Standard's Scope
SMPTE	Society of Motion Picture and Television Engineers	Cinema and broadcasting professionals	Digital Cinema Package (DCP)	2D and 3D masters for transmission to digital cinemas
			WG10E40 – 3D Home Master	3D master for transmission to the home
CEA	Consumer Electronics Association	Mass market equipment manufac-turers	WG7-R4.8 – CEA-861.1 Uncompressed video 3D Profiles	3D broadcast formats
			WG 16-R4.8 3D Eyewear	Infrared commands for active glasses
			CEA-708 Closed captioning	2D and 3D subtitles for TV
ATSC	Advanced Television Systems Committee	Digital TV, DTT and cable transmission in the USA	Ongoing work	3D TV in the USA

Acronym	Organization	Target	3D Standards	Standard's Scope
EBU	European Broadcasting Union	European broadcasters	Ongoing work	3D TV in Europe
ITU	International Telecommunication Union	Worldwide broadcasters	ITU-R BT.2160 Features of 3D TV	3D TV future planning divided in 3 generations called "Profiles"
			ITU-T H.264 Multiview Video Coding	MPEG-4/H.264 MVC (2nd generation 3D Profiles)
			ITU-T H.264 Scalable Video Coding	MPEG-4/H.264 SVC (3rd generation 3D Profiles)
DVB	Digital Video Broadcasting Project	Worldwide broadcasters	DVB-3DTV (DVB Document A154)	3D formats for broadcasting, including metadata and 3D subtitles
SCTE	Society of Cable Telecommunications Engineers	Cable TV operators (USA)	3DTV Content Encoding Profiles 3.0	3D formats for broadcasting
BDA	Blu-ray Disc Association	Blu-ray equipment manufacturers	3D spec	3D Blu-ray disc specification
ISO/IEC/ MPEG	International Standards Organization	digital broadcasting	ISO/IEC 23002-3 (MPEG-C Part 3)	3D formats, including 2D + depth and metadata
			ISO/IEC 14996-10 AVC with MVC extensions	Compression and transmission of MVC 2D + depth images
			Ongoing work	Software and methods to create MVC 2D + depth images
HDMI	HDMI Licensing	Mass market equipment manufacturers	HDMI 1.4a	Short-range transmission methods between sources and 3D TVs or projectors

Note: WG stands for *working group*; the goal of a WG is to prepare an as-yet-unpublished future standard.

Constraints

File transmission during production is subject to various constraints, above all that of preserving quality in all shooting and production phases, but also reducing the volume of data to transfer during transmission to end users. Reducing storage costs at intermediate steps calls for file compression. Lossless compression is less efficient than lossy compression and difficult compromises must be made at each step of the workflow.

This is true in 2D and 3D alike. However, with the advent of stereoscopy, the smallest compression artifacts, which are typically not identical for each of the two views, will become very visible because they introduce retinal rivalry. These annoying little details will be present to one eye and not the other, and our brains will interpret the difference as a depth anomaly.

To illustrate how precise the synchronization of the two image streams should be, note that a good half of viewers are only slightly disturbed by a 4:3 video image shown with a 16:9 aspect ratio or vice versa, where the error on pixel positions is 12.5%, but when they watch an HD-3D sports event, a wrong image shift of only five pixels between the two images makes the scene totally unbearable: It loses all its depth, gives you a headache, etc. Yet five pixels make up only 0.5% of the width of the image, or 25 times less than the aspect ratio error that is deemed simply a little bothersome by half the audience!

Various forms of transmission are involved in many places in the 3D workflow. The diagram below illustrates the main transmission

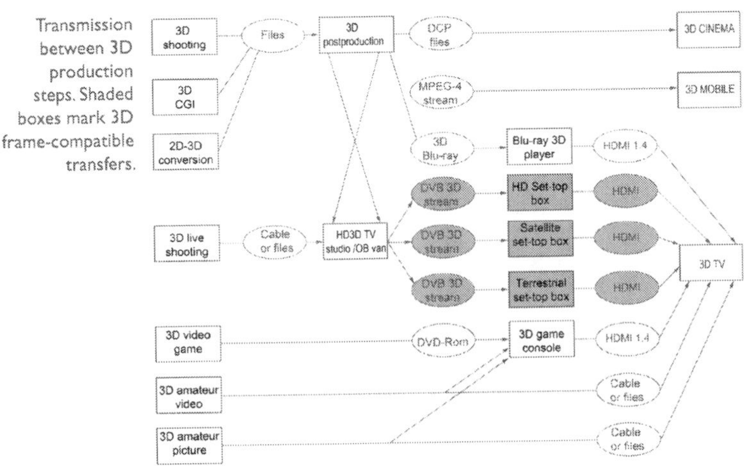

Transmission between 3D production steps. Shaded boxes mark 3D frame-compatible transfers.

types, but it is far from exhaustive. Note that the shaded areas indicate 3D images' transfers in so-called "frame compatible" formats, which reduce the signal quality to keep it compatible with the existing infrastructure. Ultimately, these transfers will become "service compatible" and reach the quality level of the rest of the chain.

From Camera to Recorder

In dual-HD shots, recording is not always done inside the camera. In the case of live transmission, the video signals must also be transported from the cameras to the OB van in real time. The video signals from a dual 720p or 1080i HD camera are transmitted by two independent HD-SDI cables, each with a throughput of 1.5 Gbits/s, in compliance with the SMPTE ST292 standard. Broadcasters are increasingly using a single 3G-SDI link at 3 Gbits/s (SMPTE ST425) to transport the two signals, which saves wiring and reduces complexity. A special signal in the data stream indicates whether a stream is 3D.

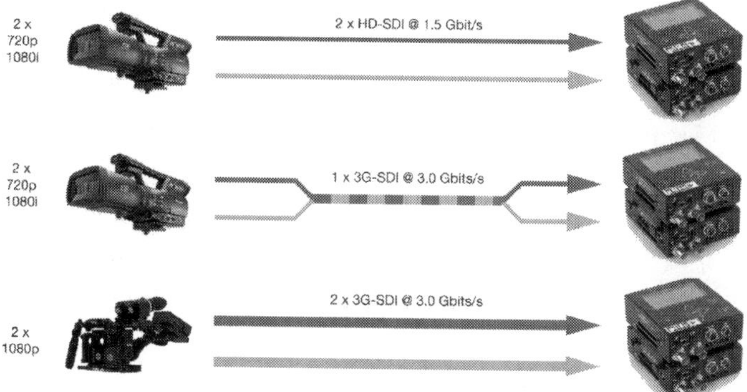

Transmission from camera to recorder requires a very large bandwidth.

For the best quality in Full HD 1920 × 1080, images cannot be transmitted in 1080i (interlaced format), but in 1080p (progressive format), which doubles the overall throughput. Two 3G-SDI cables are then required to ensure a 3 Gbits/s throughput from each camera.

More and more often, the electrical signal is converted to an optical signal inside the camera or just at its output and then transmitted by fiber. This increases the transmission range, decreases the number and weight of cables, and adds the convenience of a return path.

Pitfalls of the Camera-Recorder Link

It is obvious that everything is done in order to synchronize the two cameras' images perfectly. However, the signal path is strewn with a number of traps, especially when mirror rigs are used. First, at the entrance of the OB, and sometimes directly at the camera, you have an inverter designed to flip the image of the stereo pair that was reversed by the mirror. As a rule, this inverter takes a full frame to do its work. Therefore, you have to delay the other signal by the same amount of time. Good 3D processors take care of that, but they still have to be set correctly.

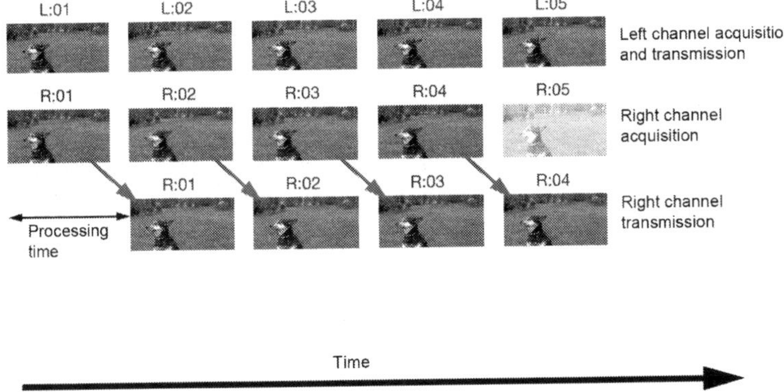

Delay caused by flipping one image from a 3D mirror rig may desynchronize the cameras.

For live broadcasts, a real-time image correction step is applied by a SIP (Stereoscopic Image Processor) or a pair of image processors calibrated to cancel unwanted disparities between the two images. For example, it will match the geometric distortions due to the lenses or to the cameras' incorrect mechanical settings. Again, it

© B. Michel

Left and right views from a mirror rig after flipping one of the views. Even with an interaxial distance set to zero, the images are not strictly identical. The image on the right is the difference between both views; it highlights here a slight difference in the zoom factor.

would be wrong to believe that the processes in these equipments were of exactly the same duration, so careful synchronization checks are required.

In the last example, a mirror rig set up with a zero interaxial distance is supposed to show two identical images after the one reversed by the mirror is flipped back. Comparing the two images, however, shows that a slight desynchronization between the two zooms gives an approximately one-pixel shift between the two images, a difference that should be corrected. To avoid such annoying mismatches' appearing in the middle of a broadcast simply because a parameter has changed in an Image Editor, signal correction is entrusted to a SIP, which will ensure an identical processing time for the two images of the 3D stream in all cases.

Compression Pitfalls

For maximum quality, the images are either uncompressed or compressed losslessly, for example with a JPEG2000 codec. But in many cases, MPEG-2 or H.264/MPEG-4 lossy compression is used in order to save storage space. These compression methods owe their efficiency to the fact that they use the redundancy between successive frames to compress the signal. The operation is done on groups of pictures, called "GOP." A GOP typically contains twelve images, but this is not always the case. At each new GOP, the compressor starts by recording a complete image. The following images are then compressed by differences within the GOP so that they occupy far less space than the first frame of the GOP.

One might think that the compressors associated with the two cameras are and will always remain synchronized, but this is not the case: The complex MPEG-2 and MPEG-4 compression algorithms may decide to start a GOP earlier than planned because they have detected a significant change in the image. Given that some of the compressors' decisions are content-based, you can be sure that after a while the GOPs coming from the two cameras will no longer start at the same image, even with two strictly identical compressors. As the compression and decompression operations introduce artifacts in the image, these artifacts are not identical and, in the case of 3D, introduce retinal rivalry problems. With this type of compression, you can prevent 3D errors due to compression only by using a dual compressor specially designed for 3D that will ensure constant GOP synchronization between left and right streams.

Shooting Data Formats

The formats used in postproduction are, in most cases, lossless or uncompressed. Images are saved either as a video stream or as separate images and organized in directories (DPX, TGA). The native format from professional cameras such as Red One or Arri Alexa is called "RAW," which means that it is an exact copy of data acquired by the camera sensor. These formats are bulky but preserve all the captured information, thus giving maximum flexibility in postproduction. They are almost always too large to be stored in the camera and must be transmitted to a dedicated recorder. The formats used internally by the usual camcorders are well known by videographers. They all include some form of lossy compression allowing long sequences to be stored on memory cards or hard drives inside the cameras.

An image's quality obviously depends on its resolution but also on the number of bits stored for each pixel. In addition, the cameras' sensors are almost never made up of three subpixels of red, green, and blue color to form one conventional RGB pixel. Many cameras have four subpixels, *e.g.*, R, G, R, B, a configuration that has the advantage of recording more brightness information. The brightness of each subpixel is encoded more or less accurately. In general, 8, 10, and even sometimes 12 or 14 bits of information are stored in a pixel in a RAW image.

The rush files' quality levels and sizes are adapted according to the type of production and budget. For example, a cinema-quality movie production may demand a RAW format, but sacrificing one level of precision and simply using compressed 12-bit ProRes444 files will yield substantial savings in equipment and storage size for marginally lower quality. In the first case, an external hard disk recorder is essential. In the second case, the same professional camera, such as the Arri Alexa, will record on its integrated memory card with the ProRes444 format, thereby cutting out the recorder, cabling, associated handling, and conversion of RAW images into images that the editing station can use immediately.

Of course, shooting in 3D tends to double storage and recording costs. Compact recorders using very low-loss compression are extremely useful. The market now offers highly efficient compact recorders, such as Cinedeck (*www.cinedeck.com*), which comes with a small screen for instant verification of the recorded footage. They owe their small size to the use of solid state drives with no moving parts, such as those used in notebook computers.

It is extremely important to check that both cameras are correctly configured to record in identical formats. It is equally important to check the file names. These are generated automatically by the camera with a serial number. It is imperative that the two cameras of a pair be set to generate identical numbering. If at all possible, change the files' base names to include the "L" and "R" prefixes. Unfortunately, few cameras offer this feature. In all cases, label memory cards and hard drives legibly with the letters "L" and "R" to distinguish the left images from the right ones.

Postproduction Formats

The file formats used in postproduction are imposed by the software. We thus shall not dwell on the subject here. However, one must check that the left and right files not only use the same file type, but also the same options (codecs, compression ratio, etc.). Variants of the same file type are usually accessible through a dialog box when saving files.

Depending on the production, files received by the postproduction studio have HD, 2K, or 4K resolution and may take the form of streams encoded with an appropriate codec or separate directories containing files (most often TIFF or DPX). As usual, as 3D doubles the volume of incoming data, "compressed stream" formats are often preferred: A great many projects are imported in XDCAM HD, R3D, or AVCHD.

Codecs

In many cases postproduction entails lots of file exchanges between studios and various remote sites, between different software options, between Mac and PC computers, etc. It is therefore advisable to select one transfer format that everyone can understand, is reliable, and fully preserves the quality. Using files with an .avi extension alone is no guarantee of compatibility!

Codecs, Containers, and Standards

A codec is a hardware or software process able to compress and decompress audio or video data according to a standardized format. A container is a file format that contains audio and/or video data streams encoded by one or more codecs. Coding standards define the formats of compressed audio and video data. For example, the file test3D.avi uses an .avi container to store its video stream, itself encoded with the XVID codec according to the MPEG-4 standard. The most commonly used containers are AVI (Microsoft), MPEG, FLV (Adobe Flash Video), and MOV (Apple QuickTime).

RGB and YUV

RGB: The three color components of a video stream are red, green, and blue. The colors visible on a movie screen or television are all formed by a combination of different amounts of these three components. These three colors are also those to which the human eye is sensitive.

YUV: Y, U, and V are three numerical values that likewise describe the color of a pixel. Also called "YcbCr," YUV encoding dates back to the early days of color television, when the luminance signal Y had to be retained for compatibility reasons with black and white television. With color, brightness, Y, is retained as is and the rest of the color information is conveyed by the two U and V values, also called "chroma."

The image's U and V layers contain fewer variations than the Y layer and thus compress very well. In fact, they are very often subsampled to reduce their size further. This is called "4:2:2" coding instead of "4:4:4" (the numbers indicate the proportions of each of the Y, U, and V components within the transmitted signal).

There are many quality-preserving codecs, but few of them offer the ability to replay HD video – even monoscopic HD – in real time. The different codecs used in .avi and .wmf files are easy to distinguish through their "four-letter codes" (the first four bytes at the beginning of the binary file). The most popular codecs in the world are listed in a database accessible on the Web at *www.codecsdb.com*. For lossless compression, I can recommend the Huffyuv and Ut video codecs among the many possible choices.

On the left, the three R, G, and B layers of an image; On the right, the three layers of the same image coded in YUV.

The Huffyuv codec (fourcc = HFYU) is a free codec for Windows developed by Ben Gould Rudiak. It uses a method similar to JPEG compression with a lossless Huffman encoder (hence its name). Each color channel is encoded separately for both RGB and YUV modes.

Ut Video codec (fourcc = ULY0) is a free codec for Windows developed by Takeshi Umezawa and implemented as a VfW (Video for Windows) codec. It compresses lossless images coded in RGB or YUV, the two most common modes. Performance, both in encoding and decoding, are excellent – typically twice as good as the Huffyuv codec – and they exist in 32-bit and 64-bit versions. However, compatibility issues are known to exist with the Edius (Grass Valley) and Vegas (Sony) editing software.

CineForm codecs: The various versions of the CineForm codec, especially those with Final Cut, are very popular on postproduction sites equipped with Apple hardware.

The CineForm Active Metadata codec is used by 3D Cinedeck recorders, the SI 2K 3D cameras from Silicon Imaging, and Neo3D software. It offers the advantage of creating a single file for both views, ensuring that they will stay matched at all times in postproduction.

Distribution Formats for Live Streaming

We face two choices when preparing files for shipment to end users: image format and type of coding. These two choices are highly dependent on how the files will be used, *i.e.*, TV broadcasts, Blu-ray 3D, on a computer connected to a projector, digital cinema, etc.

In the future, transmitting two full-HD frames will be a must. However, the changes in the entire workflow that this entails are such that distribution will have to wait many years to see it standardized, and then to have the whole distribution chain equipped from the studio to the home set-top box. Currently, the only economically acceptable solutions are called frame-compatible, meaning that 3D images are transformed to fit the existing HD channels. Quality suffers, but the extra cost of moving to 3D remains bearable. The first standardization efforts have thus focused on these frame-compatible formats so that equipment manufacturers can provide interoperable solutions as soon as possible.

Frame-Compatible Formats

A typical example of such formats using the existing distribution infrastructure is the horizontally compressed side-by-side format. This format is transported without modification across terrestrial, cable, and satellite networks and through set-top boxes up to the home viewer's 3D TV. None of the devices used in the chain is aware that it is carrying a 3D signal. The major disadvantage of these formats is that they require the standard 2D version of the same movie to be transmitted through another channel, so you have to occupy the bandwidth of two HD channels to transmit both 2D and 3D versions, and the latter's resolution is greatly penalized. Moreover, as the signal is more complex, it is compressed less easily by the encoders, and the 3D channel bandwidth is thus 10-30% larger than that of a 2D channel of the same quality.

Examples of frame-compatible formats

Frame-compatible formats that interweave lines (interlaced mode) or pixels (quincunx) are present in the HDMI standard, but they lend themselves very badly to compression if encoded. They thus run the risk of being damaged by any extra compression or decompression steps during transport. That is why they are not included in the methods for long-distance transmission.

Service-Compatible Formats

If one complies with the DVB transmission standards, it is possible to transmit a normal picture for the left eye and service data that contain information to reconstruct the right view. On the plus side, this method does not require a second channel for 2D. Indeed, a conventional television ignores service data and displays the left image as if nothing special had happened. However, a 3D TV takes service data into account, automatically switches itself to 3D mode, and displays a stereoscopic image at full resolution. On the minus side, the bandwidth increases, as service data that contain the right image can increase the size of the transmitted stream 30 to 70%, depending on the encoding method. Moreover, even if the standards are met and the transmission infrastructure remains unchanged, encoders and decoders must be replaced by more advanced models.

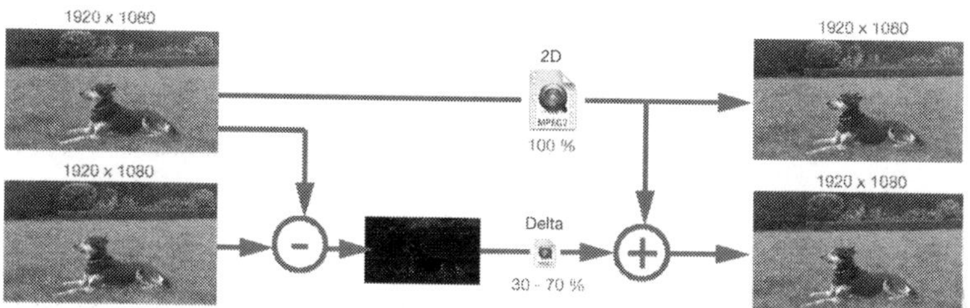

2D + Delta service-compatible transmission: The difference image takes less space than a full image. The bandwidth used is typically between 130 and 170% of a 2D feed.

Service-compatible formats are not yet fully standardized, but various initiatives exist. One of the more mature is proposed by TDVision (*www.tdvision.com*). Their coding system is known as "TDVCodec" and was proposed to be part of the MPEG-4 VMC, an extension of the standard H.264 AVC. The first set-top boxes incorporating this 2D + Delta feature appeared on the market in 2011.

Interim Formats

Pending the emergence of service-compatible formats, Dolby and Sensio already offer so-called "frame-compatible with enhancement layer" solutions. With these methods, a stereoscopic pair is transferred in a single frame, for example in over-under mode, as explained above, but additional information is included in the feed to improve the image resolution at unpacking time.

Today's frame-compatible decoders are able to manage 3D flows, but usually with half resolution per eye only, while decoders including Dolby or Sensio technologies are able to send full-screen images of near-HD quality to the 3D TV set. The enhancement layer is of acceptable size, 10-15% of the total stream size, and contains the spatial information missing in the base double image, the resolution of which is halved. This type of solution, though not standard, offers the advantage of immediate availability, which interests, among others, video on demand (VOD) providers, who are eager to start their 3D services as soon as possible, even with a lower quality than frame-compatible modes.

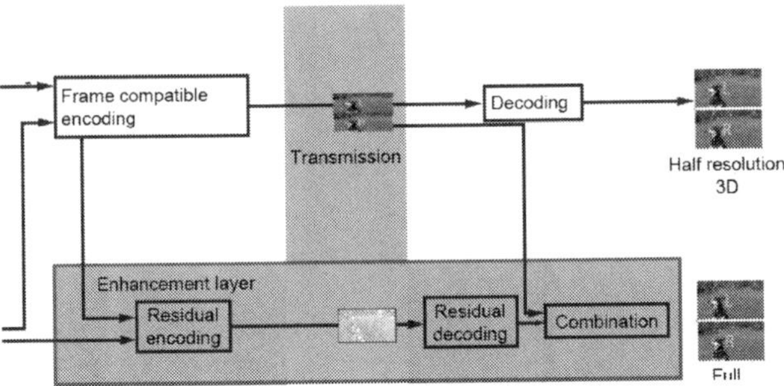

Theoretical diagram of a frame-compatible transmission with enhancement layer. Above, the classical frame-compatible transmission; below, the enhanced solution rendering full HD images for both eyes.

Sisvel Service-Compatible Format

A very interesting special case of frame-compatible mode is the Sisvel method (*www.sisveltech.com*) used by Sky Italia, which is both frame- and service-compatible. It is, however, limited to the particular case of images transmitted in 720p format inside a 1080p channel. It stems from two findings:

- a 1080p (1920 × 1080 pixels) image contains 2,073,600 pixels, while a 720p stereoscopic pair (2 × 1280 × 720 pixels) occupies 1,843,200 pixels, or 89% of the area of a 1080p image. It is therefore possible to get a 3D 720p image in a single full HD one (see figure below, left);
- a transport stream meeting the H.264 standard contains many metadata. One of them is called "cropping rectangle." This parameter was introduced because the encoders manipulating the images work on 16 × 16-pixel macroblocks and the number 1080 is not divisible by 16. So, to create a 1920 × 1080-pixel image, the encoder must actually create a 1920 × 1088-pixel image and then drop the last eight lines. The information transmitted through H.264 corresponds in fact to 1088 lines and the cropping rectangle metadata are what asks the decoder to display only the useful 1920 × 1080 rectangle (see figure below, right).

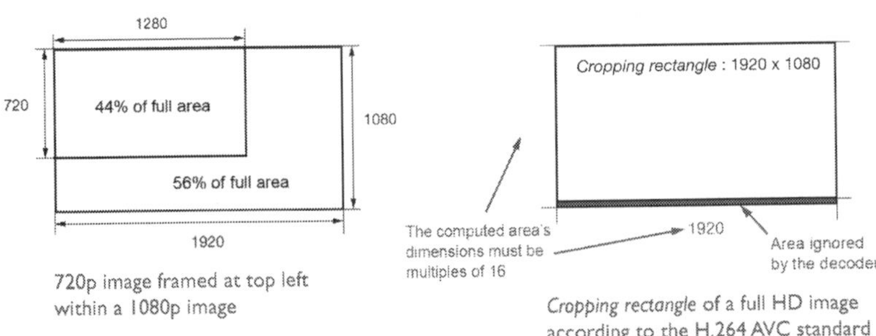

720p image framed at top left within a 1080p image

Cropping rectangle of a full HD image according to the H.264 AVC standard

Knowing this, we simply insert the left image of our 720p 3D pair in the top left of the full HD image frame, cram the right image in the rest of the frame, and then write-insert a cropping rectangle value of 1280 × 720 into the stream. A standard HDTV decoder receives the stream, decodes it, and simply displays the left image, neglecting the rest of the frame. This is indeed a service-compatible transmission, since it is 100% compatible with the existing HDTV infrastructure. Obviously, a decoder updated to detect this format is able to reconstruct easily the right image and to transmit it to a 3D TV through the HDMI link. For this detection to occur, an appropriate H.264 SEI message is inserted into the stream (SEI messages are described later in this chapter).

As the amount of pixels to be processed is not greater than in a 2D full HD transmission, the decoder does not require additional computing power and this function can be added to existing decoders easily by a simple remote update of their internal software.

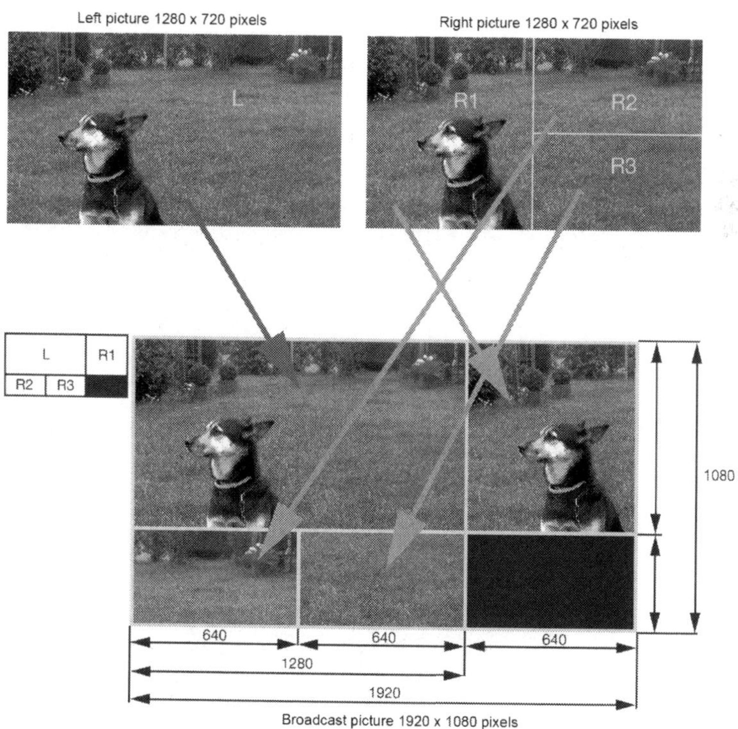

Reshuffling of parts of a 3D image by Sisvel service-compatible encoding

Transmitting TV Streams

Standard definition television (PAL or NTSC) is usually transmitted digitally with the ubiquitous MPEG-2 encoding. High definition (HD) streams are encoded in MPEG-2 or MPEG-4. As 3D TV ignores standard definition, we shall exclude it from the discussion below.

DTT, cable, and satellite TV operators follow the recommendations issued by the major international standardization bodies such as ISO, ITU, and DVB. The following section draws heavily from the *Content Encoding Profiles 3.0* standard of Cable Television Labs (**www.cablelabs.com**), one of the first to have formalized the inclusion of stereoscopic 3D in TV streams, as well as the *Commercial Requirements for 3D-TV note* issued by the DVB consortium (**www.dvb.org**).

Distribution networks are all trying to attract the largest possible number of subscribers. The content they carry must therefore be compatible with all projectors and TV sets on the market. That is why the "Commercial Requirements for 3D-TV" issued by the DVB consortium restricted the number of possible 3D formats to eight, all of them included in the HDMI 1.4 standard described later in this chapter. Better still, if a broadcaster uses only the first five formats, it will be sure that all TV sets fitted with HDMI 1.4 will be able to display them. Indeed, these five formats are necessarily supported by all makes.

3D Formats Accepted by DVB's Commercial Requirements for 3D-TV

Frame-compatible formats	Size (pixels)	Timing	Frequency (images/ second)	Layout	Presence in HDMI 1.4a
720p@50 TaB	1280 × 720	Progressive	50	Top and bottom	Mandatory
1080i@50 SbS	1920× 1080	Interlaced	50	Side by side	Mandatory
720p@ 59.94/60 TaB	1280 × 720	Progressive	59.94 ou 60	Top and bottom	Mandatory
1080i@ 59.94/60 SbS	1920 × 1080	Interlaced	59.94 ou 60	Side by side	Mandatory
1080p@ 23.97/24 Tab	1920 × 1080	Progressive	23.97 ou 24	Top and bottom	Mandatory
720p@50 SbS	1280 × 720	Progressive	50	Side by side	Optional
720p@ 59.94/60 SbS	1280 × 720	Progressive	59.94 ou 60	Side by side	Optional
1080p@ 23.97/24 SbS	1920 ×1080	Progressive	23.97 ou 24	Side by side	Optional

The frame rate for interlaced formats is the frequency of frame transmission (a frame is a half-image made from the full image's odd or even lines). Divide this value by two to get the number of full images per second.

- The first five formats are mandatorily recognized by all 3D TVs.
- The first two are recommended for distribution in Europe.

In Europe, retailers are even more restrictive; so, SES ASTRA requires that 3D streams be made in one of two formats, 720p@50 TaB or 1080i@50 SbS, and only with MPEG-4 AVC compression.

3D satellite channels are slowly multiplying in Europe: Brava3D, Sky 3D, Astra 3D, HighTV 3D, etc. To avoid confusion about the equipment's ability to broadcast 3D graphics, SES ASTRA proposes to identify the devices (TVs and set-top boxes) that are compatible with frame-compatible 3D formats with a "Made for 3D" logo.

Made for
3D

The "Made for 3D" logo proposed by SES ASTRA for 3D compatible equipment

Backward Compatibility

The main goal of frame-compatible standards is to allow transmission of 3D productions on existing networks designed for HD without having to modify them. The subtleties of the encoding formats must effectively allow:

- acquisition of 3D content, including the insertion of metadata indicating how to decode it properly at the end of the chain and the insertion of 3D subtitles. And all of this must be done without disturbing the existing transmission infrastructure;
- receiving and displaying such 3D content on a 3D TV connected to either an HD set-top box with specially updated software or a legacy decoder unaware of the very existence of 3D;
- acquisition and transmission of HD content and displaying the images on an HDTV or 3D TV via the same infrastructure, including an optionally 3D-upgraded HD set-top box;
- making clean transitions, without disturbing artifacts, between HD content and 3D content or between differently coded 3D contents, on 3D TVs as well as on HDTVs. The latter will not display 3D images in 3D, but should still be able to display them in some way, even if only to detect visually that 3D content is present;
- automatically configuring 3D TV in 3D mode if a 3D signal is present or, failing that, at least allowing the user to do this manually.

This avowed goal cannot be be fully achieved, since at least the TV set has to be adapted. If the decoder is updated to understand the 3D metadata such as the 3D encoding method and 3D subtitles, it will automatically control the mode switch on the TV using a suitable HDMI cable. With these minimal infrastructure adaptations, 3D TV is becoming a reality. To achieve this goal the transmission methods described below may sometimes give an impression of unnecessary complexity; but it is precisely this complexity that enables 3D to be integrated, at a reasonable cost, in the complex transmission chain that stretches from the TV studio to the home.

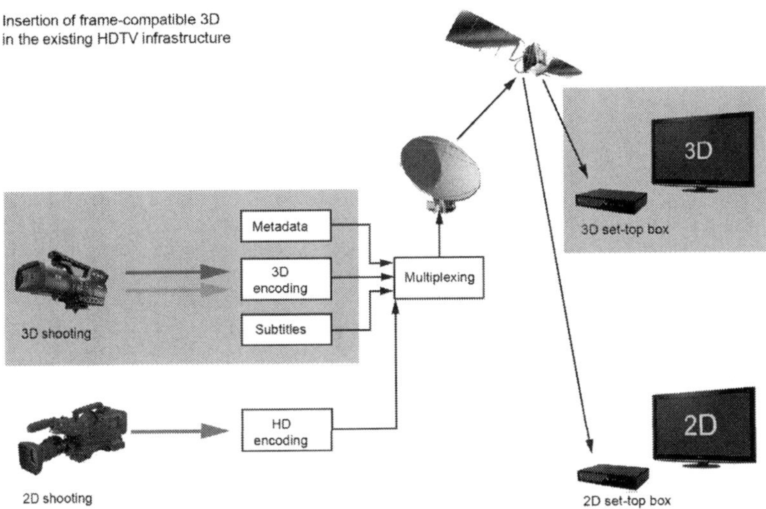

Frame-compatible 3D is broadcast without modifying the HDTV transmission workflow.

HDTV Encoding

Audio coding is usually stereo, with two audio channels (left and right) encoded at 192 Kbits/s. When multi-channel sound is available, it is encoded in AC-3 5.1 (left, right, rear left, rear right, center, and bass) with an ideal bit rate of 448 Kbits/s. In all cases, sampling is done at 48,000 samples/second.

Video sampling is done at a frequency of 23.97 Hz for cinema content and 25 or 29.97 Hz for full HD (1920 × 1080-pixel) content. Some HD video streams called "720p" are encoded at 50 or 59.94 Hz (or more rarely at 23.97, 29.97, or 25 Hz), but their resolution is 1280 × 720 pixels only.

Encoding Rate

The maximum bit rate for video encoding is set by the transport stream's bandwidth, minus what is used by the audio. You will have to be careful, because with some movies with 5.1 sound you may have multiple audio channels in a single transport stream. This can occur if dialogs are to be delivered in several languages in a given country. In such cases, audio transmission could significantly reduce the bandwidth available for the image part.

This throughput depends on the infrastructure used (cable, satellite transponder, and terrestrial digital TV transmitter). For example, here are the rates recommended for U.S. cable operators' transport streams for MPEG-4 HD channels: 5.625 Mbits/s, 7.500 Mbits/s, 9.375 Mbits/s, 11.250 Mbits/s, 15.000 Mbits/s, and 18.750 Mbits/s. This last value represents the maximum usable in practice; it is rarely used and is very nearly a satellite transponder's maximum throughput of 19 Mbits/s.

MPEG-2

Describing the MPEG-2 encoding method in detail is beyond the scope of this work, but it is worthwhile recalling that the goal is to reduce the amount of information carried in a video stream by removing redundancies between consecutive frames. The video stream consists of groups of pictures (GOP), generally 12 or 15 frames. A GOP contains full-frame images (called I-frames), images encoded by the difference with the previous I-frame (called P-frames), and images encoded by the differences with the previous and next images (called B-frames).

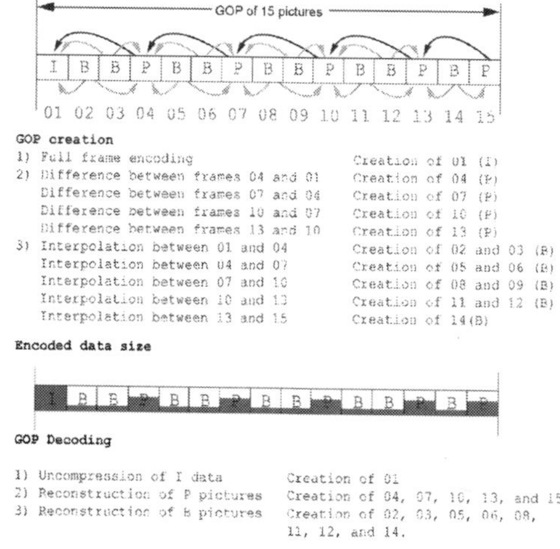

Example of MPEG-2 coding showing the creation and decoding order of I-, P-, and B-frames.

3D HDTV Encoding

Stereoscopic content is currently transmitted almost exclusively in frame-compatible mode and this situation will persist until the infrastructure evolves. This means encoding images in a frame-compatible format and adding one or more audio channels and all metadata required, and doing all that while remaining below the 19 Mbits/s limit.

The DVB consortium and Cable Television Labs' specifications clarify the groundrules:

- Resolution and quality of left and right images after encoding, transmission, and decoding must remain identical. Compressing one of the two views in a way that will degrade its quality significantly more than the other is out of the question.
- Images may be compressed with MPEG-2 or H.264 (also known as "MPEG-4 AVC") methods.

Over-Under Mode

Over-under (also called top-and-bottom) mode is the most interesting of the frame-compatible modes, because even if it reduces the amount of information by a factor of two, it does

Transmission of a 1080p 3D image in over-under mode

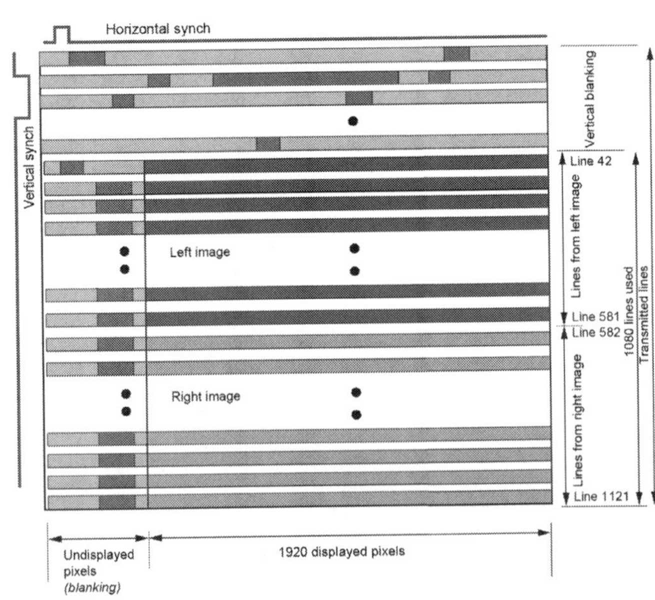

not reduce the 3D quality: As the 3D information is given only by horizontal pixel shifts, the loss of vertical resolution does not interfere. The over-under mode has the following characteristics:

- This method always has its two sub-images synchronized in time, with the left image being positioned on top.
- With over-under and side-by-side modes, the images are never mirror- reversed either horizontally or vertically.
- This mode will use only full non-interlaced frames, so the only acceptable transmission modes are exclusively 720p HD or 1080p full-HD.
- In 720p mode, the left image occupies lines 26 to 385 and the right image lines 386 to 745.
- In 1080p mode, the left image occupies lines 42 to 581 and the right image lines 582 to 1121.
- With over-under mode, images are compressed by an algorithm that reduces the vertical resolution only. If the left and right sub-images are identical, the algorithm will give the same result for both.

Side-by-Side Mode

Side-by-side is also a frame-compatible mode, which is why it reduces the amount of information by a factor of two, but this time at the expense of horizontal resolution. This lessens the perceived depth accuracy, but has the advantage of demanding fewer resources inside the decoder. It is indeed necessary to store one line at a time to recreate both views from the transmitted signal, while in over-under mode, the entire image has to be memorized before recreating the two views. The side-by-side mode characteristics are as follows:

- This mode always has two sub-images synchronized at the line level, the left image being on the left.
- Interlaced modes may be used, so the allowed HD formats are 720p and 1080p (progressive) or 1080i (interlaced).
- In side-by-side mode, images are compressed with an algorithm reducing only the horizontal resolution. If the left and right images are identical, the decimating algorithm has to give exactly the same result on each.
- In 1080 side-by-side modes, the left image occupies pixels 0 to 959 and the right image occupies pixels 960 to 1919.

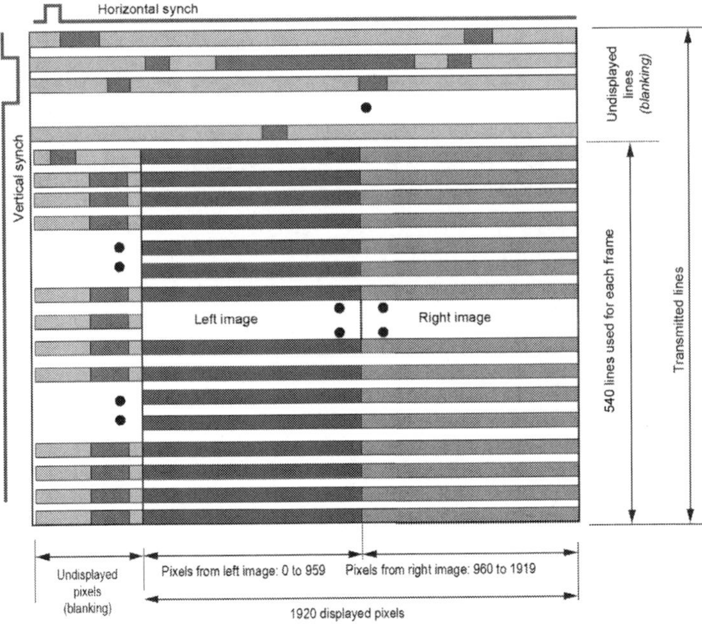

Transmission of a 3D 1080i frame in side-by-side mode. Two such frames are transmitted for each image, one for odd lines, the other for even lines.

Letterbox and Pillarbox

Many 3D content have an aspect ratio wider than 16:9. Consequently, they are displayed in a so-called "letterbox" format that includes black bands above and below the image.

Note
Black lines displayed on screen are called "blank lines" for historical reasons dating back to the analog era!

In some cases, content with an aspect ratio narrower than 16:9 is transmitted; this is typically the case of old TV shows recorded in 4:3. These images are displayed in a format called "pillarbox," where black bands complete the picture on the left and right sides of the image. This format is not very appealing in 3D, but must still be taken into account.

- Any "letterbox" content displayed in 1080 lines and transmitted in over-under format has to be arranged so that exactly 540 lines separate the elements of both images and both horizontal black bars. You cannot append two images to each other in the middle of the screen without black bars if they do not occupy exactly the whole frame before compression. In the following example (on top), we see that lines 62 and 602 correspond.

3D Image with a 1.85:1 aspect ratio in over-under "letterbox" mode in a 1920 × 1080 container

- Any «pillarbox» content displayed in 1080 lines and transmitted in side-by-side format will be arranged centered, with just as many black pixels left and right. You cannot append two images to each other along the screen's vertical symmetry axis without black bars if they do not occupy exactly the whole frame before compression. In the following example (bottom), we see that pixels 20 and 980 and again 939 and 1899 match exactly.

3D Image with a 4:3 aspect ratio in side-by-side mode with black pillars in a 1920 × 1080 container.

Description of the SEI 0x2D Metadata Block in an MPEG-4 AVC-Encoded 3D Stream (block length is 48 bits)

Name	Size (in bits)	Value	Meaning
frame_packing_arrangement_id	1	0	Indicates to the decoder that a 3D stream description exists
frame_packing_arrangement_cancel_flag	1	0	Always 0
frame_packing_arrangement_type	7	3 or 4	3 = side-by-side, 4 = over-under
quincunx_sampling_flag	1	0	Quincunx coding not used
content_interpretation_type	6	1	1 = Left-Right, 8 = Right-Left (value 8 is forbidden as images are always placed Left/Right)
spatial_flipped_flag	1	0	No mirror-reversed image
frame0_flipped_flag	1	0	No mirror-reversed image
field_views_flag	1	0	Images are not interlaced line by line
current_frame_is_frame0_flag	1	0	Always 0
frame0_self_containing_flag	1	0	Images not temporally interlaced
frame1_self_containing_flag	1	0	Images not temporally interlaced
frame0_grid_position_x	4	0, 4 or 8	Usually 0; 4 and 8 are possible options
frame0_grid_position_y	4	0, 4 or 8	Usually 0; 4 and 8 are possible options
frame1_grid_position_x	4	0, 4 or 8	Usually 0; 4 and 8 are possible options
frame1_grid_position_y	4	0, 4 or 8	Usually 0; 4 and 8 are possible options
frame_packing_arrangement_reserved_byte	8	0	Always 0
frame_packing_arrangement_repetition_period	1	0	Always 0
frame_packing_arrangement_extension_flag	1	0	Always 0

Metadata

A TV stream can be viewed from any point in time, so it is important that the essential data characterizing it are issued several times per second. A broadcast video transmission has neither beginning nor end and is called a "stream" to differentiate it from files, which are entities of well defined size. A digital transport stream complies with the MPEG-2 TS standard and can contain images compressed according to the MPEG-2 or MPEG-4 AVC standard. Although a complete description of MPEG-2 TS (transport stream) encoding is beyond the scope of this book, be aware that if decoders are to recognize 3D images as such and decode them correctly, a certain number of data must be present in the stream. These data describing data are called "metadata" and are repeated at regular intervals.

The most important metadatum describes which method is used to insert two images in a single frame (in frame-compatible mode, of course).

With MPEG-4 AVC coding, we use blocks called "SEI metadata," the largest of which is the type 0x2D (45 decimal) metadata block described below. SEI stands for "Supplemental Enhancement Information."

MPEG-2 will disappear one day from the usual 3D transmission methods in favor of the more elaborate MPEG-4/H.264 coding. Nevertheless, many infrastructures, including U.S. cable operators, still use MPEG-2 for distribution; it is also the encoding standard of all DVDs.

"User Data – S3D Video Format" Block for MPEG-2-Encoded 3D Streams (block length is 96 bits)

Name	Size (in bits)	Value	Meaning
user_data_start_code	32	0x1B2	(= 434 decimal): beginning of user_data block
S3D_video_format_signaling_identifier	32	JP3D	Indicates "3D images present;" this code was chosen to be as different as possible from other codes
S3D_video_format_length	8	3	Size of following datum is 3 bytes
reserved_bit	1	1	Value ignored by the decoder
S3D_video_format_type	7	3, 4, 8	3 = side-by-side, 4 = over-under, 8 = 2D ; other values reserved for future use
reserved_data	16	0x04FF	(= 1279 decimal): value ignored by the decoder

MPEG-2 requires that transition from a 2D sequence to a 3D one or a 3D mode change between two 3D sequences occur only at the end of a GOP; the new sequence must start with a new GOP starting with a full I-frame image. With MPEG-2 encoding, the corresponding metadatum is called "S3D_video_format_signaling()", which is part of the "user_data()" metadata block. Cable Labs specifications recommend transmitting this block once per frame throughout the 3D sequence.

Set-top Box

Video on demand (VOD) operators were early adopters of 3D, for the additional costs of the new format could be compensated easily by a small increase in paid subscriptions.

Standard Set-Top Boxes

The existing home set-top boxes decode and transmit frame-compatible formats such as side-by-side and over-under. Without any modification, they decode HD signals and transmit them to HDMI-connected 3D TV sets without being aware of the images' 3D nature. The home setup's compatibility depends only on the TV set and its HDMI connection.

Frame-Compatible 3D Set-Top Boxes

The disadvantage of this compatibility is that the HDMI signals notifying the TV of the video stream's 3D nature are not generated by the decoder. The viewer must therefore switch manually to 3D mode with the TV set's remote controller. The standard decoders fitted with specific 3D firmware are able to detect 3D metadata and insert the corresponding messages into the HDMI output stream. Fortunately, managing this information and forwarding it to the HDMI output does not require additional computing power. A simple remote update of the set-top box by the operator can solve this problem and make the 2D/3D switch automatic (if the TV set is configured for automatic 3D, of course).

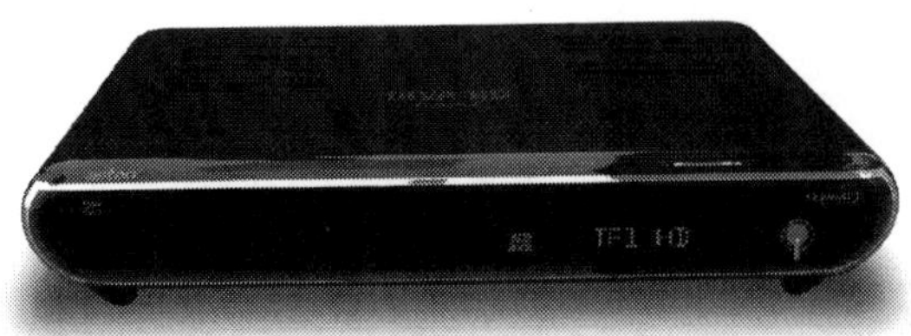

© Aston

This Aston DIVA HD Easy set-top box is fitted with a frame-compatible decoder compatible with most 3D channels.

Service-Compatible 3D Set-Top Boxes

However, future transmission standards known as "service compatible," which will double the quality of 3D images at the cost of higher bandwidth, will require new, more powerful decoders. We will then reach the same quality as a 3D Blu-ray: 50 or 60 full HD (1920 × 1080-pixel) frames per second. To be able to decode and retransmit twice the number of pixels to the TV set, not only will such decoders have be more powerful, but they will have to use components that do not exist right now. Even standards specifying the details of the decoder's internal electronics are not yet fully published. The first of these standards to become reality will probably be a particular implementation of MPEG-4 MVC (Multiview Video Coding), which is compatible with existing HD encoders and decoders. Of course, new decoders will be required to decode the 3D part of the stream and to send the result to a 3D TV. Additional costs are likely to stay rather low, because much of the hardware and software required will be similar to the decoding part of a Blu-ray player; an economy of scale will develop *de facto* through the use of common components.

HDMI 1.4 and 3D TV

The adoption of interoperable standards is the key to new technology's success. As the most critical link in the entire transmission chain is the one between a 3D source (digital TV decoder, Blu-ray player, or game console) and the TV set, it was the focus of the first standardization efforts to be made, and which produced their first results in 2010.

For the home distribution of stereoscopic movies, the interface between the 3D TV and its 3D sources and between decoders and 3D Blu-ray players is standardized under the name "HDMI 1.4." The HDMI consortium (HDMI Licensing, LLC) consists of Hitachi, Panasonic, Philips, Silicon Imaging, Sony, Technicolor, and Toshiba. It published the 3D extension of the HDMI standard under number 1.4 in 2010.

TV and Blu-ray players manufacturers have thus rushed to adopt this standard, which is the only one to have garnered broad consensus. Only HDMI-compatible products are authorized to bear the official HDMI logo.

Official HDMI interface logo

The three HDMI connector flavors

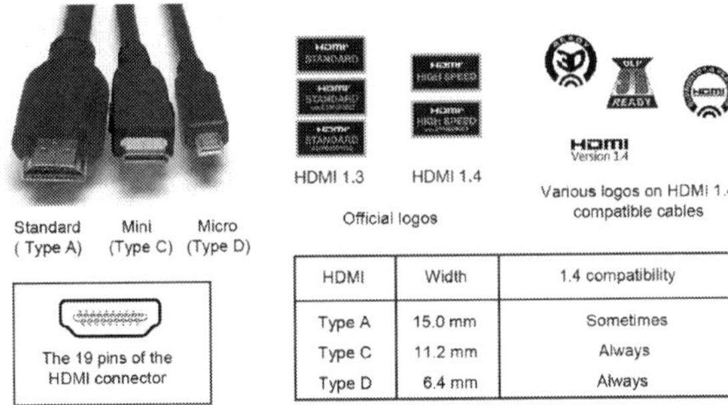

HDMI	Width	1.4 compatibility
Type A	15.0 mm	Sometimes
Type C	11.2 mm	Always
Type D	6.4 mm	Always

HDMI Cables

In addition to supporting 3D, HDMI 1.4 also supports the Ethernet protocol - in this case called "HDMI Ethernet Channel" -, thereby enabling compatible devices to exchange and share data in a simpler way at speeds of up to 100 Mbits/s. HDMI-compatible devices will be able to access the Internet without requiring any additional network cable, which will often do away with the need for an Ethernet switch as well.

A new "Audio Return Channel" feature has also been added: The sound from the TV will be redirected back to the audio/video amplifier through the same HDMI cable connecting the TV to the amp. Finally, thanks to its high throughput potential, HDMI 1.4 supports larger resolutions, such as 3840 × 2160 at 24, 25, and 30 Hz, and even 4096 × 2160 at 24 Hz.

The HDMI cables of interest for our purposes are therefore divided into several categories:

- HDMI standard cables for 720p and 1080i HD;
- HDMI High-Speed cables for 1080p and 3D;
- HDMI High-Speed cables with Ethernet.

The HDMI connectors used by these three categories of cable are available in three sizes, depending on the type of device to be connected, *i.e.*, standard, mini, and micro. The Micro HDMI Type-D Connector is 50% smaller than the Mini C-type version already defined in HDMI 1.3. It supports resolutions up to 1080p for mobile devices such as 3D cameras.

HDMI Transfer Protocol

The HDMI exchange between an audiovisual content source and a display consists of two data blocks. These two blocks are called "HDMI Vendor Specific Data Block" and "HDMI Vendor

Specific InfoFrame." We shall detail here only the data blocks on the coding of stereoscopic images, because they determine the technical limitations of TV sets and source devices. The other parts of the HDMI specification, which deal with audio, encryption, image size, etc., are available directly on the HDMI consortium website (*www.hdmi.org*).

HDMI Vendor Specific InfoFrame (VSI)

The VSI data block begins with a 24-bit-long optional recording identifier with the value 0x000C03, indicating that this InfoFrame is defined by the HDMI consortium. The rest of the block is defined below. But if the source device sends data corresponding to the 3D format declared in the HDMI 1.4 specification, it must include this block, which contains the information coding method used for 3D, at least once every two video frames, or at least twelve times per second. Reading and interpretation of this data packet by the receiving device are optional (for example, in the case of an older HDMI 1.3 TV set, the receiving TV set will ignore the 3D information).

The VIC (Video Identification Code) parameter, which describes the type of video, is transmitted in a data block called "AVI InfoFrame" (in a part of the HDMI standard not described here). This VIC parameter is very important, for it sets the format used to transmit video data. It can take many different values, the most useful being 4, 19, and 32, which correspond to the 720p-60 Hz, 720p-50Hz, and 1080p-24Hz formats, respectively, used by DVDs and Blu-rays. All other values of the AVI InfoFrame data block remain valid, regardless of whether this VSI (Vendor Specific InfoFrame HDMI) packet has been transmitted or not. The useful content of the VSI packet is described hereunder.

HDMI VSI Header

Byte \ Bit	7	6	5	4	3	2	1	0
HB0	Packet type = 0x81							
HB1	Version = 0x01							
HB2	0	0	0	Length in bytes of following data (= Nv)				

The 5-bit "Length in bytes of following data (= Nv)" field defines the length in bytes of the whole VSI data block, which has a maximum size of 32 bytes. This size does not include the 3-byte header.

HDMI VSI Data

<table>
<tr><th>Byte \ Bit</th><th>7</th><th>6</th><th>5</th><th>4</th><th>3</th><th>2</th><th>1</th><th>0</th></tr>
<tr><td>PB0</td><td colspan="8">Checksum</td></tr>
<tr><td>PB1</td><td colspan="8" rowspan="3">InfoFrame indicator(0x000C03)
(least significant bit first)</td></tr>
<tr><td>PB2</td></tr>
<tr><td>PB3</td></tr>
<tr><td>PB4</td><td colspan="3">HDMI 3D video format</td><td>Reserved (0)</td><td>Reserved (0)</td><td>Reserved (0)</td><td>Reserved (0)</td><td>Reserved (0)</td></tr>
<tr><td>(PB5)</td><td colspan="4">3D_Structure</td><td>3D_ Meta_ Present</td><td>Reserved (0)</td><td>Reserved (0)</td><td>Reserved (0)</td></tr>
<tr><td>(PB6)</td><td colspan="4">3D_Ext_Data</td><td colspan="4">Reserved (0)</td></tr>
<tr><td>(PB7)</td><td colspan="3">3D_Metadata_type</td><td colspan="5">3D_Metadata_Length (= N)</td></tr>
<tr><td>(PB8)</td><td colspan="8">3D_Metadata_1</td></tr>
<tr><td>...</td><td colspan="8">...</td></tr>
<tr><td>PB (Nv)</td><td colspan="8">Reserved (0)</td></tr>
</table>

The three useful bits of PB4 (HDMI 3D Video Format) define the type of 3D encoding as explained hereunder. The possible values for this field are 0, 1, and 2. Values 3-7 are reserved for future use.

The Three Useful Bits of PB4 "HDMI 3D Video Format" (bits 7, 6, and 5)

Value	Description of accepted values for the 3D video format
000	No special HDMI 3D format in this packet
001	"Frame packing," default 3D transmission method
010	Another 3D format is present. The following 3D_Structure is described in the PB5 and following bytes.
011 à 111	Reserved for future use

Default 3D transmission method

To understand the next data block, one must first understand how normal 2D images are transmitted through the HDMI interface. With both 1080p and 720p formats, each image is always transmitted sequentially, one at a time, as a series of consecutive lines, separated by black lines that are never displayed (and curiously called "blanking"). The default transmission of 3D images (when PB4 bits 7, 6, and 5 = 001) simply extends this method and sends the left and right images one after each other, separated by a group of black lines. This mode is called "frame packing." It is a very simple way to transmit images between one source, such as a 3D Blu-ray player, and a receiver, usually a 3D TV. Its disadvantage is to more than double the transmitted data's size. However, it is extremely simple and offers the advantage of being backward compatible, that is to say that old TV sets that comply with earlier versions of the HDMI standard accept the signal and use only the part that they understand, that is, the left image. They pay no attention to the remaining data and display the received images correctly in 2D.

HDMI transmission of a 2D (on the left) and 3D (on the right) image

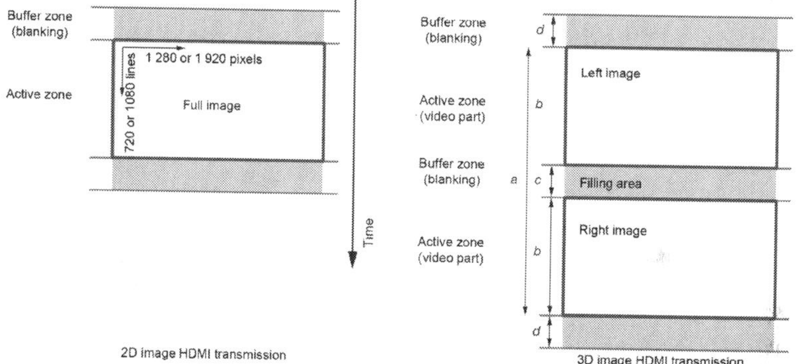

Image Formats And Corresponding Video Identification Code (VIC) Values – (a, b, c, and d refer to the figure above.)

Video Identification Code (VIC)	Format description	pixels/ line	lines/ image	# of lines a	# of lines b	# of lines c	# of lines d
32	1080p/23.98 Hz or 24 Hz	1920	1080	2205	1080	45	45
4	720p/59.94 Hz or 60 Hz	1280	720	1470	720	30	30
19	720p/50 Hz	1280	720	1470	720	30	30

When images' data are in 3D, the number of transmitted lines per frame is doubled, but the overall structure is retained for compatibility with HDMI 1.3, which ignores the 3D extensions but should still be able to display one of the two transmitted images. In the case of HDMI, it's always the left image (the first one) that is transmitted and the right image that is ignored by old equipment. The blanking area between the two images of a pair is usually black and is called the «active zone.» This area corresponds to the length of a conventional blanking (inactive area between two 2D images) to prevent artifacts if a HDMI 1.3 device should display one or more extra lines because of a transmission error. This area should be coded the same way as the adjacent video areas and contain a uniform color. Using zero, corresponding to black, is recommended. This area should be ignored by the receiving device.

Other 3D Transmission Modes

If the value of the PB4 "HDMI Video Format" field is set to 010, the "3D_Structure" field is used to define a known method of 3D images transmission different from the default method. Other types of 3D format, as yet unknown, will be handled by future 3D TV sets. They will be defined by the values 011 to 111 of the "HDMI Video Format" field.

The "3D_Structure" field (the first half of the PB5 byte within the VSI block) declares which 3D transmission mode is used. It is defined by the source device after checking that the receiver accepts it, a fact that will be indicated by the HDMI "Vendor Specific Data Block" (VSDB), which will be described later. Besides the default method (frame packing), six other methods are defined. These methods are described below.

The 3D_Structure Field

Value of the 4 bits 3D_Structure field	3D display method
0	Frame packing – default method
1	Field alternative
2	Line alternative
3	Side-by-Side (Full)
4	L + depth
5	L + depth + graphics + graphics-depth
6	Reserved (= 0)
7	Reserved (= 0)
8	Side-by-Side (Half) – Sub-sampling methods are described in 3D_Detail_X
9 to 15	Reserved (= 0)

The «3D_Ext_Data» field (the first half of the PB6 byte in the VSI block) defines the subsampling used in the case of the Side-by-Side (Half) method (3D_Structure = 8). It explains how the pixels are divided between the left and right images in this case; it is ignored in all other cases.

Sub-sampling is required because the resolution of each image must be halved. With simple subsampling, all the pixels in every other column are ignored. With the quincunx method, the sub-sampling is shifted by one pixel every other line, which helps mitigate the loss of precision on vertical stripes and other similar textures. However, the risk of creating moiré-like artifacts during image recreation is increased.

The 3D_Ext_Data Field

4 bits value of the 3D_Detail_X field	Used sub-sampling method	
0	All eight methods are supported	
1	Horizontal subsampling	All four methods are supported
2		Odd left, odd right
3		Odd left, even right
4		Even left, odd right
5		Even left, even right
6	Quincunx subsampling	All four methods are supported
7		Odd left, odd right
8		Odd left, even right
9		Even left, odd right
10		Even left, even right
11 to 15	Reserved	

Table of 3D Transmission Modes

The different modes of 3D transmission are not applied to all resolutions automatically. In practice, the transmission format depends on the pair of values VIC (image format) and 3D_Structure (3D transmission mode 3D). The mode that is used is, of course, conditioned by the receiving device's ability to interpret it. The table on the next page is not exhaustive.

The Various Possible 3D Transmission Modes

3D_ Structure value	3D Mode	VIC code	Description	Pixels	Lines	Trans-mitted active pixels	Trans-mitted active lines
0	Frame packing	31	1080p – 50 Hz	1920	1080	1920	2205
0	Frame packing	16	1080p – 60 Hz	1920	1080	1920	2205
0	Frame packing	5	1080i – 60 Hz	1920	1080	1920	2228
0	Frame packing	20	1080i – 50 Hz	1920	1080	1920	2228
1	Field alternative	5	1080i – 60 Hz	1920	1080	1920	2272
1	Field alternative	20	1080i – 50 Hz	1920	1080	1920	2272
2	Line alternative	16	1080p – 60 Hz	1920	1080	1920	2160
2	Line alternative	31	1080p – 50 Hz	1920	1080	1920	1260
3	Side-by-Side (Full)	16	1080p – 60 Hz	1920	1080	3840	1080
3	Side-by-Side (Full)	31	1080p – 50 Hz	1920	1080	3840	1080
4	L+Z	19	720p – 50 Hz	1280	720	1280	2190
5	L+Z+GFX+G-Z	19	720p – 50 Hz	1280	720	1280	2250
8	Side-by-Side (Half)	16	1080p – 60 Hz	1920	1080	1920	1080
8	Side-by-Side (Half)	2 or 3	480p – 60 Hz	720	480	720	480

Timing of HDMI transmission of a field interlaced image: 2D image on the left, 3D image on the right.

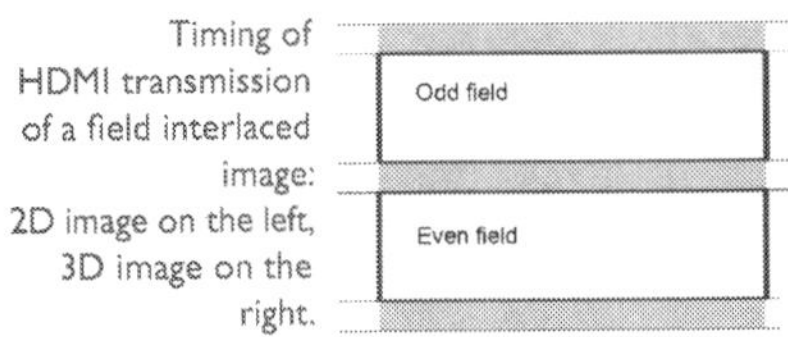

2D interlaced image HDMI transmission

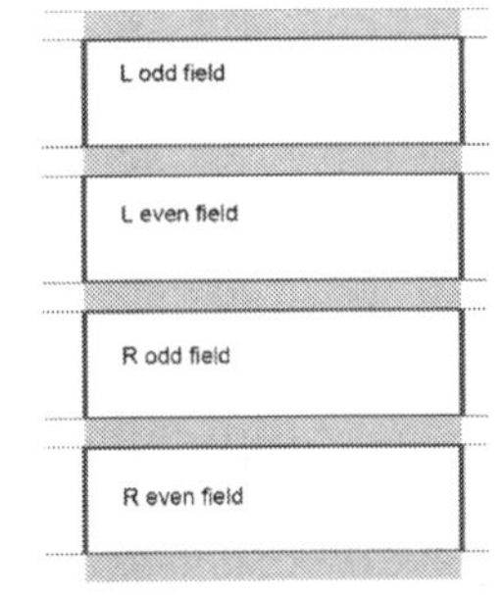

3D interlaced imageHDMI transmission

Field Alternative Interlaced 3D Transmission Mode

This mode is used if 3D_Structure = 1. It is typically used by terrestrial TV broadcasting, satellite, and cable.

Line Alternative Interlaced 3D Transmission Mode

This mode is used if 3D_Structure = 2. It is more interesting than the frame packing mode if the TV displays the left and right images simultaneously and line by line. It does not have to wait to receive a complete frame before starting to display it.

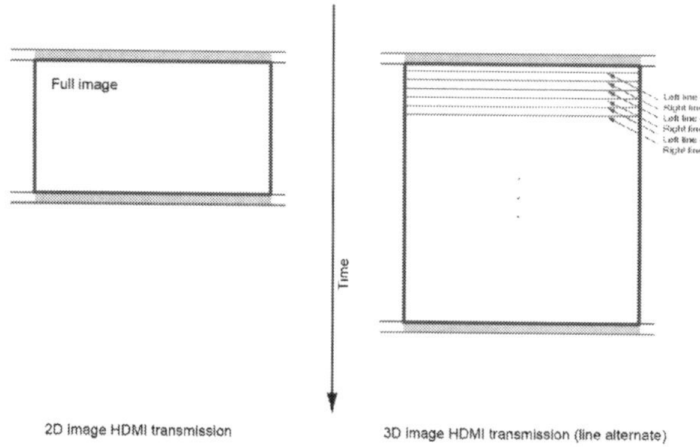

Timing of HDMI transmission of a line-interlaced image: 2D image on the left, 3D image on the right.

Side-by-Side (Full) 3D Transmission Modes

This mode is used if 3D_Structure = 3. It is as interesting as the line alternative mode. The only difference lies in how the lines are arranged: They are side by side instead of stacked.

Comparison of HDMI transmission of a 2D image (left) and a 3D side-by-side image (right)

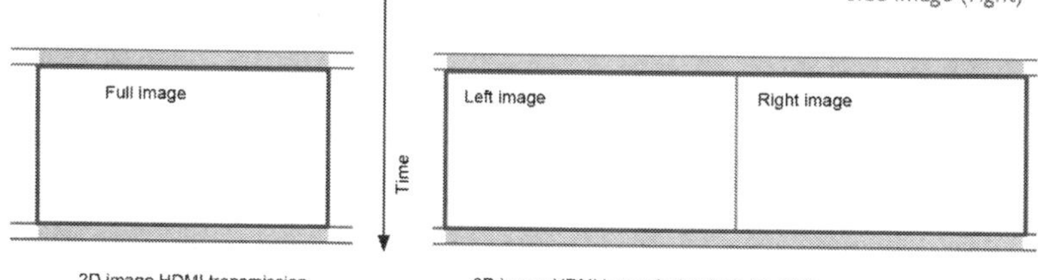

L + Depth Transmission Mode

Also called "2D + Z," this is perhaps the mode of the future. Now rarely used, it nevertheless has several advantages. With "2D + Z," regular image pixels are transmitted, but each pixel is associated with a depth value (Depth or "Z," usually used to name the third axis of the image after X and Y for the horizontal and vertical axes). These depth values are stored in a memory area called by analogy the depth map or Z-Map. Here, too, the display device must change the color and distance of each point in a 3D image. In future, autostereoscopic displays will require a large number of different views that exceeds the current HDMI interface's throughput capability. These multiple views are then generated by calculation from 2D + Z data.

Remark
Note that algorithms generating the various views are not defined in the standard and are left to the screen manufacturer's discretion. The depth map is transmitted as an ordinary image, as the depth values occupy the same number of bits as the color information of a normal pixel.

This assumes a high computing power in the TV set, which is hardly the case today. The 2D + Z methods have, however, an important advantage, that of allowing the viewer to adjust the intensity of 3D effects.

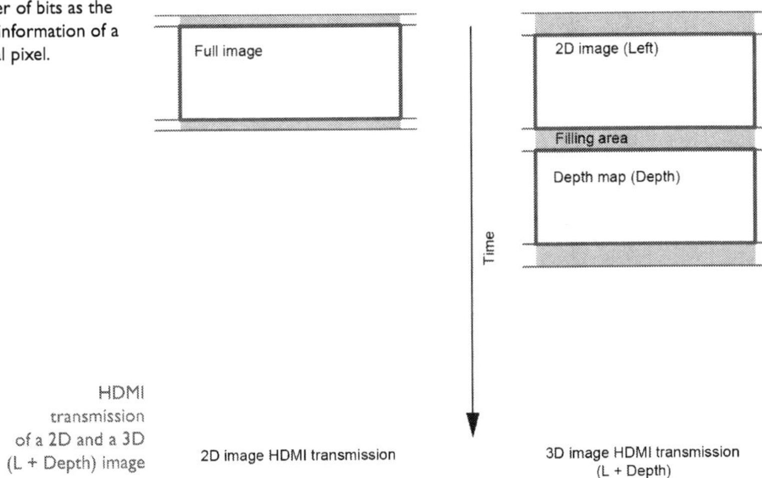

HDMI transmission of a 2D and a 3D (L + Depth) image

L + Depth + GFX + G-Depth Transmission Mode

This mode, which is a little more complex than the previous one, is an extension of the latter in which the image consists of two layers called "L" for the 2D left image and "GFX" for a graphics layer overlaid on the image. This zone typically contains text or logos. Each of these two layers is a viewable image accompanied by its own depth map (Z or Depth for the L layer and G-Z or G-Depth for the GFX layer). Again, the display device will be in charge of turning the color and distance of each point in each of the two layers into a single stereoscopic image. This method should give better resolution to texts and logos overlaid on a base image. Indeed, if the texts were embedded in the image beforehand,

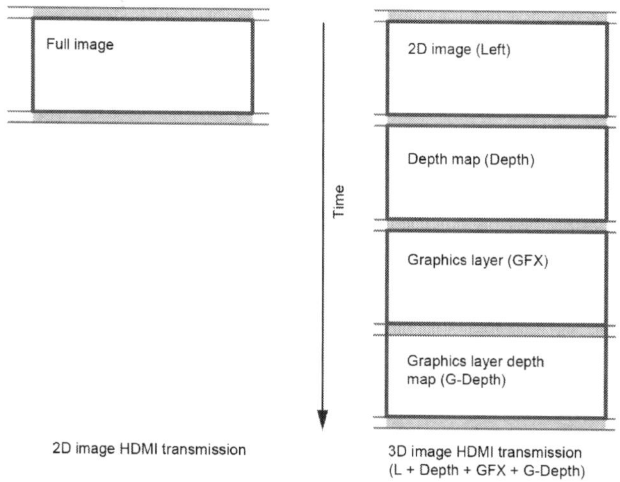

HDMI transmission of a 2D image (on the left) and a 3D (L + depth + GFX + Gdepth) image (on the right)

resizing done by the TV to change their depth might introduce artifacts along the edges, which would degrade the overall quality. Apart from this segmentation into two layers, the transmission method is identical to the previous one.

Side-by-Side (Half) Transmission Modes

These methods are frequently used in these early days of 3D TV because they allow a 3D image to be processed as if it were a single normal resolution image, with very few changes in the production workflow. However, the tradeoff is high, with a 50% loss in horizontal resolution. This is not a solution for the future, but a transitional one. As these formats halve the image's horizontal resolution, which means that the TV set has to regenerate the missing pixels by interpolation.

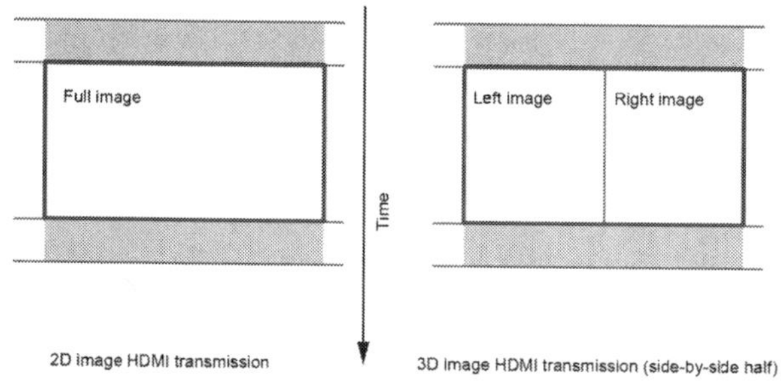

HDMI transmission of a 2D image (on the left) and a 3D side-by-side (half) image (on the right)

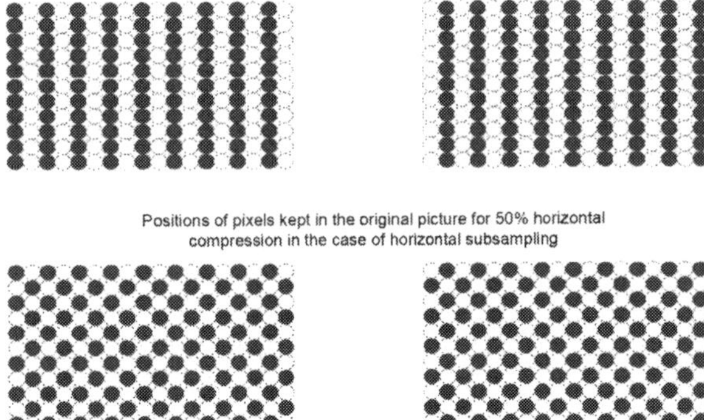

Positions of pixels kept in the original picture for 50% horizontal compression in the case of horizontal subsampling

Positions of pixels kept in the original picture for 50% horizontal compression in the case of staggered or quincunx subsampling

The two subsampling methods for HDMI side-by-side transmission

HDMI Vendor Specific Data Block (VSDB)

The VSDB data block is defined by the receiving device – typically a TV set – and specifies its display capabilities, which the sending device will consider to send data in one of the formats it accepts. The VSDB block size is between 6 and 32 bytes (N = 5 to 31). If a receiver supports one or more of the functions described in the extensions below, it must indicate that in a VSDB of sufficient size. The sending device must be able to understand a VSDB of any length, as the VSDB size may increase in subsequent versions of the HDMI standard. After receiving the VSDB, the transmitter sends the video data in one of the formats accepted by the receiving device.

Bytes 6 to N are optional in the standard, but mandatory to define 3D formats. Information in bytes PB0 to PB5 are present for all HDMI receivers and do not concern 3D. We are not interested in the optional bytes related to audio and video latency, either, but since they are present or absent, depending on the value of bits 6 and 7 of byte PB8, we need them to know the positions of the following bytes.

Byte PB13, renumbered PB9 if latency information is not present, is the most important one. If its bit 7 = 1, then 3D video is present (PB13 = 0x80).

Byte PB14 gives the lengths of the extension block – hence the length of the remaining of the VSDB – and the 3D block.

The remaining bytes – there are HDMI_3D_LEN of them – contain detailed information on the 3D formats supported. These bytes are not all necessarily present. They are called "3D_Structure_ALL"

Description of the Vendor Specific Data Block (VDSB)

Byte \ Bit	7	6	5	4	3	2	1	0
PB0	Tag (= 3)			Length (= N)				
PB1	InfoFrame identifier (0x000C03) (with least significant bit first)							
PB2								
PB3								
PB4	A				B			
PB5	C				D			
PB6	Sup-ports_AI	DC_ 48 bits	DC_ 36 bits	DC_ 30 bits	DC_ Y444	Reserved (0)	Reserved (0)	DVI_ Dual
PB7	Max_TMDS_Clock							
PB8	Latency_Fields_ Present		HDMI_ Video _Present	Reserved (0)	see HDMI 1.4			
PB9 to 12	Audio and video latency information							
PB13	3D_ Present	Reserved (0)	Reserved (0)	Reserved (0)	Reserved (0)	Reserved (0)	Reserved (0)	Reserved (0)
PB14	HDMI_X__LEN			HDMI_3D_LEN				
PB15	3D_Structure_ALL (only if HDMI_XX_LEN >0 and HDMI_3D_LEN)							
...	3D_MASK (only if HDMI_XX_LEN >0 and HDMI_3D_LEN)							
...	VIC_order_X (only if HDMI_XX_LEN >0 and HDMI_3D_LEN)							
...	3D_Structure_X (only if HDMI_XX_LEN >0 and HDMI_3D_LEN)							
...	3D_Detail_X (only if HDMI_XX_LEN >0 and HDMI_3D_LEN)							
...	(...)							
...N	Reserved (0)							

(2 bytes), "3D_MASK" (2 bytes), "VIC_order_X" (4 bits), "3D_Structure_X" (4 bits), and "3D_Detail_X" (4 bits). They are used to specify the ability to receive 3D modes other than the three basic formats.

If the receiver accepts at least one 2D format at 60 Hz, it must also accept 1080p-24 Hz and 720p-60 Hz formats. If the receiver accepts at least one 2D format at 50 Hz, it must also accept the 1080p-24 Hz and 720p-50 Hz formats. In both cases, the receiver, if it declares 3D_Present = 1, should accept at least the three basic formats described above. In this case, HDMI_3D_LEN is zero and it is not necessary to add extra bytes to define the display capabilities of the receiver. However, many 3D televisions use different technologies that require different data configurations. The TV set must then signal to the source device in which format it wants to receive data. It does so through the 3D_Structure_ALL bytes and those that following them.

The HDMI 1.4 standard is still in its first version (HDMI 1.4a), but it has already listed the most common methods for displaying 3D. If a TV set uses a method other than the three basic ones, it must write in bytes PB15 to PB (n) a list of image formats supported (given by their VIC numbers) and the 3D display method used for each one. A table of up to 16 entries is provided, so a 3D TV can offer 16 specific 3D formats for 16 different image formats. This makes it possible to cover all PAL, NTSC, and HD formats, if necessary.

The HDMI 1.4 standard is not yet mature enough to present a comprehensive selection of 3D display options, but the main ones are already listed. A typical television proposes only one 3D display method for each image format, but this is not a requirement. Methods accepted by the receiving device are each marked by a bit in a 16-bit word called "3D_Structure_ALL." So, for each display format in the list, the device must assign the value 1 to the bit corresponding to each supported 3D mode.

Generally, 3D TVs use the same 3D display method for all supported image formats and cover the usual range of SD and HD TV signals with 4:3 and 16:9 aspect ratios, 50 and 60 Hz frequencies, plus the 1080p-24 Hz mode. (The latter is one of the most important because it is the preferred mode for 3D Blu-ray, since cinema content is currently shot natively at 24 frames/second only.)

HDMI 1.4 takes an additional complexity into account if 3D images are down-sampled (Side-by-Side (Half), 3D_Structure_ALL_8=1) to transmit two frames in the same number of pixels as a single image. As noted earlier, these formats obviously halve the horizontal resolution of the image, since for each view they transmit

3D Display Methods in the HDMI 1.4 Standard

Bit of 3D_Structure_ALL	3D display methods
0	Frame packing – default method
1	Field alternative
2	Line alternative
3	Side-by-Side (Full)
4	L + depth
5	L + depth + graphics + graphics-depth
6	Reserved (= 0)
7	Reserved (= 0)
8	Side-by-Side (Half) – Sub-sampling methods are described in 3D_Detail_X
9 to 15	Reserved (= 0)

* In future versions of the HDMI 1.4 standard, this table will have to be enlarged to include additional methods.

only every other pixel. The TV will recreate the missing pixels by interpolation. It is also necessary to report what subsampling method is used. It is the role of the "3D_Detail_X" structure (4 bits), which is also attached to each entry in the list.

Sub-sampling Methods for Frame-Compatible Hdmi Transmission

4-bit value of the 3D_Detail_X field	Horizontal subsampling method used	
0000	The eight methods are supported	
0001	Horizontal subsampling	The four methods are supported
0010		Odd left, odd right
0011		Odd left, even right
0100		Even left, odd right
0101		Even left, even right
0111	Quincunx subsampling	The four methods are supported
0111		Odd left, odd right
1000		Odd left, even right
1001		Even left, odd right
1010		Even left, even right
1011 to 1111	Reserved	

If a HDMI receiver supports down-sampling (3D_Structure_ALL_8=1), it must support at least the 0010 mode, and then use the odd pixels of each left and right image to display a 3D image line.

Patents and Intellectual Property

TDVision Systems has a patent portfolio for which it sells a license. This portfolio covers the 2D + Delta coding, also known as the "TDVCodec." "TDVCodec" is a key component of the MVC codec, the generic name of the "MPEG-ISO-14496-10:2008-Amendment 1" codec, which is itself an extension of AVC (ITU-H.264 Advanced Video Codec), adopted by the Blu-ray Disc Association, among others, as a format for 3D Blu-ray encoding. The patents cover the areas of acquisition, encoding, decoding, and displaying 2D + Delta stereoscopic content. Exclusive rights to the TDVision patent portfolio are negotiated by Sisvel (*www.sisvel.com*) under the name "3D Licensing Program."

Blu-ray 3D Discs

The Blu-ray disc is a medium of choice for 3D movies thanks to its full HD resolution and high capacity, so that compression is minimal, thereby ensuring excellent picture and sound quality. A Blu-ray single layer disc holds up to 25 gigabytes of data and a dual layer disc up to 50 GB. BDXL Blu-rays with three or four layers have existed in the Blu-ray specification since 2010 and offer 100 GB and 128 GB, respectively, of storage.

The maximum bit rate of a Blu-ray Profile 1 disc was initially set at 36 Mbits/s, then increased to 48 Mbits/s with Profile 2. With Profile 5, HDMI 1.4-connectivity appeared, as well as a throughput increase to 72 Mbits/s.

BDA *(Blu-ray Disc Association)*

The BDA (Universal City, USA, *www.blu-raydisc.com*) was founded in 2005. This association of Blu-ray hardware manufacturers includes all the big names in consumer electronics: Dolby, Intel, LG, Panasonic, Philips, Samsung, Sony, Technicolor, and over 160 others. It publishes technical recommendations such as *Blu-ray Disc Format*. The name "Blu-ray Disc" and associated logo are registered trademarks of the Blu-ray Disc Association.

© Blu-ray Disc Association

The official logo identifying a Blu-ray disc

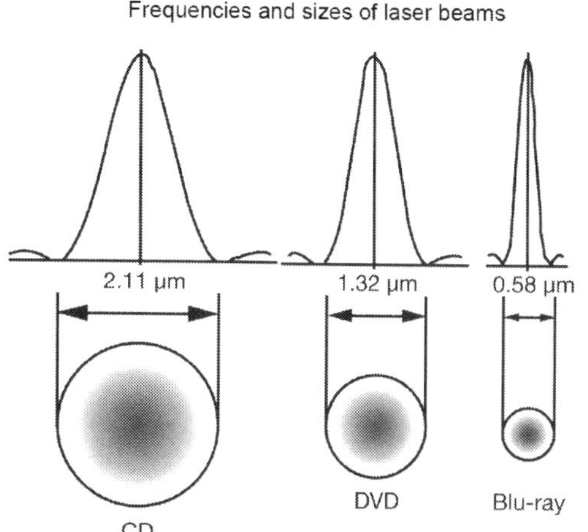

Comparison of optical disks' reading beams

Note that, despite their identical sizes, the storage space on a Blu-ray disc is greater than that of CDs and DVDs, which are earlier generations of the same concept: A 12-cm-in-diameter optical disk is covered by a spiral track on which a digital code is engraved by a laser. The higher the laser's frequency, the smaller the markings, thereby increasing the density. Inside Blu-ray players, reading lasers are at the limit of the ultraviolet end of the visible spectrum. These lasers are able to focus their rays at four different depths within the disk, thus multiplying by four the area available for storing information.

Capacities of Various Types of Optical Disk

Disk Type	Laser wavelength	Color	Number of layers	Dual face	Capacity per layer (GB)	Max Capacity (GB)
CD	780 nm	Infrared	1	No	0.7	0.7
DVD	650 nm or 635 nm	Red	1 or 2	Possible	4.7	9.4
Blu-ray	405 nm	Blue	1 or 2	No	25	50
BDXLBlu-ray	405 nm	Blue	3 or 4	No	33.4 or 32	100 or 128

In general, players can read all formats preceding them in the table, but not those following them. The double-sided DVD must be turned over manually to read the second side.

As far as coding is concerned, the Blu-ray standard allows the use of MPEG-2 (as on DVDs) and of VC-1 (a Microsoft Windows Media codec), in addition to H.264 MVC. They do not interest us, however, because they cannot be used for 3D movies.

3D images have twice as many pixels as HD images and therefore the data rate on a Blu-ray 3D is higher. H.264 MVC coding efficiently compresses the second image using the many redundancies between it and the base image. The data increase thus about 50% in size instead of 100%. The maximum throughput of a standard Blu-ray does not exceed 40 Mbits/s, but for 3D Blu-rays this maximum had to be increased to 72 Mbits/s.

The BDA, the association of Blu-ray hardware and software manufacturers, standardized Blu-rays' handling of 3D in December 2009. The 3D standard BDMV V2.4 explains how to encode menus and subtitles at varying depths. This allows displaying graphic elements overlaid on the image in a way that ensures that they always stay in front of the current scene.

Wherever possible, the 3D Blu-ray disc is compatible with existing equipment. Thus, on a computer or PlayStation 3, a software update is enough to ensure the Blu-ray reader is compatible with the new format, as long as it is able to read a disk at twice the standard speed. However, some older Blu-ray players have no update possibilities and thus will be unsuitable for 3D Blu-ray playback.

A 3D Blu-ray disc (BD3D) remains readable in 2D on an HDTV set thanks to H.264 MVC coding. Blu-rays use the 720p-60Hz, 720p-50Hz, and 1080p-24Hz formats, the latter being by far the most used, since the native frequency of motion pictures is 24 fps. Stereoscopic movies are transmitted to the TV or projector through HDMI in frame packing mode, a method that does not sacrifice any resolution, unlike the frame-compatible modes used in broadcasting.

The Three Image Formats on 3D Blu-ray Discs

Possible formats for 3D Blu-ray	Resolution	Image frequency
720p-60 Hz	1280 × 720	59.94 Hz
720p-50 Hz	1280 × 720	50 Hz
1080p-24 Hz	1920 × 1080	23.976 Hz

Important Warning

If you shoot a 3D film with the intention of distributing it on Blu-ray discs, keep in mind that 3D Blu-rays support only 24 frames per second at full resolution: 1080p-24 Hz 3D is the only full HD format accepted by the standard. Do not bother to shoot at 25 or 30 fps, on pain of having to do frequency transcoding, which is always highly detrimental to quality. However, if your movie contains a lot of fast-action scenes, let yourself be tempted by the 720p-60Hz format, which will give you smoother movements, at the cost of reduced resolution.

BD3D Disc Playback on PCs

On PCs, several applications, such as Stereoscopic Player, Corel WinDVD, and PowerDVD Ultra, can play 3D Blu-rays on any properly configured 3D display. Obviously, sufficient graphics processing power must be on board. Nvidia or ATI 3D graphics boards approved for 3D and a recent multi-core processor (Intel Core i5 or i7) are recommended.

A view of the Blu-ray player software CyberLink PowerDVD Ultra; The 3D configuration dialog box is visible in the middle of the screen.

Blu-ray 3D Professional Encoding

The MPEG-4 VMC encoder delivers a stream that is fortunately compatible with the Blu-ray players from before the advent of the Profile 5 drive. On a Profile 4 or earlier drive, images are displayed in HD, without any resolution loss, but not in 3D. The size of the MPEG-4 MVC-encoded 3D files is divided roughly into two-thirds for the base layer, which contains the complete left picture, and one-third for the dependent layer, which contains the information necessary to reconstruct the right image from the left one.

To encode a 720p or 1080p film to Blu-Ray 3D, you will have to encode it with a BD3D-compliant MPEG-4 MVC codec. The encoder must be given the uncompressed left and right images. For example, the MVC encoder from NetBlender accepts QuickTime, AVI, and YUV formats. The most important parameter you will have to specify to the encoder is the bit rate (in Mbits/s).

The encoder produces three output files. One contains the base layer and the dependent layer (my3Dmovie.264). The second file contains the base layer with the same name as the double layer, followed by 0 before the extension (my3Dmovie0.264). The third file contains the dependent layer only, with its name completed by a 1 (my3Dmovie1.264). Depending on the authoring program you use, you will have to provide either the combined file or the two separate files. This last solution offers more flexibility in bit-rate fine tuning during the multiplexing phase. However, the combined file is directly readable by many Blu-Ray applications, which allows for very quick testing of the encoding quality without having to burn a test disc – and in 3D, of course!

Where can you find an MPEG-4 MVC encoder for Blu-ray? Various companies offer them or have announced a coming roll-out. They include Cinema Craft, Sony, Sonic Solutions, and NetBlender with its DoStudio MVC encoder. Once it has been converted into MPEG-4 MVC, the movie is ready for transfer to the disc with a suitable authoring program.

Blu-ray 3D Professional Authoring

The content encapsulation process is called authoring. The choice of Blu-ray authoring software is still limited, even though the situation is improving. 3D Blu-ray authoring software is even more rare and mostly confined to the professional field.

The top two Blu-ray authoring applications are DoStudio from NetBlender and Blu-print from Sony. The NetBlender DoStudio Full 3D Bundle (*www.netblender.com*) software suite costs about $20,000. Sony's Blu-print production suite, for its part, is at the high end of the professional category. It includes Blu-print, Z-Depth for 3D subtitle placements and the 3D encoder DualStream MVC (*www.sonycreativesoftware.com/bluprint*).

Sony's Blu-print authoring system user interface

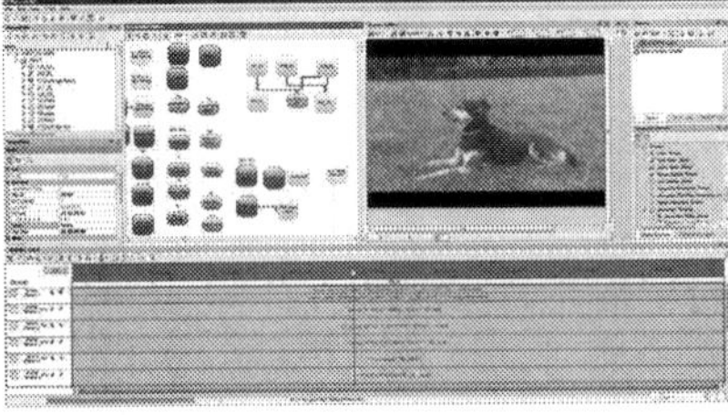

At the time of writing, Sony was selling Blu-print for around $80,000, a temporary offer including a 40% discount. Add about 10% of the price for annual maintenance.

Blu-ray authoring is more complex than one might think at first: You must select all the movie sequencess and their soundtracks and subtitles, usually along with many variations of formats and languages. Next, you have to create a fairly complex file tree, cut the movie into chapters, and prepare additional bonus tracks: cut scenes, making-of, and movie trailers. A graphically attractive menus hierarchy and a possible legal notice, which must be seen before anything else, must be developed. Then all this must finally be encoded on a single disk. Do not forget a series of more technical acts, such as setting the number of layers to use, determining the transition method from one layer to another, adding possible anti-copying encryption, activating region codes, etc. Authoring a good quality Blu-ray typically requires three weeks of hard work and strong programming skills to write all the scripts that manage the interaction with the menus.

Several points should be checked for each stream to be encoded on the disk:

- The total 3D video + audio + subtitles throughput must not exceed the allowed maximum;
- the lengths of video and audio tracks are identical; and
- the subtitles' and images' time codes must be synchronized.

Different points must be checked for menu items:

- All images have a transparency component in addition to RGB color, so they are coded with 32 bits/pixel.
- Menu items are not unnecessarily large; one-third of the screen is enough for almost any menu.

The authoring steps are as follows:

- Create the playlist by concatenating all video streams: 2D and 3D streams may be mixed in any order;
- define the usual disk information (disk ID, publisher ID, etc.);
- launch the multiplex process that will create a disk image for testing;
- check the disk image on a software player or a Sony PlayStation 3 with 3D mode activated;
- create menus and submenus with their main navigation logic, animations, buttons, and other effects, assigning each its own depth (Z value);
- compile the project and test it on a software player;
- optionally add AACS (Advanced Access Content System) encryption to protect the disk from illegal copying; and

Note
If, at this stage, the result looks correct, multiplexing will no longer be repeated, unless changes are made to the audio, vide, or subtitles stream or to the disk information.

- as soon as the encoding passes this step, retest everything on a reference drive, such as the Sony PlayStation 3, to ensure that the menus do not use more memory than expected in the player. Check the 3D for the last time, especially the depths of menus and other graphic elements that are overlaid on the image. They must always appear in the foreground in front of the scene.

Before large-scale distribution, it is always necessary to check that the Blu-ray disc runs properly on several reference players, such as Sony's PlayStation 3, Panasonic's DMP-BDT350, and Sony's BDP-S470.

Blu-ray 3D Amateur Authoring

Starting from release 10, the editing software Sony Vegas Pro lets you not only see 3D previews during editing, but also burn 3D Blu-rays with a true MVC compressor. While this does not make it a full-size authoring software package on a par with the "Pro" versions mentioned above, it does support its claim to be the cheapest solution for creating 3D Blu-rays on the market.

Sony's editing software Vegas Pro 12 includes 3D Blu-ray creation with MVC coding.

Half-Resolution Blu-Ray 3D Authoring

Besides the above methods, there are various ways for amateurs to produce 3D Blu-rays on a budget, albeit with some quality loss. One solution is simply to use authoring software designed for DVDs, but with an HD option. In that case, the software is used to encode a frame-compatible HD picture, for example by compressing the images laterally in side-by-side or over-under format. The authoring software is then used without any 3D-specific features. The movie will be encoded to Blu-ray in AVCHD, just like any movie. We still have to trick the software by injecting stereoscopic frame-compatible content. This content is divided into two parts: menus and videos. The videos will be prepared in side-by-side format with your usual editing software, the left and right images being squeezed 50% laterally, and then put one beside the other. The tricky part is making menus that will match the same size-by-side format without preventing interaction, *e.g.*, menu items that change color when the cursor passes in front of them, selection areas, etc. Fortunately, some possibilities of circumventing the problems are available at the price of a few constraints: You have to create clickable areas arranged in a single horizontal column and spanning the entire width of the screen. In this case, the interaction will work just as well in 3D as in 2D.

For example, let's illustrate the various steps of creating a simple menu with three choices: Trailers, Film, and Bonus. The result we want to see on our 3D TV – with 3D mode enabled from the start – is shown in the figure below.

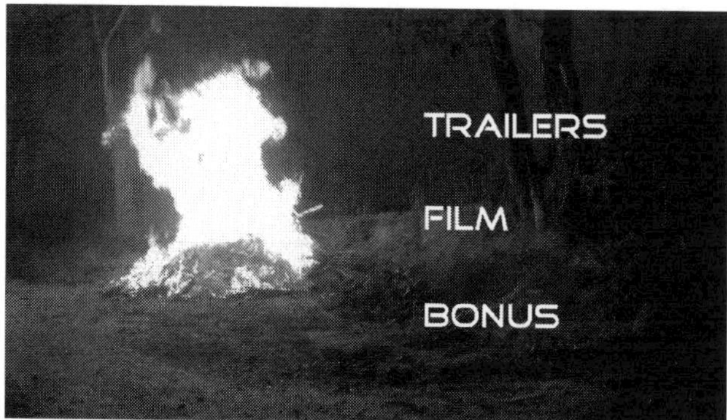

The menu of the 3D Blu-ray disc is visible in 3D but is located in the screen plane.

1. Let's select a picture from the movie in frame-compatible side-by-side mode, thus with a 2 × 960 × 1080-pixel size. We have to make sure that we have good contrast between the background and menu, and especially that the image has no out-of-screen effect. All the pixels of the menu background must be behind the screen plane, as the menu – which will be in the screen plane – will appear before the background image.

3D Background image for the 3D Blu-ray menu in frame-compatible side-by-side mode

2. Now we create the images of the three buttons. These will have a transparent background and a rectangular format covering the width of the screen. Here, we choose 150 × 1920 pixels.
3. Let's duplicate these buttons and squeeze them laterally by 50%. This gives the same size buttons (150 × 1920 pixels), but the content is anamorphic to the same extent as the background image. The button images are then saved in .png format, which preserves their transparency.

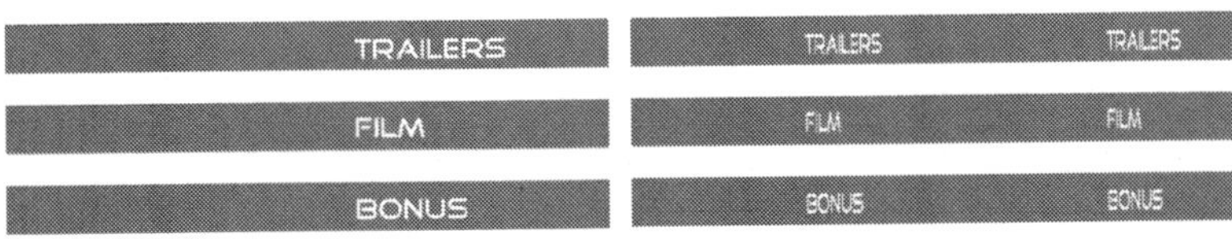

The size of the three buttons is 150 x 1920 pixels (grey areas are transparent).

Here, the same three buttons have been duplicated and squeezed horizontally.

4. The last step is to configure the authoring software to use the three buttons to launch the corresponding actions, and then burn the disc.

A view of Roxio DVDit with the finished menu, ready to be burned on a 3D Blu-ray disc in side-by-side format

The advantages of this method are simplicity, low cost, and no need for dedicated 3D Blu-ray software. In contrast, the disadvantages include the halving of resolution, the need to switch on the 3D side-by-side mode manually on some 3D TVs, and the constraint on the buttons' arrangement (they have to be in a single full-screen column).

Creating a Frame-Compatible 3D Blu-ray

To create such a disk we use the side-by-side mode so as to fool the program that manages the interaction and creates menus. Whatever the X,Y position of a click on the 3D screen displaying the side-by-side image, the program will consider the X value to be correct. And this is true regardless of whether the action is seen as located on the left or right side of the screen, since, thanks to our horizontal buttons, only the vertical position of the click matters. For something similar in over-under format, the layout would require the buttons to be on a single line instead of in a single column, which would impose very short texts on the buttons.

Remember that half-resolution full HD is nevertheless three times sharper than a regular DVD! As long as H.264 MVC encoders and 3D Blu-ray authoring programs' costs remain high, the "half resolution" method's cost/performance ratio will remain unbeatable!

Blu-ray Authoring Software Compatible with Half-Resolution

There are numerous software applications consistent with the above method. However, the most prominent for burning Blu-ray discs are Adobe's Encore, Sonic's DVDit Pro HD, and CyberLink's PowerProducer for a more accessible solution.

Adobe Encore is part of the Creative Suite Production Premium from Adobe (approximately €2,500) and is the logical complement to Adobe Premiere. It manages, *inter alia*, the 16:9 format in 1920 × 1080 pixels, the format that we are interested in to encode stereoscopic films. Video formats managed at the import step cover most common cases, such as MPEG-2, MPEG-4, H.264, MOV, AVI, and WMV. It accepts the MP3, AAC, AC3, and DTS audio formats. On the output side, Encore can burn a disk or save a disk image on a hard drive so that it can then be tested and burned on to a disk later. It lets you choose the BD-R, BD-RE, and BD-ROM Blu-ray formats, region code, copy protection scheme, and number of layers (single or double). Among the features that make it a professional product, Encore accepts masters produced in 4K and ensures that projects shot at 24 frames per second are fitted with the correct metadata at burning time. Its main drawback is its inclusion in Adobe Premiere, which makes it expensive for those using another editing software (*www.adobe.com*).

Roxio Creator Pro is the professional version of Roxio's DVD authoring software (about $109). Its interface is very similar to Adobe Encore's and it supports the same video and audio formats as inputs. Output options are also similar, with choice of region code, protection, type of Blu-ray disc, and number of layers. Its independence from the editing software used makes it an excellent choice for those not editing with Adobe Premiere (*www.roxio.com*). It also has the advantage of generating a Blu-ray or DVD from a single project without repeating the authoring phase and often tedious menus creation. The Blu-ray discs can be generated in 720p or 1080p (1080i is possible but not recommended for 3D!) at speeds of up to 40 Mbits/s; adding copy protection is possible and surround sound is managed in 5.1 and 7.1 up to 640 Kbits/s.

Note
This program calls itself "3D compatible" because it generates anaglyph videos and even includes an automatic 2D/3D conversion module. These options are not likely to produce a quality stereoscopic movie. Trust rather your usual HD editing software to create a video in frame-compatible full HD format and use Roxio Creator Pro only for creating menus and for burning the disk (*www.roxio.fr*).

Roxio Creator Pro handles AVCHD and burns 3D Blu-ray discs, as long as you add the optional "HD/Blu-ray Disc Plug-In for Roxio Creator" module ($19.99). This additional module is not free because it includes the price of the licence fees that all Blu-ray technology developers are charged. The HD/Blu-ray Disc Plug-in includes twenty high-resolution menu styles covering a wide range of themes, from family activities to sports and more. Movies generated by Roxio Creator are playable on all standard Blu-ray players, including Sony's Playstation 3.

CyberLink PowerDirector Ultra is aimed at the general public and its price is very affordable (about $85). It is also regularly included with the purchase of various Blu-ray disc writers. It combines a general-purpose movie editor with a menu generator and disc burner. Release 10 and up include 3D editing features, so it is easy to add 3D titles, 3D particles, 3D effects, 3D menus, and more. Plus, there's a broad range of 3D format output and burning options - including on 3- and 4-layer Blu-ray discs –, and even options for uploading 3D videos directly to YouTube.

The generated menus are simple and certainly less complex than what is possible with Encore and Creator Pro HD, yet powerful enough in most cases. It also offers an automatic menu generator: You just drop (by cutting and pasting) various video files into the application, and a menu is generated automatically. All this makes it a good choice for taking one's first steps in authoring 3D Blu-rays (*www.cyberlink.com*), despite sometimes frustrating ergonomics.

Transmission to Theaters

Distributing 3D content to movie theaters is fortunately much simpler than creating a 3D Blu-ray. Indeed, it suffices to encode the two video streams, soundtrack, and subtitles in the preferred format for cinema distribution, namely, the DCP (Digital Cinema Package).

3D DCP Format

A DCP is not a file but a directory containing several files. The DCP format was proposed by DCI (Digital Cinema Initiatives, *www.dcimovies.com*), a consortium including all the major Hollywood studios, and then formally adopted and published by the SMPTE (Society of Motion Picture and Television Engineers, *www.smpte.org*). The DCI Specification, Version 1.2, is available for free on the DCI website.

DCP-Encoding Hardware and Software

The DCP format is quite complex, but fortunately various software applications are able to create it for us from the original streams. In the DCP, the images are encoded using the JPEG2000 compression method. There are both hardware and software solutions for DCP encoding. The leader of the DCP hardware encoder market is Doremi Cinema (*www.doremicinema.com*) with its Rapid and Rapid-2x workstations. Many DCP encoding programs also exist, and some are even available for free!

© Doremi Technologies

The Doremi Cinema DCP encoding workstation

The main disadvantage of the software method is its slowness due to the JPEG2000 encoding of each frame of the movie, a computing power-hungry process. Fortunately, the process can also be accelerated by means of specific JPEG2000 compression boards, such as the Im-XL from Image Matters (*www.image.matters.pro*) or the Pristine from intoPIX (*www.intopix.com*).

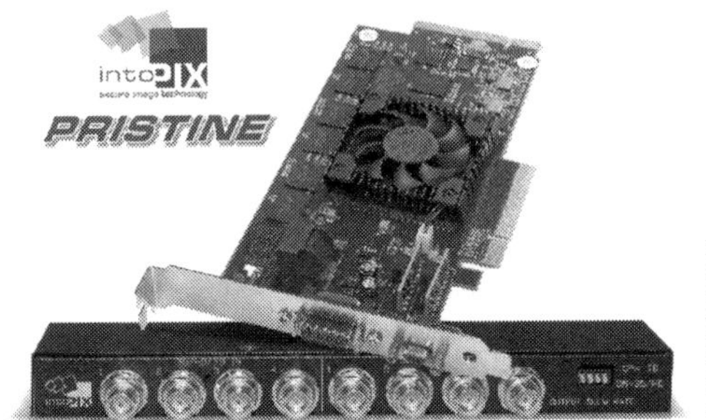

© www.intopix.com

The JPEG2000 compression-decompression hardware board from intoPIX, the ideal solution to accelerate DCP encoding

One of the best known encoding software packages is undoubtedly *easyDCP Creator+*, developed by Fraunhofer Institute in Germany (*www.easy-dcp.de*). Available for about €3500, easyDCP Creator+ can generate encrypted 3D DCP, for complete protection against illegal copying. It also generates keys (KDM) used to grant reading rights to specific projectors for specific periods.

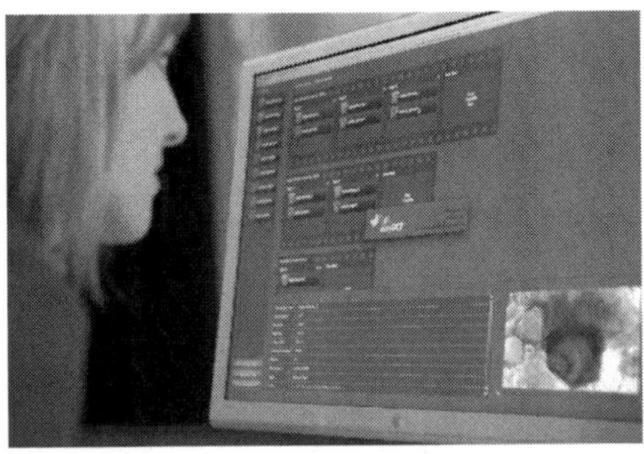

The easyDCP Creator+ software generates encrypted DCPs and the corresponding KDM keys.

DCPC, Everybody's DCP Encoder

DCPC is a 2D or 3D DCP creation software package (*cinema.terminal-entry.de*).

It comes in two versions: a free one, without copy-protection encryption, and a "Pro" one, which includes such encryption.

DCPC Free Version

Apart from protection against copying, the free version of DCPC is a comprehensive tool to create 3D DCP compatible with most cinema servers. It has been tested successfully on the servers from Doremi, XDC, Dolby, and Sony. DCPC runs on both 32- and 64-bit Windows XP, Windows Vista, and Windows 7. The interface is multilingual: English, German, and French are supported. Here we describe how to encode an unencrypted DCP from a 3D movie using DCPC.

1. Download and install DCPC, then install the ImageMagick utility that DCPC uses for JPEG2000 encoding. If you do not have separate images at the output of your editing tool, you can use VirtualDub or another utility to generate the separate views from streams in .wmv, .avi, or .mov formats.
2. Prepare your left and right video streams as suites of images in .tif, .bmp, or .dpx format, separated and numbered consecutively in two directories christened "L" and "R." The frame rate must be 24 fps. Even though DCPC accepts speeds of 24, 25, 30, and 50 fps, most digital cinema projectors are compatible only with 24 fps for 3D.

   The images must fit within a 1920 × 1080- or 2048 × 1080-pixel window, so if the aspect ratio of your movie differs from 16:9, it will appear with appropriate black bands to achieve the desired dimensions. To meet the DCP standard, at least one dimension of the image must nevertheless fill the frame. In practice, the image will be either 1920 or 2048 pixels wide, with the height depending on the film's aspect ratio. For example, the "Flat 1.85" format will use 1998 × 1080 pixels.

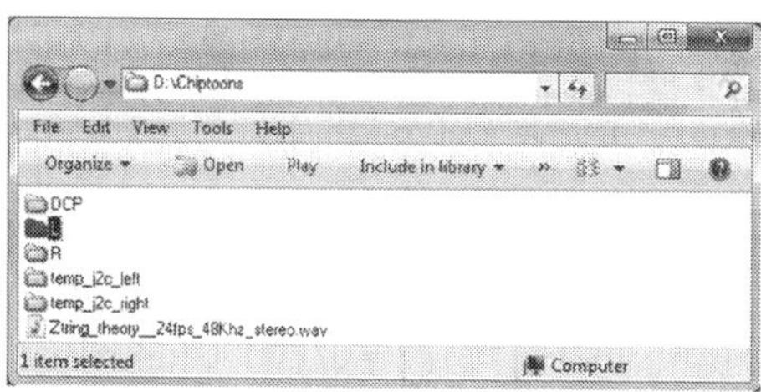

DCP creation: The directory tree shows the source (L and R) and destination (DCP) directories, plus two temporary directories created by DCPC to store the intermediate JPEG2000 images. At the bottom, you see the stereo sound file with the sound converted to the "24 bits signed" format.

3. Prepare your audio streams in .wav files (PCM, 48 kHz, 24 bits per sample, mono or stereo). If your sound is in 5.1 surround, you must put your six audio channels in a single .wav file or in pairs in three stereo files. The sound will be saved in the .wav PCM 24-bits signed format specified by the DCP standard.

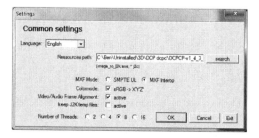

The DCPC settings dialog box

4. Tell DCPC the path to its support programs such as ImageMagick. Be careful not to insert any spaces in directory names, as this tends to confuse most scripts using ImageMagick.
5. Choose the type of film. Note that the 3D Still Image option can be selected to generate a DCP with a single image, which can be useful for displaying a message during an intermission (the display time is definable; as a rule, ten seconds is used so that the projectionist has enough time to press the "Pause" button).

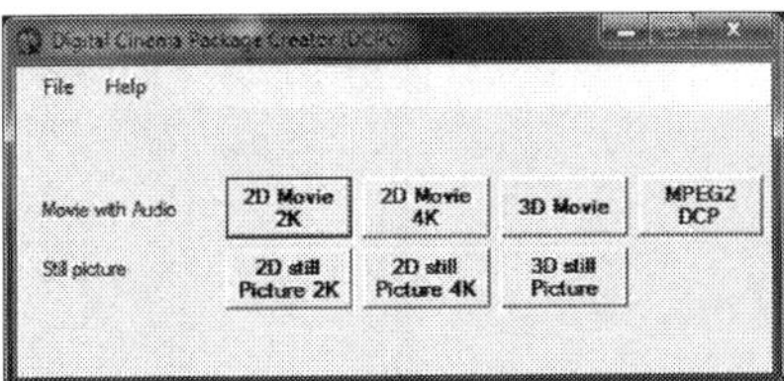

This window let you choose the type of 2D or 3D movie DCPC will encode; you may also create still panels of predefined duration.

6. Tell DCPC where your audio and image files are located.

   For a movie with stereo sound, specify a single file in the first text box. For a 5.1 or 7.1 surround-sound movie, the channels are divided into either six or eight (respectively) mono-

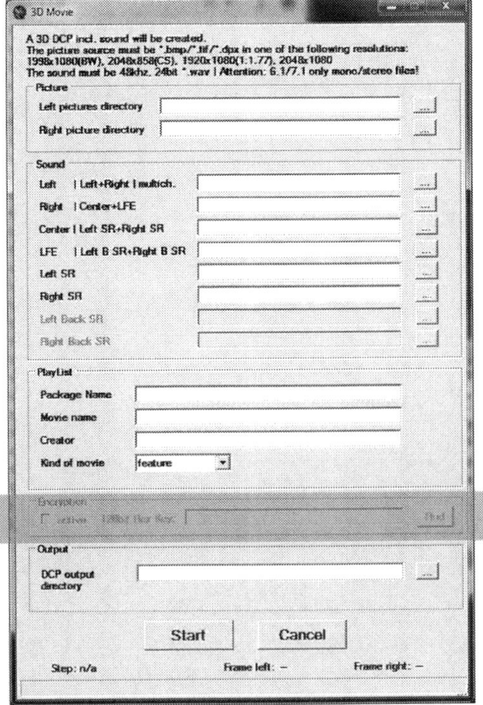

On this screen you give the names of the directories where the left and right images are located, and then the locations of the sound files (here, a single stereo file). Under "PlayList" you give the names that will be displayed in the projection booth. Under "Output" you give the directory name where the new DCP will be created. The highlighted zone is active in the Pro version only.

sonic files or three or four (respectively) stereo files, as follows:

- left and right;
- center and bass;
- left surround and right surround;
- rear left surround and rear right surround (for 7.1).

7. Click Start, check the settings in the confirmation dialog box, and then wait a few hours...! Do not forget to prepare a large enough temporary disk space on the work drive. In addition to the uncompressed source data, DCPC typically uses twice the size of these data during the process. For a 1920 × 1080 movie at 24 fps, the disk space required, including for the source images, is approximately:

N minutes × 24 × 60 × 2 × 3 × 6.22 Mb = 53.7 Gb per minute of movie,

or 3,200 Gbytes per hour of movie.

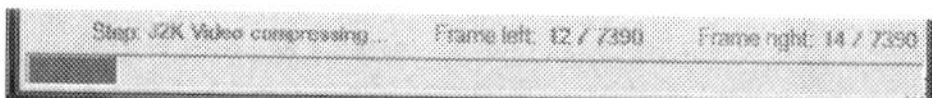

The DCPC status bar indicates the number of the frame being compressed. A simple interpolation lets you estimate the overall computation time required.

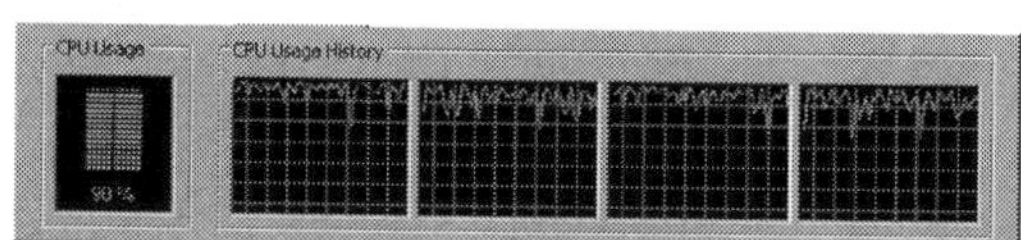

The four cores of the PC are used by four branches of the DCPC program; each one compresses one-quarter of the images into JPEG2000 format.

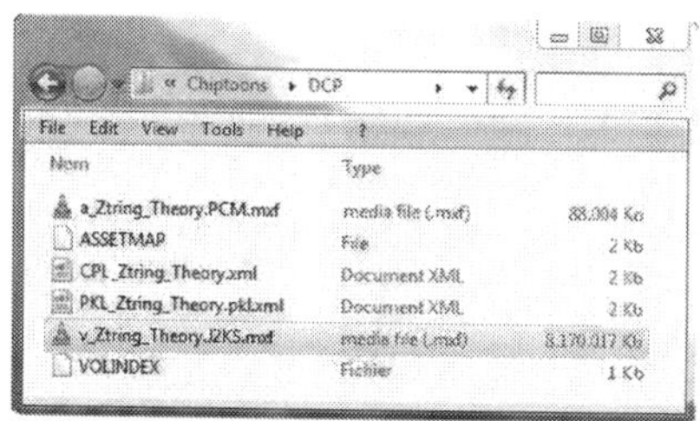

Files in the DCPC directory, ready to be transferred to theaters.

A DCP is not a file, but a directory including several files, to wit:

- a short text file indicating the number of volumes the movie is made of; usually, only one (VOLINDEX);
- a list of audio and video files (ASSETMAP);
- audio content (xxx.PCM.mxf);
- video content (xxx.J2KS.mxf);
- a packing list containing the name of the movie, its type, and its author (xxx.pkl.xml); and
- a composition playlist indicating how to reproduce the content: 3D or not, number of images per second, etc. (CPLxxx.xml).

The two .xml files have binary content. All the other files are simple text files (it is sometimes useful to look inside them to find out the basic movie parameters). A 3D DCP differs from a 2D DCP by a few parameters included in the composition playlist. Here is an excerpt of a 3D movie CPL; we see that the <FrameRate> parameter is set to 48 for a 24 fps 3D movie:

```
<?xml version="1.0" encoding="UTF-8"?>

<CompositionPlaylist xmlns="http://www.digicine.com/
PROTO-ASDCP-CPL-20040511#">

....

<msp-cpl:MainStereoscopicPicture ....>

 <Id>urn:uuid:65acac74-72ff-43f1-92b1-607fc1355a8b</Id>

 <EditRate>24 1</EditRate>

 <IntrinsicDuration>7390</IntrinsicDuration>

 <EntryPoint>0</EntryPoint>

 <Duration>7390</Duration>

 <Hash>QiHxmPLaPp0yadFz/DB1uJFiBmc=</Hash>

 <FrameRate>48 1</FrameRate>

 <ScreenAspectRatio>1.78</ScreenAspectRatio>

 </msp-cpl:MainStereoscopicPicture>

...

 </CompositionPlaylist>
```

DCPC Pro Pay Version

DCPC Pro, the pay version of DCPC, adds encryption features (€650) and a 7.1 sound module (€125). Support for subtitles and a KDM generator are also provided.

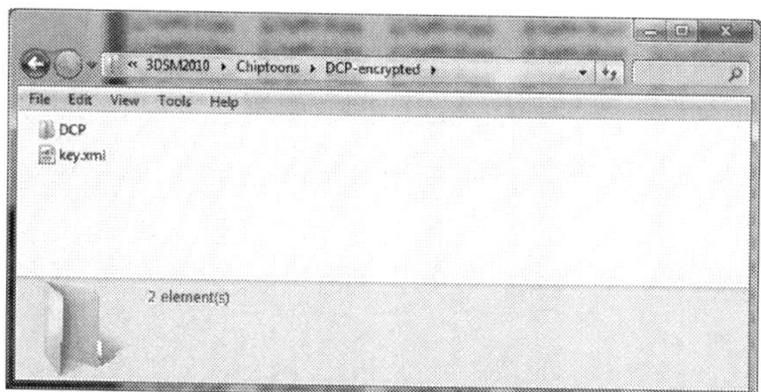

A key.xml file generated by DCPC Pro includes the encryption key necessary to generate the KDM (Key Delivery Message), the small file for decrypting the movie in the theater.

Anaglyph Distribution

In view of the foregoing, it is clear that distributing a movie or still images in anaglyph is an obsolete solution. It is nevertheless still used for several reasons: primarily because it is the only method for printing 3D images. In addition, images encoded in anaglyph can be viewed on all existing electronic media, even the oldest ones. The only constraint is the use of red-cyan glasses, which are unsightly and give unpleasant views of all red and high-contrast subjects, but are oh so cheap!

If you plan to distribute your stereoscopic videos in anaglyph format, avoid codecs with high compression ratios and check carefully how they work, for most codecs tend to degrade color information significantly, which is highly damaging to the 3D's quality in the case of anaglyphs.

To produce an anaglyph video, just keep the R component (red) of the left movie and the G and B components (green and blue) of the right movie and combine them into a single film. Editing software with 3D features, such as Magix Movie Edit Pro, offers this possibility in its 3D output options.

Stereoscopic Display 8

How to watch 3D: on a monitor or projected on a big screen?

What are the advantages and disadvantages of polarization?

How to choose a 3D projector?

Is a DisplayPort connector mandatory for 3D displays?

Are active glasses better?

Can I print 3D pictures?

Is watching a 3D movie on a PC easy?

3D Presentation Methods

Presentation methods fall into three main categories: direct screens, projectors, and static media such as paper, plastic, and, of course, film. It is clear that electronic screens are the most universal media: They can display both static and animated images and cover a size range from a few centimeters to tens of meters. Paper and other fixed media offer the advantage of being usable everywhere without expensive equipment and are distributable at very low cost.

As for the traditional distribution of 3D movies in the form of silver film, it has become anecdotal. Even if companies such as Technicolor continue this trend to satisfy a demand in the short term, it is doomed to extinction in the medium term for economic and ecological reasons, but also because of the lack of consistent quality.

3D TV

3D with sizes between 80 cm and 150 cm meet the needs of most users: watching TV at home, displaying information and advertising, and playing video games. Modern flat-screen TVs can display full HD-resolution images that are almost identical to movie theater quality, with high brightness, which is an essential element for 3D systems, and acceptable prices. Manufacturers quickly realized that the electronics needed to convert an HDTV set into a 3D Ready TV called only for a minimal extra cost. A few more functions have to be performed by the electronic circuits: You have to show twice as many frames per second and add a way to control infrared active glasses. The corresponding cost increase amounts to barely a few dollars. On the other hand, the increase in the selling price is more on the order of 10 to 40%, and the associated profit margin justifies the advertising efforts for 3D TV, even in these times when available content is still very scarce. 3D TVs are usually provided with active glasses, but in recent months, more and more passive sets that use the same cheap polarized glasses as offered in theaters have been proposed.

Manufacturers also swiftly agreed on the interface of choice for 3D content, namely, the HDMI 1.4 connector described in the previous chapter. This small connector can receive images from a digital set-top box, a Blu-ray or DVD player, or a game console.

Active or Passive 3D TV?

A whole range of both active and passive 3D TVs, mainly in the 40- to 65-inch size range (100 to 160 cm diagonal), have hit the market since 2010. All have an aspect ratio of 16:9. All the major brands - Panasonic, Samsung, LG, Sony, etc. - have their own 3D products.

A Panasonic 3D TV set

© B. Michel

Instead of displaying images faster and sending slightly different pictures to each eye alternately thanks to expensive active glasses, passive 3D TVs use inexpensive polarized glasses, like the RealD ones used in movie theaters. A passive 3D TV has a special filter that polarizes the image's odd lines in one direction and even lines in another direction. This means that only half of the pixels are visible to each eye through the glasses, which practically halves the resolution.

Pros and cons of passive 3D TV. On the plus side, passive glasses are light and inexpensive, and the viewer's vision is not disturbed by the fast 120 Hz flickering of active glasses. On the minus side, the image quality is lower, as only half of the pixels are visible to each eye. However, this is not as bad as one might imagine, because the human brain is a marvelous tool able to reconstruct fine details from information coming from the two views.

Passive may become a good solution if 1) many people are usually watching, and/or 2) the heavier expensive active glasses risk being stolen or broken. Active may be the first choice if you are usually just a few people carefully watching high-quality 3D Blu-rays in a dark environment. Of course, everybody hopes that high-resolution autostereoscopic 3D TV sets will become available in the near future, but today the only real choice we have is active versus passive.

Choosing a TV set is always the result of a compromise: Active vs. passive is part of the equation, but budget and size must also be taken into account. Do not forget that the last factor is very important, for a 3D show can be truly immersive only if the apparent size of the screen is large enough. So, when it comes to screens, the rule is usually the larger, the better!

Active Glasses

Active glasses have liquid crystal shutters blocking vision alternately for each eye in synchrony with the projector or TV. Synchronization is usually done by infrared transmission, such as with NVIDIA and XpanD glasses. However, there are solutions using radio frequency synchronization, for example the "NVIDIA 3D Vision Pro" system. The latter method is somewhat more expensive, but has advantages in environments where other infrared signals, such as wireless headphones for simultaneous translation at conferences, other IR commands from the stands neighbors in an exhibition or a museum, or other 3D screens in a large control room, and so on, can interfere with the glasses' signals.

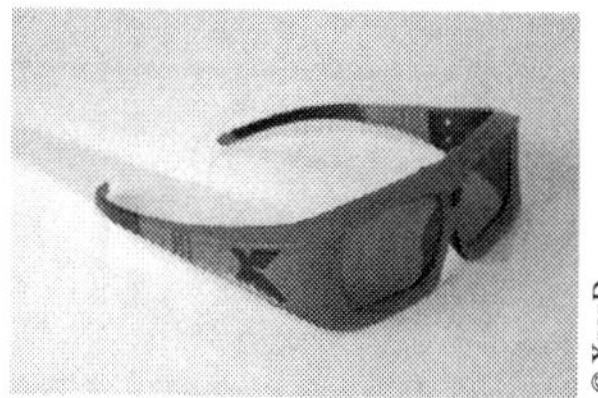

© XpanD

The X101 infrared-controled active glasses are manufactured by XpanD.

In the case of movie projection, the IR emitter connected to the projector is generally located next to the projector and sends its signals towards the screen. A normal projection screen is reflective enough to send the infrared signal back to the viewers› glasses without being disrupted by the film projection. With televisions and computer monitors, the infrared emitter is placed near the screen facing the audience and connected to the PC via a USB cable.

The main drawback of the shutter glasses system is its cost: about one hundred dollar per pair! Once the number of viewers becomes large, the glasses total cost can quickly exceed that of the projector itself. In theaters, active glasses are distributed at the entrance and collected by staff when the audience leaves the room. They then have to be checked and cleaned before returning to service. The cost of all these operations is far from negligible.

Active glasses are, however, the most effective solution in terms of light output. Loss of light is barely above 50%, as each lens is opaque just a little more than half the time. Colorimetry is well preserved because, in active mode, the glasses are a neutral gray color.

The different providers of 3D TVs and cinema equipment, as well as NVIDIA for computer screens, have each developed what they considered to be the optimal solution, but these "solutions" are

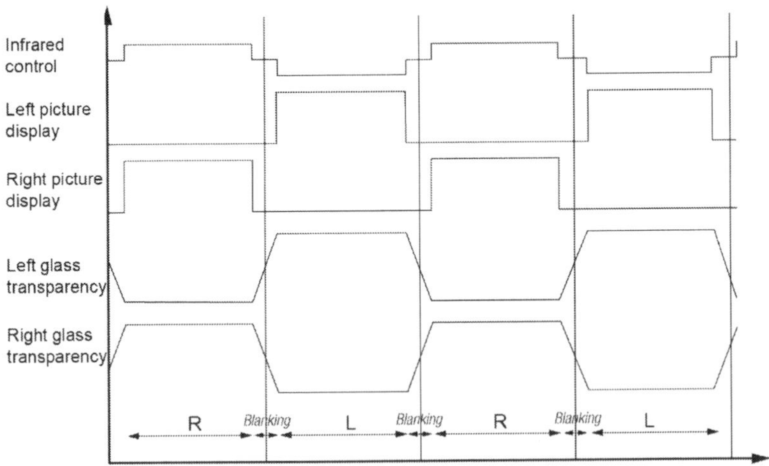

Timing diagram of active glasses display. R and L are the display times for the right and left images; blanking is the interval during which no image is displayed (the drawing is not to scale; blanking may last a very short time).

not compatible with each other. This incompatibility problem may be resolved soon, as XpanD with its model X-103 and the Japanese glasses manufacturer Sanwa are now offering "universal" glasses able to handle signals from most 3D TVs on the market. However, nothing will move without the manufacturers' good will and the establishment of an international standard. A step in this direction was made by XpanD and Panasonic, who launched the M-3DI (m-3di.com) in 2011, and the CEA (*www.ce.org*), whose R4-WG16 working group is developing a future standard for active glasses with infrared control that should be called "CEA-2038 IR-Synchronized Active Eyewear Standard." However, pending these standards' publication and wide adoption, it is always wise to try before you buy!

In summary, we can say that active glasses offer a high-quality solution but with a high price tag if the viewer count is high. As far as home use is concerned, it is probably the best compromise for from one to six viewers. For public shows, the active solution is better qualitywise, but the extra cost over passive methods will have to be borne by either the operator or the viewers.

Digital Projection

Very large images cannot be displayed on active flat screens. This type of display effectively faces numerous technological barriers when the panel size exceeds 150 cm. Some outstanding achievements reached several meters, but these feats are not reproducible on an industrial scale. Digital projectors naturally have to take over from there and can display images that are up to 30 meters wide. Various solutions are used with single or dual projectors complemented by active or passive glasses, according to needs and screen size.

Single Projector and Active Glasses

The easiest projection solutions to implement use only a single projector and synchronized active glasses. Thus, there is no setting, no matching, and consequently no risk of losing this match. High-end-projection home cinemas and movie theaters all use the single projector solution. The projector must be capable of

Temporal diagram of the triple flash method used in 3D digital cinemas with active glasses

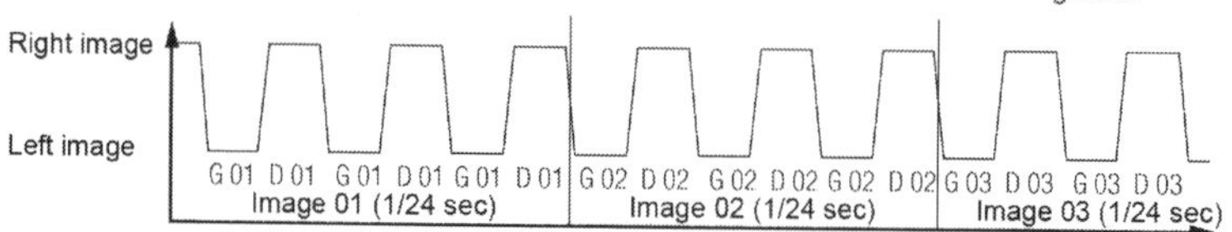

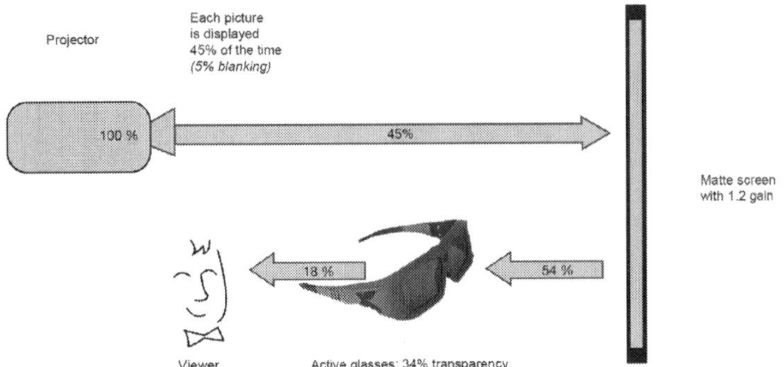

Brightness balance of an active glasses system.

displaying successive images at twice the speed of an ordinary projector, that is, at 120 instead of 60 Hz. A growing number of manufacturers provide such projectors, including Acer, DepthQ, Mitsubishi, NEC, and Viewsonic. In the case of cinema projectors, the 24 fps rate is too low to avoid a distracting flicker. The display standards require that each image is then displayed three times, a method called "triple flash." Thus, the images are alternated at a rate of 144 fps and the flicker becomes imperceptible.

Single Projector and Dolby Passive Glasses

The Dolby 3D glasses system, formerly called "Infitec" from the name of the company that discovered the process before giving it to Dolby, is based on separating the image into different colors, much like the anaglyph. But the difference is huge: Here, each filter lets red, green, and blue pass through, thus transmitting all colors to each eye, using a method called wavelength multiplex visualization. Each filter effectively lets three slices from the visible spectrum corresponding to different parts of the three regions where the eye is most sensitive, *i.e.*, red, green, and blue, pass through.

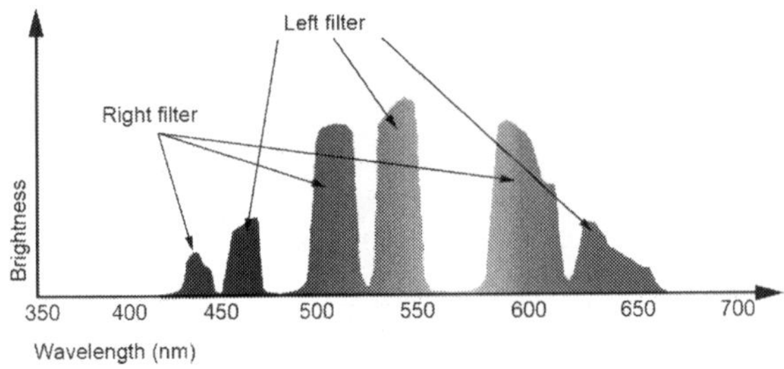

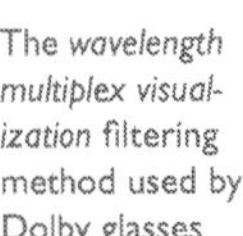

The *wavelength multiplex visualization* filtering method used by Dolby glasses

Through the Dolby glasses, the left eye perceives three spectral bands centered on 629 nm (red), 532 nm (green), and 446 nm (blue). The right eye receives slightly higher frequencies, and thus slightly shorter wavelengths, namely, 615 nm (red), 518 nm (green), and 432 nm (blue). Unfortunately, the Dolby dichroic filters are very dark and let only about 7% of the original light emitted by the projector reach the viewer's eye.

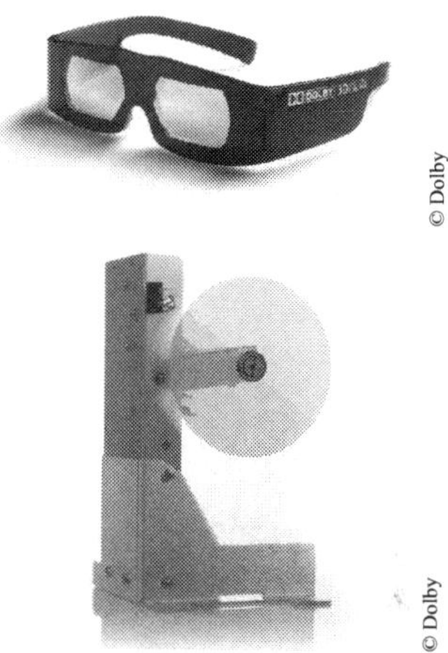
© Dolby
© Dolby
Dolby 3D glasses and the rotating wheel inserted inside the projector

Dolby glasses are passive filters, without any onboard electronics, but they are structurally complex, with many overlapping filter layers (some say 50 layers for one eye, 23 for the other), which makes them quite expensive. The filter used in the projector is a wheel rotating at high speed and synchronized with the alternating projected left and right images. This filter is inserted between the projector lamp and the electronic modulator of the projection head, thus preserving the contrast and reducing the light intensity that heats up the electronics. The timing of the Dolby display system in theaters is the same Triple Flash method used by active glasses.

The cost of the Dolby system is not negligible: about $8,000 for the rotating modulator, plus the price of the glasses, which, given their complexity, are not cheap. Dolby glasses, like active glasses, have to be recovered after use, cleaned, and reused many times. Their cost and low brightness are offset by the high reliability of the system (no electronics in the glasses) and its good color rendering. This system shares one big advantage with active glasses: It works with an ordinary screen and does not call for a metallic screen such as polarized systems require. Among its drawbacks, it should be noted that for good color rendering, movies planned for Dolby require specific color adaptation treatment at encoding time.

Single-Projector Polarized Projection

While active glasses work best, passive solutions are much cheaper. The separation of images by polarization dates back nearly a century and remains more advantageous when the number of pairs of glasses at stake is high. Indeed, at about $2 apiece, the least expensive pair of polarized glasses costs only 2% of the price of a pair of active glasses. Therefore, installing a polarized projection system makes sense if the number of spectators is greater than a dozen people.

RealD circular polarization glasses and the corresponding filter placed in front of the projector lens

Many amateur photo and video clubs and even some cinemas (mainly the very large ones) use a pair of polarized projectors.

The technique is simple: The left and right images are polarized in two complementary ways at the projector. The viewer wears glasses with matching polarizing filters letting only the desired image go through to each eye. While the principle is simple, it still entails a major constraint: It requires a metal screen that reflects light without changing its polarization. The polarizing filters block just over half of the light in the projector, and again half of what remains at the glasses level. One will therefore understand that the light output is worse than with active methods, and this lack of light must be offset by a more powerful projector. Fortunately, the mandatory presence of a metallic screen has a beneficial effect here, for metallic screens are more directional in the way they reflect light into the room. Their gain is said to larger, which means that viewers close to the projection axis will receive more light than in front of an ordinary cloth screen. However, the spectators who are at the edges of the room will receive much less light and, what is more, it will be spread unevenly across the screen. The moral is: In theaters with polarized system, sit as much as possible in the middle of a row to benefit from the highest brightness and best light distribution.

Linear or Circular Polarization

There are two types of polarization: linear and circular. Linear filters are identical to those of polarizing sunglasses, but inclined differently: Instead of their both being oriented horizontally (0°), the left and right filters are placed at a 90-degree angle to each other, usually at 45° and 135°. The filters placed in the two images' projection paths are tilted the same way. The disadvantage of this method is that if a viewer nods, the glasses will no longer be angled correctly and s/he will see "ghost" parts of the left image passes through the right lens and vice versa.

Circular polarized filters are left or right polarized; they have the advantage of being insensitive to head tilt. This advantage has a cost, however as circular polarizers are more difficult to produce and thus a bit more expensive.

On the projection side, we use either two projectors, each with a fixed filter, or a single projector at double speed and a dynamic filter that changes polarization in sync with the image display sequencing.

Ghostbusted Version

Ghosting or "crosstalk" is one of the classic flaws of polarized systems. As filters are not perfect, they let through a small percentage of the image that they are not supposed to, giving the unpleasant impression of seeing double. Movie studios were still recently preparing special versions of their movies for polarized theaters. In these versions, a small percentage (usually 2%) of the left image's brightness is subtracted from the right image and vice versa. Thus, at projection time, the flaw is added to the precorrection applied by the studio to give the viewer a correct image. Creating such "ghostbusted" versions of movies has now disappeared in favor of a specialized algorithm implemented directly in the projector's electronics. It is therefore now possible to use a standard stereoscopic version in a polarized projection room simply by configuring the projector so that it activates the integrated deghosting method.

Projection on Two Polarized Projectors

This consists of two projectors, each displaying one image for a single eye, fitted with polarizing filters oriented perpendicularly to each other. The two images are projected onto the same silver screen, which reflects the polarized light back to the spectators wearing glasses matching the projector's filters so as to block the image not intended for the left or right eye. The simplicity of the solution makes it reliable and cheap. However, a number of disadvantages must be taken into account, such as significant loss of light, the presence of ghosts, or the projectors' sometimes difficult mechanical pairing.

Polarizing Filters

Polarizing filters absorb more than half of the light that passes through them; their light output is between 42 and 46%. We deduce from this that they must withstand severe thermal stress. Glass filters are thus better able to withstand such stress than plastic ones. As the light output of projectors may become greater for large screens, one must adjust the filters' sizes accordingly. Moreover, the filters' mechanical mounting must allow them to cool properly. To minimize heating, the filters should be positioned as far as possible from the projector according to their size. For example, if only a 7-by-7 cm square on a 10-by-10 cm polarizing filter is used, the illuminated area will sustain twice as much heat as if the entire surface of the filter was used (since the same illumination will go through 49 cm^2 instead of 100 cm^2). It is easy to check the filter's temperature after projection using a thermometer; if it exceeds 70°C, the lifetime of the filter will be reduced. A simple low-voltage small fan is enough to increase the system's reliability at little cost.

Dual Projector Setup

The dual projector is an economical solution, but as soon as you start designing your own setup, you are forced to draw up a list of its constraints. The first consequence of the double projection is the need for two separate, synchronized video sources. The obvious solution is to use a PC with a dual DVI or HDMI video output. The display drivers and software available on the market allow you to send to each of the two screens the image for the corresponding eye.

Remark
Take care nevertheless to avoid too high gains if viewers are not well centered along the screen axis, as the most off-center portions of the audience will see excessive brightness variations.

To build a 3D projection system, the following suffices:

- a computer with a powerful graphics card with dual DVI or HDMI outputs. Many PCs labeled "gamer" have the computing and graphics power needed to display 3D movies. The computer must include hard drives large enough to store lots of content locally, keeping in mind that 3D movies occupy twice as much disk space as their 2D counterparts;
- two identical projectors with full HD resolution and a brightness sufficient to offset the polarizing filters' losses. At equivalent prices, preference should be given to the model with the highest contrast. DLP projectors are often the best in this regard, with values exceeding 3,000:1;
- a pair of polarizing filters. Two solutions are possible here: linear or circular polarization. Circular filters are more expensive, but slightly better in removing ghosts and insensitive to the tilt of the viewer's head. However, linear filters and the associated goggles are more readily available and cheaper. Therefore, it will be solution that is adopted most often;
- a silver screen of appropriate size. Careful! Even if the suppliers market them as rollable and washable, these screens are fragile and handling them too frequently is not advisable. An important parameter of the screen is its gain. In general, silver screens have a high gain, which increases brightness;
- passive 3D glasses, polarized in the same way as the filters used for the projector, *i.e.*, linearly or circularly. Again, circular polarization will increase costs; and

Budget

As a guide, the setup (without the computer) for a 20-seat room with a 3-meter-wide screen will cost about:

- two 3,500-lumen full HD projectors: 2 x $900;
- a pair of linear polarizing filters: 2 x $100;
- a 3.4 x 1.9 m silver screen with 2.5 gain (canvas and frame): $1,000;
- a home theater amplifier with six speakers: $500;
- 20 pairs of linear passive glasses: $100; and
- transportation, small hardware, connectors, and miscellaneous accessories: $400.

Total: $4,000

- a rigid attachment for securing the two projectors to each other and to the wall or ceiling facing the screen. You should never underestimate this point because the quality of 3D obtained will depend on how precisely the projectors are aligned. As a rule, the projectors are mounted one above the other so that only a vertical adjustment of the image is necessary to achieve alignment. The same assembly should also keep the two polarizing filters in front of the projectors in place. Look out for the setup's ventilation: Do not block the projectors' air vents – both inlets and outlets – and make sure that the filters are given plenty of cooling.

Dual Projector Calibration

Once the dual projector has been installed, you still have to adjust it. A wall-, floor-, or ceiling-mounted installation is always preferable, if the premises allow it. In this way, the geometric settings will remain correct longer.

You begin by setting the frame of the left projector so that it fills the screen. Then you do the same with the right projector after turning off the first. After that is done, with both projec-

Example of a total disparity calibration image

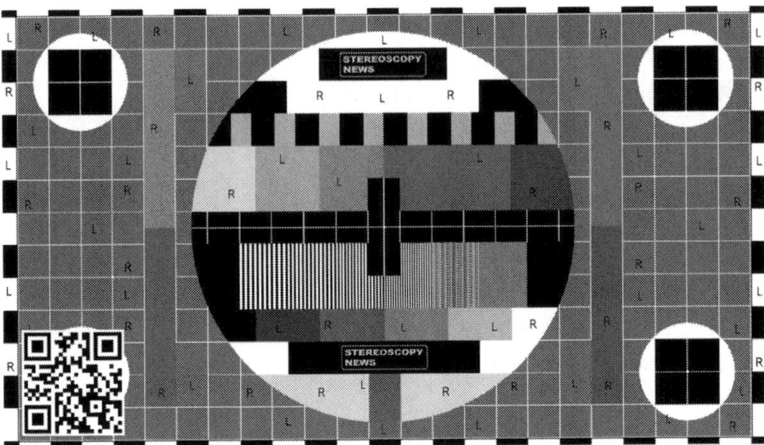

The same image when both views are perfectly superimposed

tors powered on and displaying identical images, you adjust the projectors' mechanical positions and internal settings until the images are perfectly superimposed.

Fine adjustment of the two projectors' geometry is essential but not sufficient to obtain a perfectly satisfactory effect. The brightness, contrast, and color must also be balanced. For this valuable assistance is provided by a so-called "total disparity" test image. The two images of such a pair are superimposed to reveal a set of lines and colored areas that are perfectly continuous when viewed without glasses. In addition, no pixel in the left image will contain anything other than black if the corresponding pixel of the image is colored, and vice versa. So once they are superimposed on the projection screen, the two images of the total disparity pair should form a seamless image. The slightest difference in color or brightness is clearly visible. The areas projected by each projector are identified by the letters L and R to distinguish them easily and to allow adjustment of the corresponding projector.

Polarization Systems

As noted above, it is possible to use the polarized system with a single projector. In this case, you need a projector capable of displaying at double the usual speed, as with active glasses. You must also add an alternating synchronized filter to polarize the successive images one way and then another way. In cinemas, the rhythm of alternation is the already mentioned 144 Hz Triple Flash.

Active Polarizers

The synchronization filter may be electronic or mechanical. Two suppliers share the bulk of this market: RealD, which uses an electronic filter called "Z-Screen" in front of the projector, and MasterImage, which employs a rotating disk that is likewise placed in front of the projector lens.

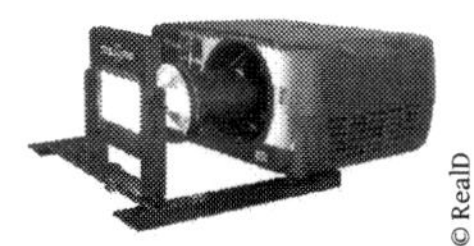
© RealD

The Z-Screen polarizer from RealD has no moving parts

In addition to their low cost, the passive glasses used by these systems hugely simplify theater operation. The cheap glasses can be simply sold to the viewers, who can keep them for future projections. This eliminates the steps of collecting, checking, and cleaning glasses. RealD recommends the use of a silver screen with 2.3 or 2.4 gain. With this type of screen, the apparent brightness for a well positioned viewer is greatly improved, making it possible to recover a portion of the light lost in the various filters. The higher the gain, the narrower the optimal viewing area.

© Masterimage

The rotating polarizer from MasterImage is usually mounted on a trolley so it is easy to move from one theater to the next

In very large theaters, it will not be possible to use a high gain screen without penalizing spectators located on the sides.

The viewing position for a 2.4 gain screen must remain within a 17° angle of the direction of projection (compared with 50° for a 1.4 gain screen). The minimum gain for an off-center viewer is then halved. For large venues, RealD offers the RealD XL system, which recovers some of the light blocked by the filter placed in front of the projector and returns it to the screen. This increases the complexity and difficulty of installation, but brightness and efficiency are improved.
Another way to reflect more light towards the audience and expand the comfortable viewing area is to use a cylindrical screen instead of a flat one. The curvature is usually 5%: the screen shows a depth of curvature that is 5% of its width. For a 12-meter-wide movie screen, the screen's maximum depth of curvature would be 60 cm. Curved screens are highly recommended for 3D cinema.

Passive Polarizers on Sony 4K

The Sony 4K projector is a rather special case. It is effectively a single projector that emits two different images simultaneously thanks to a specific optical setup.

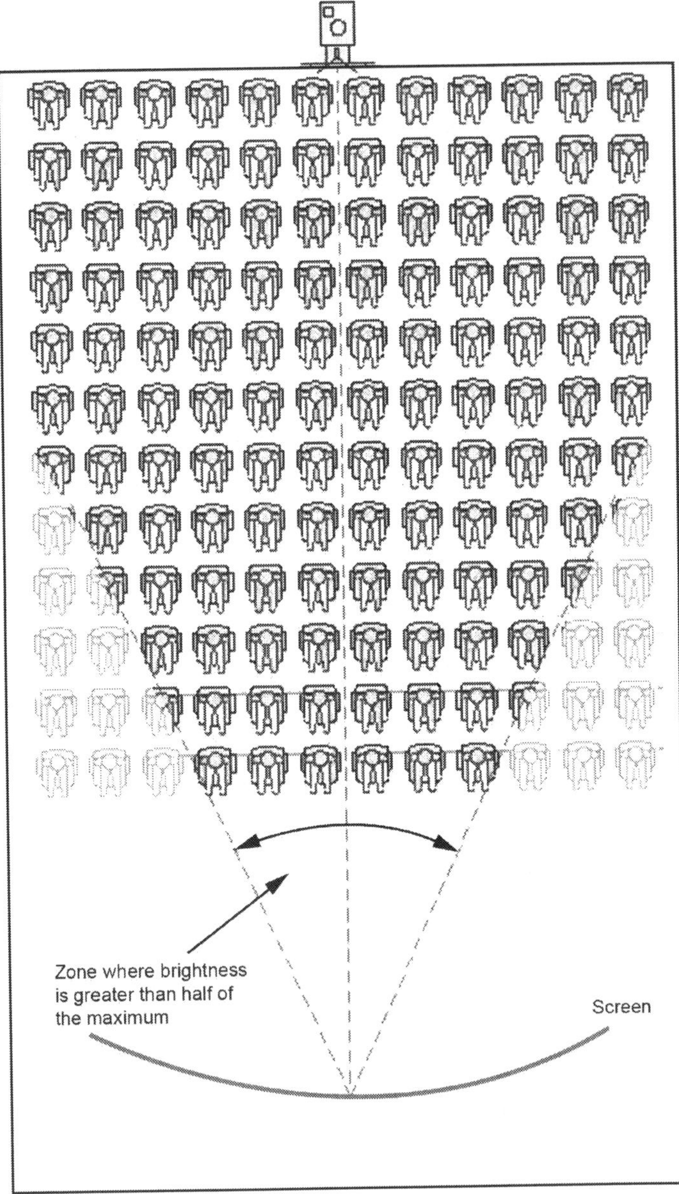

Optimal vision zone in a 3D cinema

As can be seen in the diagram **below**, the Sony 4K projector, with its 4096 × 2160 pixels resolution, is used to project the left and right images simultaneously, one above the other. A set of lenses and prisms then projects these two images onto the same location on the screen. Each lens is provided with a RealD circular polarized filter and the whole setup works like a dual projector setup. Of course, the optical system must be calibrated so that the two images overlap perfectly, as with dual projectors. The Sony 4K's advantage over dual projector systems is the use of a single lamp, so no difference in brightness is caused by the bulbs' aging. The images do not have to be alternated at high frequency, as in conventional single projector systems, and this increases the perceived brightness by avoiding the downtimes when the

Sony 4K 4096 x 2160 pixels picture

Unused area

2048 x 1080 pixels

2048 x 1080 pixels

Unused area

Layout of the two 2K images in the frame of a Sony 4K projector

The 3D optical adapter of the Sony 4K projector

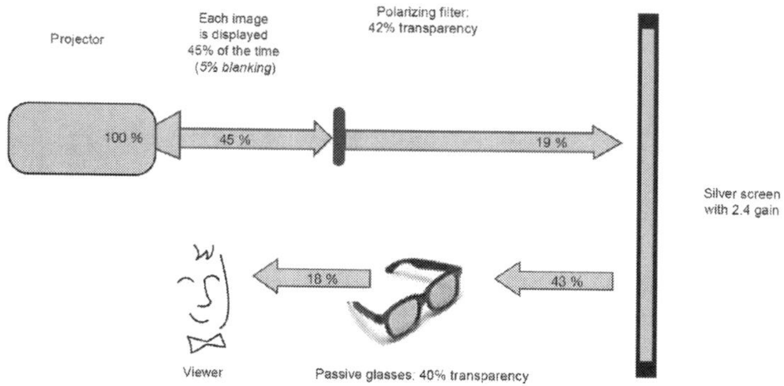

Brightness balance of polarized projection

Screen Gain

Screen gain is a measure of the screen's reflectivity. A high gain screen does not create additional light; it simply shifts more of it toward the center and less towards the edges. The gain is measured relative to a reference (magnesium carbonate) according to a standard known as *British Standard BS5550*. A matte white screen gain is 0.8-1.0. A screen usually has an average gain of 1.2-1.4. The metallic high-gain screens used for stereoscopic projections have gains between 1.8 and 2.5. Above 1.8, the risk of observing "hot spots," that is, reflections of the projector's lamp, becomes important. One can find now on the market 2.2-gain matte white projection screens specifically designed for the Dolby system; it's the screen's high gain offers a welcome increase in brightness while the screen remains compatible with conventional 2D projection.

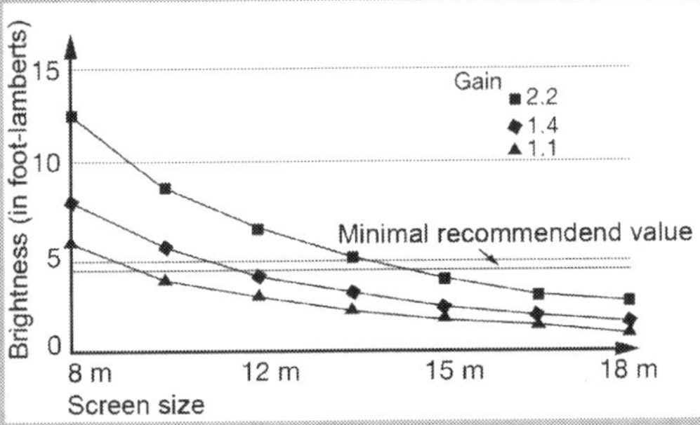

Ratio between gain and brightness with a silver screen (here in the case of a 20,000-lumen projector)

projector has to be shut off to black between two images. However, only the DCI-compliant cinema servers from Doremi and Sony are currently able to feed these projectors for 3D projection.

Lasers and Brightness

The movie projectors of the future will be illuminated by laser sources, which are more efficient, longer-lasting, and brighter than the xenon lamps in use today.

Laser sources are a combination of three lasers with frequencies of 615, 546, and 455 nanometers that generate the three components – red, green, and blue – that combine to give all the colors of the spectrum. The range of possible colors, called the "color gamut," is even wider than that of xenon projectors and exceeds the official DCI standards.

Laser Sources

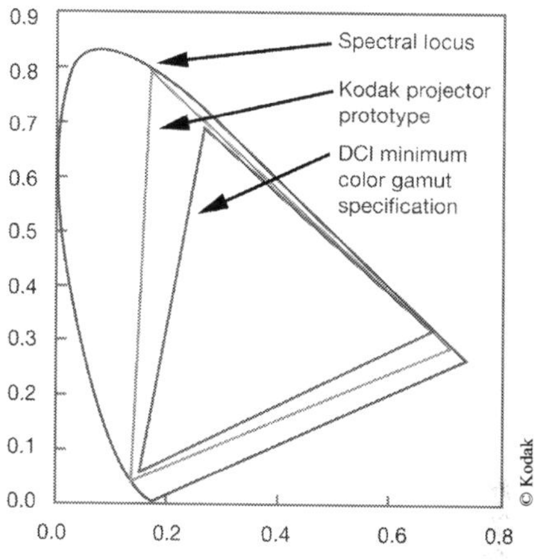

Color gamut accessible to Kodak's prototype laser projector

Laser diodes are light sources that have only one very specific color. They cover a narrow spectral band and overflow very little. Consequently, a laser emits neither infrared nor ultraviolet waves, unlike conventional xenon lamps, in which only 5% of the energy is converted into useful light. Furthermore, lasers emit their light in a specific direction, so they do not need reflectors to direct light toward the projection head, giving yet another efficiency gain. As laser light scatters little, blacks remain black, while the light areas of the screen can be very bright. Laser projectors therefore offer a remarkable contrast ratio of up to 20,000:1.

The small size of a laser source and its directivity offer yet another advantage: The projectors' lenses will no longer need to be as large. It will be possible to work comfortably with f/6 lens opening instead of the f/2.4 used now, reducing the lens's diameter by 60% and its length by 80%. When you know that the laser source can easily be deported and the light sent to the projector via an optical fiber, it is clear that the reduction in projector size will be spectacular!

Best of all, lasers have a very long lifespan, which means that, in practice, it will never be necessary to change the light sources during the projector's life. In contrast, a large xenon lamp costs well over $500 and lasts only a thousand hours or so, while replacing it is always risky, for the bulbs are fragile and filled with a high-pressure gas. A protective suit, mask, and gloves are required to replace them. Because of its short life, a movie projector bulb must be changed several times a year.

Stereoscopy, for its part, currently suffers from a lack of light that lasers can fill. Laser brightnesses of between 7 and 8 foot-lamberts are said to be reached without problem, where conventional projectors with 6 kW xenon lamps are struggling to provide 3.5 foot-lamberts. However, the laser offers yet another advantage: Its light is naturally polarized. It is therefore simple to provide,

Prototype laser projector for 2D and 3D cinema

with an even number of laser diodes in the light source, a source whose polarization can change for every other image. 3D thus becomes available with simple passive glasses, without any polarizer mounted in front of the projector, and therefore without losing half the brightness at this step. The overall brightness gain would be 65% compared with a conventional passive 3D projector.

Laser sources are thus advantageous with regard to all important issues: color rendering, cost, efficiency, brightness, and adaptation to stereoscopic passive glasses. Why, then, are they not yet fitted to all projectors?

Simply because a final problem barred the way for the engineers in charge, that of eliminating the specific flicker (a.k.a. speckle) observed when a laser illuminates a rough surface. These flashes are caused by phase variations due to the roughness of the projection screen surface. Solutions have been found from 2010 by Kodak and Laser Light Engines, the two companies at the forefront of the field, have found solutions to this problem in the past few years (from 2010 on). Part of the problem is solved by switching from high frequency polarization to another in the laser source. As this polarization change is done at the laser level, it will not cost anything.

The main technical barriers seem now to have been overcome mastered and we can therefore expect to see the first laser movie projectors in the near future.

Advantages of Laser Projectors

The mainstreaming of laser sources should lead to a progressive decrease in their cost price. Their numerous advantages will therefore offset their cost more quickly. These advantages are:

- 30 to 50% reduction in operating costs (electricity, cooling);
- no more lamp replacements (lamp lifetime > 30,000 hours);
- constant brightness and color over the projector's life;
- increased brightness and contrast; and
- no need for a polarizer for 3D projection with passive glasses.

The last step before commercialization will be to overcome a legal obstacle, for projection lasers are currently put on the same footing as the light shows that illuminate rock concert stages and

LIPA's logo

are subject to severe restrictions. The use in movie projection is different, as the laser beams are spread over a large area, the screen, rather than being concentrated as they are in light shows. That is why manufacturers of laser sources created LIPA (Laser Illuminated Projection Association, *www.lipainfo.org*), a lobby pushing for adequate U.S. regulation, in 2010.

No doubt other countries will follow suit once this goal is achieved. LIPA's founding members are Barco, Christie Digital, Disney, IMAX, Kodak, Laser Light Engines, NEC, Necsel, and Sony.

Choosing a 3D Projection System

What is the best choice among all possible technologies? The main criteria are the budget, the size and type of audience, and quality. Active solutions are the best, but they are the most expensive per viewer and therefore used mainly for small meetings. The type of content to project, ease of handling glasses (disposable or not), and number of shows will also influence the choice. The table below helps tease out the problem.

3D projection solutions compared

3D Display	Glasses	Screen (m)	# viewers	Source	Cost (K$)	Quality	Ease of use
LCD monitor	Active	0.4-0.5	1-3	TV/PC	<1	++	++
Control monitor	Active	0.4-0.7	1-3	Server	5-10	+++	++
TV3D	Active	0.7-1.4	1-5	TV	2-6	+++	+++
Projector	Active	2-3	5-15	TV/PC	4-10	++	+
Projector	Passive	2-3*	10-30	TV/PC	4-15	++	++
Double projector	Passive	2.5-4*	10-100	TV/PC	4-30	++	++
D-Cinema 2K Projector	Passive	10-17*	100-500	DCI Server	60-90	+++	++++
D-Cinema 2K Projector	Active	10-22	100-800	DCI Server	70-130	++++	++
D-Cinema 4K Projector	Passive	15-30*	300-1 000	DCI Server (Sony or Doremi)	>130	++++	++

Note: Screens marked (*) require a special silver canvas that can be removed if 2D movies are also to be shown.

Theater Size

Once the method is selected, one has to choose the ideal projector and its screen according to the room characteristics. The projector's position with respect to the screen is determined by the theater's geometry, which will also determine the choice of lens throw. The required brightness will dictate the projector's power. We must also take into account the fact that the brightness may drop by almost 40% during the lifetime of a projection lamp (600 to 2,500 hours). If we want a minimum brightness at any time, we have to take a "lamp-wear margin" into account.

For optimal 3D projection, in a well darkened room, brightness should ideally be between 4.5 and 7 foot-lamberts, or 17-24 nits. For an optimal 2D projection, the values are usually much higher, from 11 to 17 foot-lamberts or 38-58 nits.

Brightness required for movie theater projection

Screen	Brightness (foot-lamberts)	Brightness (nits)
SMPTE D-Cinema standard	11-17	38-58
SMPTE 35 mm cinema standard	16	55
Minimum allowed for 3D	4.5	15
Optimum 3D	7	24
Optimum 3D with lamp-wear margin	10	34

Note: 1 foot-lambert = 3.426 nits = 3.426 lumens/square meter/steradian.
1 nit = 0.291 foot-lambert = 1 lumen/square meter/steradian.

Brightness

If you know a projector's characteristics, you can determine the zoom value needed to achieve the correct image size and calculate the brightness. But this is not a trivial calculation, and computational errors can have serious consequences! Fortunately, there is a comprehensive online calculator on Projector Central (*www.projectorcentral.com*) that lets you find the brightness achieved with a typical configuration. You then have to check whether this light level is sufficient to meet your requirements.

The input parameters of the calculation are:

- projector model (giving the lamp's power and extreme zoom values);
- screen size;
- screen gain;
- distance between projector and screen (throw distance); and
- aspect ratio: 4:3, 16:9, or another ratio.

The online calculator determines instantly the screen's brightness under normal circumstances for flat (2D) projection. To account for the stereoscopic system loss, the resulting brightness, given in nits (1 nit = 1 candela/m²), must be multiplied by the efficiency factor of the 3D system used. This factor depends on the type of filter, type of glasses, and number of projectors (1 or 2).

Efficiency factors for various 3D projection systems

3D glasses system	# of projectors	Screen Gain	efficiency in percentage of standard projector nominal brightness (estimated)
Active XpanD, NVIDIA	1	1.2	18%
Active XpanD, NVIDIA	1	2.2	32%
Passive Dolby	1	1.2	7%
Passive Dolby	1	2.2	12%
Polarized, dynamic filters (MasterImage, RealD, others)	1	2.4*	18%
Polarized RealD XL	1	2.4*	25%
Polarized, passive filters	2	2.4*	36 %

Note: Screens marked * are silver coated.

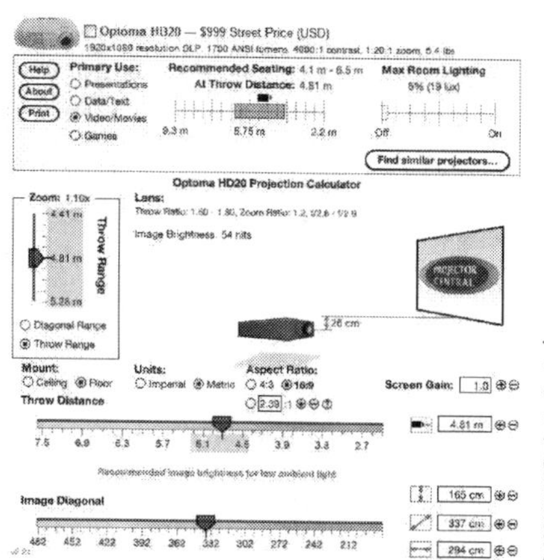

The online brigthness calculator from *www.projectorcentral.com*

It is possible that in seeking a solution, you can find yourself faced with different brightness units expressed in nits or in foot-lamberts. There is also an online converter that specializes in brightness units.

- Projection parameters computer: *www.projectorcentral.com/projection-calculator-pro.cfm.*
- Brigthness units conversion: *www.unitconversion.org/luminance/foot-lamberts-to-nits-conversion.html.*

Canvas Screen

As noted above, the room size and projector power may sometimes require choosing a high-gain screen or a silver screen. If choosing a medium-gain canvas is possible, it is better to adopt that solution because of its larger positioning leeway for projector placement in front of the screen. For silver screens, we must also take their fragility into account. If you plan to move your canvas screen frequently, avoid metallic ones as much as possible, and if you really must move one, always have an appropriate transportation cylinder on hand. Never bend a metallic screen, as the folds will leave a mark in the canvas that may be impossible to remove!

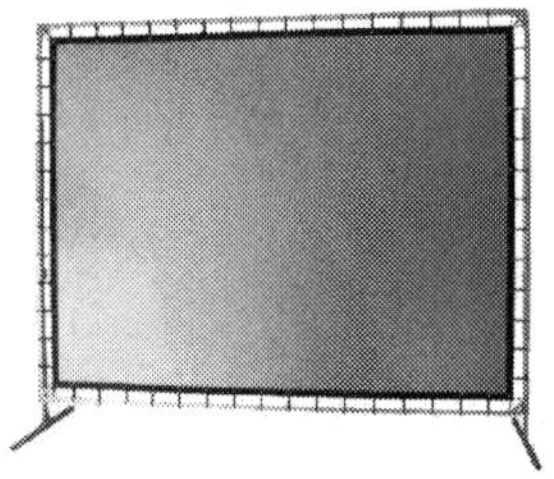
© B. Michel

A 3-meter silver screen with its frame and tensioning rubber bands

Frame: Do not forget the screen's frame! A nice screen is a taut screen. It is therefore necessary to fix the screen to a rigid frame with rubber links that will maintain a uniform tension across the surface and avoid folds that are impossible to get rid of.

Curvature: The larger the screen, the greater the required brightness. In the larger sizes, it is more elegant to increase apparent screen brightness by bending it cylindrically toward the audience rather than to increase projector power. This search for maximum brightness is especially necessary in Dolby theaters, but all systems should benefit from added light. Please note that the projector focus settings may limit the acceptable curvature; it is always best to consult the screen supplier before ordering a curved screen frame.

Perforations: For 3-meter-wide screens and up the speakers must be placed behind the screen. You then must choose a perforated cloth to prevent treble sound distortion. For small screens, microperforations will be preferred, as the holes are less visible from the front row.

Immersive display: While a curved screen offers advantages in terms of brightness, it also gives a better sense of immersion in the image. So why not go further and install a spherical screen as in professional simulators? There are spherical screens for personal use in sizes ranging from 1 to 5 meters that are compatible with active glasses and a 120 Hz projector.

Aspect Ratio and Masking

In movie theaters, the unused portion of the screen is usually hidden to allow maximum contrast between the white and black parts of the image. Masking is done by motorized black curtains or by mechanical or electronic masking within the projector. Masking can change frequently because movie sizes and thus aspect ratios are not standardized.

© ThinkOutOfBox

A semispherical TOOB used for single-user 3D projection

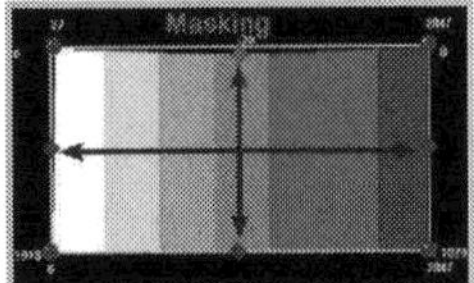

The masking user interface of a Barco D-Cinema projector

Scope on Flat

Scope on flat is a special masking technique that consists in hiding a "Flat 1.85" area (1.85:1 aspect ratio) that is much larger than the film that is projected in CinemaScope size (aspect ratio of 2.29:1) in order to allow the image to invade the area outside the frame at a few select moments. The movies *G-Force* from Walt Disney Pictures and *Dredd 3D* from Lionsgate use this technique. In a D-Cinema 2K projector, the maximum size of the projected image is 2048 × 1080 pixels. *G-Force* is a Scope film, so its size should be 1998 × 858 pixels, but the masking imposed by the studio is 1998 × 1080 pixels, allowing out-of-screen effects to invade the top and bottom black bands for about 1% of the movie's duration, which makes these moments more impressive. The masking procedure is clearly explained in the required preshow test procedure to prevent the projectionist from erroneously masking the black bands that are actually part of the movie.

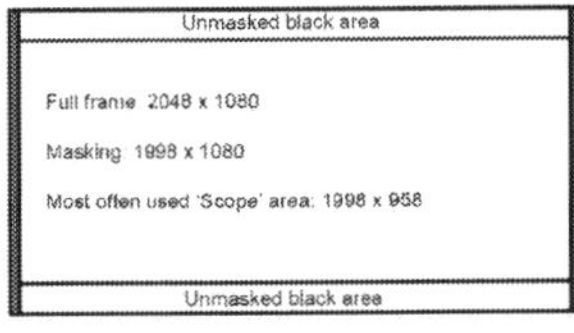

'Scope' format used for 99% of the film

Out-of-frame effect using the top and bottom 'scope on flat' areas

Scope on flat masking allows for characters and props to jump out of the screen frame, strongly reinforcing in-your-face moments.

Cinema Projection

Cinema operators wishing to convert their theaters to 3D face many dilemmas. Each case is different, but the questions to ask are always the same. Let's review them briefly.

Digital Transition

Few theaters receive public money to finance their conversion to digital. The possible solutions are self-financing, sponsorship contracts with the studios (known as "VPF" or Virtual Print Fee), and third-party financing. In the latter case, a technical and financial intermediary deals with all legal and financial problems in return for a monthly rent. In the U.S., third parties are generally the big studios themselves. In Europe, Dcinex (*www.dcinex.com*), Europa Cinemas (*www.europa-cinemas.org*), and Arts Alliance Media (*www.artsalliancemedia.com*) are among the largest third parties. It is generally believed that the number of digital theaters in the world surpassed the number of 35 mm theaters at the end of 2011. The conversion rate will continue to grow for several years and the general belief is that the worldwide movie industry should be fully digitized before 2020.

Equipment Cost

Converting a theater to digital can cost from $60,000 to more than twice that amount depending on size, the amount of work required, and whether or not a 35-mm projector is to be kept in the projection booth. Maintenance costs are not negligible, because even though the reliability of digital equipment is as good as with old 35-mm projectors, its maintenance requires skills that projectionists do not have.

Additional Cost of 3D

The additional cost of 3D is between 10 and 20% of the digital equipment's price, depending on the options chosen. Nevertheless, 3D movies mean higher incomes, which alone often justify the digital conversion. And if the theater is already equipped with a digital projector, converting it to 3D will generally pay for itself very quickly, especially if it is the only one (or at least the first one) in the neighborhood to switch to 3D. The cost of consumables is higher in 3D theaters: Higher-power lamps are needed, so replacement costs and power consumption are slightly higher. The glasses' management can be expensive, especially in the case of active and Dolby glasses, but will remain proportionate to attendance. The servers can store fewer movies, as stereoscopic files occupy more space on the disks. A larger server is thus required to provide the same versatility as in a 2D theater.

Choosing a 3D System

As discussed above, the choice is not simple and physical and economic conditions vary from case to case. Obviously, in the case of cinemas, the operating cost becomes a very important criterion, as does the choice of screen: A silver screen will require a lot of handling if the room is not dedicated exclusively to 3D and this will be reflected in maintenance. However, in a theater dedicated solely to 3D, this will not be a problem.

The Projectionist's Role

In addition to loading films and encryption keys in servers and verifying content (colorimetry, aspect ratio, sound, and masking), the 3D projectionist must ensure that 3D is indeed present and is not reversed, a time-consuming (when done conscientiously) but very important task. Training projectionists to use digital equipment is essential, but in the case of 3D, it will require an indispensable complement.

Control Monitors for Postproduction

The tenth commandment of 3D states: "*Always check thy 3D shots most verily in 3D.*" This is true as of the shooting stage. It is therefore important to check whether takes are valid and 3D is correct as close as possible to the camera. The cameraman will use a small monitor to check his frame, but for the rest of the team, director of photography, lighting technicians, and others, a monitor as large as possible should be available on the site. Setting up a complete screening room in the studio is not always easy, but becomes impossible outdoors. A large monitor is then a good compromise, but its quality must be excellent.

Transvideo

Transvideo (*transvideo.eu*) control monitors are by far the most widely used for real-time camera control in 3D production. The CineMonitorHD 3D from Transvideo, which comes in three sizes (6, 10, and 15 inches), connects to the outputs of a pair of professional cameras through HD-SDI, and accepts 720p and 1080i modes. It is capable of reversing one of the two images horizontally or vertically before displaying it to reflect the possible mounting of cameras on a mirror rig. Features such as zoom, reference grids to check alignments, and an anaglyph or shutter glasses display make it a very valuable control tool. Its sturdy metal construction and 6-inch (15-cm) screen allow it to be attached directly to the camera stand or rig.

A Transvideo control monitor

© B. Michel

Transvideo monitors have a 4:3 aspect ratio, making it possible to display a 16:9 image at the top of the screen and technical information at the bottom. They display audio or video signal waveforms, timecode from the HD-SDI input stream, and other information. Quality does not come cheap: The smallest CineMonitorHD 3D monitor sells for around $7,000.

Polarized Screens

Polarized professional monitors are perfect for 3D checking on location: wide viewing angle, good brightness, low price for a large glasses inventory, portability, all the ingredients!

With their very wide viewing angle, the StereoMirror double screens from the American manufacturer Planar (*www.planar3d.com*) allow several people to watch the footage in full HD resolution simultaneously without flicker or compromising on resolution or color. Cheap passive glasses allow a large number of people to check the images at the same time. Although there is a 17-inch version, only larger models offer Full HD resolution: 1920 × 1080 is accepted by the 22-inch one and even 1920 × 1200 by the 26-inch version. Input connectors are two DVI connectors, one for each view.

The Korean manufacturer RedRover markets 8- to 24-inch polarized double screens under the True3Di brand (*www.true3di.com*).

Other polarized screens are available, but most have only a vertically halved resolution, because every other line is used to display one of the two views. However, this is not really a drawback if one visualizes an HD 3D signal in 1080i interlaced mode, which transmits 540 lines to each eye. In this case, the polarized screen displays effectively all the transmitted pixels. The JVC 46-inch professional monitors (including the GD-463D10E) are typical examples of this category.

3D TV

3D TVs are also a good solution for onsite visualization. Plasma displays, with their high brightness and low crosstalk, are the most popular. As with polarized monitors, larger sizes are better to appreciate fine details. However, their lower cost will be increased by that of the many pairs of glasses needed on a shooting stage. Models from 46 to 65 inches in size offer the best compromises between size and mobility. The HDMI 1.4 connection allows for

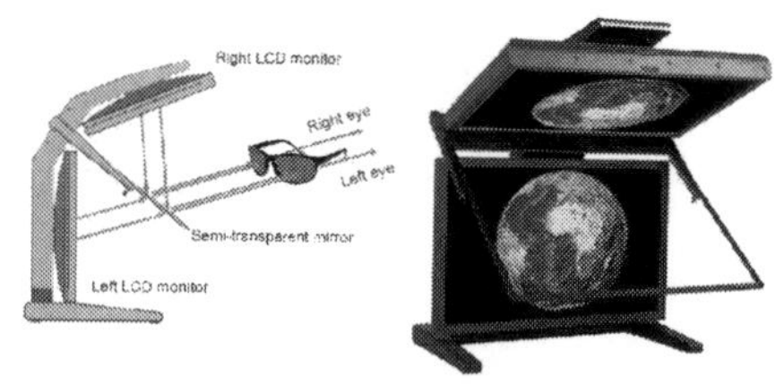

Operating principle of the Planar StereoMirror dual monitor

full HD 3D display. Be careful, however. It is not because a screen is using active shutter glasses that it always offers full resolution: In some transfer modes, resolution may be reduced by half (in 1080i mode carrying interlaced frames, for example).

Computer Screens

The computer is a tool of choice to benefit from new technologies. 3D display performance benefits from the performance of today's graphics adapters. Several methods of display are possible from a PC. The best known are dual polarized projection and the 120 Hz monitor with active glasses, but polarized monitors also exist. Several graphics card drivers and other software applications will take charge of displaying photos and videos in the many existing 3D formats for you. The best known application is the Stereoscopic Player software from the company 3dtv.at, of which several variants are bundled with 3D hardware for PCs, such as graphics cards, monitors, and active glasses.

Active Glasses

Active glasses for computer screens must be synchronized with display. The most common method is infrared (IR) control, used among others by NVIDIA in its 3D Vision and 3D Vision 2 systems. This is possible only with screens offering a refresh rate of 120 Hz or more. Various OEMs, such as Samsung and Panasonic, offer them.

In a number of places, such as trade shows and exhibitions, museums and multi-screen control rooms, infrared is subject to interference. For example, two 3D screens located side by side may scramble each other›s signals. We must then switch to radio-frequency-controlled systems, such as the NVIDIA 3D Vision Pro.

Passive Glasses

Polarized monitors with passive glasses are another option for environments where infrared is not an option and cases where a large number of glasses has to be distributed. There is a handful of polarized PC monitors, such as the Zalman 3D LCD monitors (*www.zalman.co.kr*), which are available in 22 and 24 inches and use circular polarization glasses. Their resolution is 1920 × 1080 full HD and their input connector uses the DVI format. The effective vertical resolution is halved when displaying 3D images because the horizontal polarizing filters use every other line of the display for each of the two images.

Video Walls

A video game using the 3D Vision Surround technology from NVIDIA

© www.nvidia.com

There are not many multi-screen applications, especially because of the constraints that they place on the hardware. However, proven solutions do exist. One such is the NVIDIA 3D Surround, which allows a single PC to control three 3D screens mounted next to each other. The PC must have three compatible video outputs, either through two NVIDIA graphics cards connected in SLI mode (which also requires an SLI-compatible motherboard, see *www.slizone.com* for details) or through a single GeForce GTX 680 or an equivalent model. Various existing software packages support this mode of operation, such as certain games. The resolution of the composite image is then three times full HD, or 5760 × 1080 pixels.

To create large 3D video walls, we must synchronize multiple PC displays. The solution then entails locking («genlock») the different graphics cards. The most mature solution is currently offered by NVIDIA with the optional NVIDIA Quadro G-Sync II, which is added to each Quadro FX-type graphics board. The images are then perfectly matched and glasses controlled by a single PC will display 3D video correctly on all interconnected screens.

Quadro FX graphics board

Cable connection with the G-Sync II board

Quadro G-sync II synchronization board

© B. Michel

Multi-PC NVIDIA Quadro G-Sync II synchronization board

Drivers

Efficient drivers managing hardware components are the foundations of any computer system. There are three for 3D displays, each with its own approach to the question:

- GeForce 3D Vision from NVIDIA (*www.nvidia.com*) ;
- 3D display driver from iZ3D (*www.iz3d.com*) ;
- TriDef Experience package from DDD (*www.ddd.com*).

© www.nvidia.com

© iZ3D Inc

© DDD Group plc

Logos of the various 3D systems used in PCs.

GeForce 3D Vision

The NVIDIA 3D Vision driver works only with graphics cards from the same manufacturer and the corresponding active glasses. It is free, but incompatible with graphics cards from other manufacturers (such as ATI) and does not handle double projection.

3D Display Driver

This driver is independent from the graphics card and therefore works with graphic cards from both ATI and NVIDIA. It is oriented above all towards gaming applications. It handles particularly well (but not only) 3D polarized monitors from iZ3D and Zalman, which have two DVI inputs corresponding to the left and right views. We can therefore use it as well to drive a polarized dual projector, but it can also display anaglyph images on a regular screen. It is provided free by AMD for use on rear projection 3D TVs. The driver is free for anaglyph modes and is bundled with screens sold by iZ3D. It costs $50 for other display types.

TriDef Experience Package

DDD (*Digital Dynamic Depth*) is not an OEM for hardware, but it provides a driver for Hyundai 3D monitors, 2D/3D conversion solutions, and software for the photographic industry. For $50, the TriDef Experience package solution includes a driver, the TriDef Media Player, and photo software. The TriDef Media Player is able to display any 2D or 3D movie and also offers an automatic real-time 2D-to-3D conversion mode. The included photo software also offers this possibility for still photos. The package is often bundled with Samsung 3D devices.

Essential 3D Software

StereoMovie Maker

This program and its counterpart StereoPhoto Maker are distributed for free by the author, the Japanese Masuji Suto (***stereo.jpn.org***). StereoMovie Maker lets you synchronize the two video sequences of a stereoscopic pair, even in HD formats; it automatically fixes unwanted disparities, and then displays and saves the result in many different formats. It offers unbeatable value for money, but its main limitation is the size of the files handled, which may not exceed 2 GB. It uses NVIDIA 3D Vision and iZ3D display drivers, as well as a wide range of anaglyph modes.

The free StereoMovie Maker software cancels unwanted disparities automatically.

StereoPhoto Maker

This freeware, from the same author as StereoMovie Maker, is a veritable toolbox for manipulating 3D still pictures. It works with all versions of Windows and on Intel Macs (with the Windows emulator). Starting with a stereoscopic image pair, it does all the necessary corrections, and then displays and saves the corrected image in a variety of 3D formats. A batch mode allows batch processing of image pairs, making it not only an experimental tool, but also an effective tool for daily work. SereoPhoto Maker handles the same display modes as StereoMovie Maker.

Free StereoPhoto Maker software after automatic adjustment of a picture shot by the Fuji 3D W3camera

Stereoscopic Player

Many software packages are able to display 3D movies on a PC, but none does it as well as Stereoscopic Player, an application developed by Peter Wimmer (*www.3dtv.at*). It supports virtually all 3D encoding and display modes. A free version limited to 5 minutes is available; a single-seat private license may be purchased for €39.

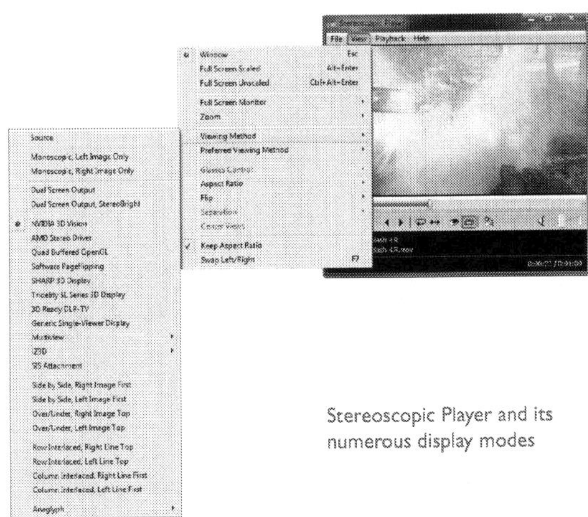

Stereoscopic Player and its numerous display modes

DisplayPort: The future of 3D Connectivity

VGA and DVI connectors have been linking computers to their screens for many years. However, increasing the resolution and number of frames per second pushes this type of connection to the limits of its possibilities. With the advent of digital stereoscopy and haptic interfaces, display systems were suddenly asked to more than double performance. The VESA association, which includes 95% of computer electronics, graphics cards, and computer monitor manufacturers, decided to standardize a new system for high-speed image transmission called DisplayPort (ANSI/CEA-2017- A). The standard is now mature since the advent of release 1.2.

© www.displayport.org

The official logo of the DisplayPort system that will soon be found on all computers

DisplayPort is an open, royalty-free standard that has already been adopted by most PC and monitor manufacturers, including Apple, Dell, Hewlett-Packard, NVIDIA, Intel, and many others. This standard is expected to last for a long time. It offers the possibility of transferring signals with a bandwidth of 21.6 Gbits/s through a regular cable, making it possible to connect multiple high-resolution screens supporting stereoscopy and high color fidelity. DisplayPort connectors are available in two versions, namely, Standard (DP) and mini (mDP).

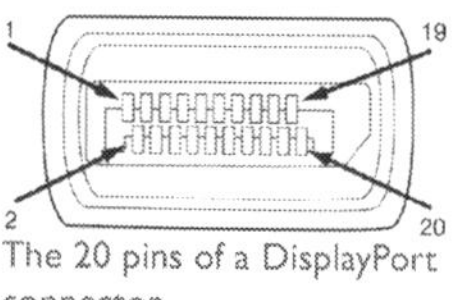

The 20 pins of a DisplayPort connector

The DisplayPort protocol was adopted by the Consumer Electronics Association (CEA) as part of the PDMI (Portable Digital Media Interface) interface, a standard established in February 2010 for connection between display screens and portable media devices.

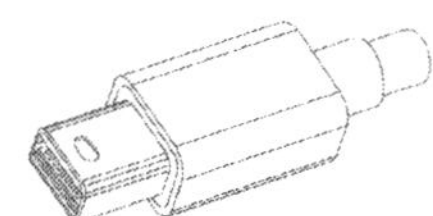
The mDP mini connector

According to its promoters, DisplayPort should be universally available on all computers, monitors, and devices in 2013.

The PDMI connector

DisplayPort was designed from the outset to support signals from 3D Blu-ray players. The maximum cable length of 15 meters allows easy connection of single and dual projectors in venues ranging from conference rooms to home theater installations. The latest information on the standard is available on a dedicated website: *www.displayport.org.*

HDMI and DVI compared with DisplayPort

HDMI and DVI	DisplayPort
Synchronous pixel transfer	Packet serial communication
3 data pairs for digital video components	8B/10B encoding with embedded clock
Pixels' timing synchronized with the display, a legacy from the analog age	Fixed transfer rate according to needs (1.6, 2.7, or 5.4 Gbps)
Variable clock frequency	Use of 1, 2, or 4 differential pairs (called *lanes*) according to needs
No memory inside the display device	Mandatory memory inside the display device
I2C/CEC-type backchannel	Bidirectional auxiliary data channel
R G B Clock Backchannel	Lane 1 Lane 2 Lane 3 Lane 4 Bidirectional AUX channel

Comparison of most common video interfaces

Properties	VGA	DVI	HDMI	DisplayPort
Lockable connectors	√	√	Option	Option
Integrated audio connection	No	No	√	Option
USB size connectors	No	No	√	√
3D Blu-ray support	No	Option	√	√
Embedded power supply	No	No	No	√
Native optical fiber cable compatibility	No	No	No	√
2560 × 1600 resolution and above	No	Option (double DVI)	Option (Cat 2 cable)	√
120 Hz Refresh rate	No	No	Option	√
2× (2560 × 1600) 120 Hz stereoscopy	No	No	No	√
10-bit color definition	No	No	Option	√
15-meter cables	No	No	No	√
Bidirectional auxiliary channel	No	No	No	√
Ultra-flat screen compatibility	No	No	No	√
Extensibility	No	No	Limited	√

Compatibility

In addition to video signals, DisplayPort (DP) cables carry a large enough power supply to power a variety of converters. DP-HDMI, DP-DVI, and DP-VGA adapters are readily available. The ability to transfer DisplayPort signals over optical fiber also makes the system compatible with transmissions beyond 15 meters, as in large conference rooms and stadiums.

Audiovisual Data

DisplayPort supports up to 8 channels of audio at 192 kHz with 24 bits/sample, synchronized with the video to 1 ms.

DP 1.2 supports 1080p signals up to 240 Hz and a color definition of 6 to 16 bits/component. It handles RGB, YCbCr 4:4:4, and YCbCr 4:2:2 color spaces.

Stereoscopic Data

3D transport protocols are included in the DisplayPort specification. DP 1.1's throughput of 10.8 Gbits/s is now increased to 21.6 Gbits/s in release 1.2. That is enough to carry stereoscopic images in all existing modes: all various interlaced modes and the 720p60, 1080p24, and even 1080p60 progressive modes. The standard is able to handle future resolutions up to the double size of 2560 × 1600 pixels per eye for 120 Hz stereo (120 frames per second for each eye).

The data packet controlling the stereoscopic protocol is called «MSA» (Main Stream Attribute). This packet is transmitted at every frame during the vertical blanking interval. Bits 1 and 2 of the MSA packet indicate the type of image carried (see table below).

Bits 1 and 2 of the MSA packet of the DisplayPort interface

Bits 2 and 1 values	Meaning
00	No stereo video transported
01	For progressive video, the next (coming) video frame is for the right eye For interlaced video, the top field is for the right eye and bottom field for the left eye
10	Reserved
11	For progressive video, the next (coming) frame is for the left eye For interlaced video, the top field is for the left eye and bottom field for the right eye

Auxiliary Channel

The DisplayPort auxiliary data channel can carry data in both directions at a rate of 720 Mbits/s. It is therefore possible to include information from USB 2.0 device, such as sound from a microphone or images from the webcam embedded in the screen, inside the connection.

Security

DisplayPort supports security protocols as defined in HDMI 1.3 and 1.4 and therefore offers the same copy protection guarantees.

Sequential Display on CRT Screens

Although now obsolete, the 3D sequential video display method was very popular in the 1990s. Active glasses synchronized with the CRT (CRT) refresh rate by a cable or through an infrared interface. In both cases, a physical connection with the electronic screen was necessary. This connection has been standardized under the "VESA connector" name.

If a stereoscopic video is displayed sequentially by alternating fields, the VESA connector indicates which frame is currently displayed. It was described and standardized in *VESA Enhanced Video Connector and Plug and Display Standards*, published by VESA (Video Electronics Standards Association, Milpitas, California, USA) in 1997.

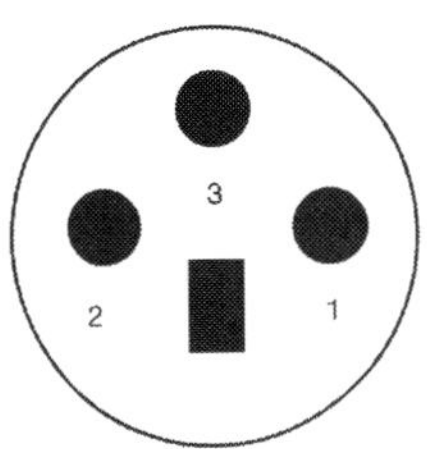

Pin	Signal
1	+ 5 volts DC
2	Ground
3	Stereo Sync

IEC-1076-4-105 or equivalent connector

The VESA connector uses the 3-pins mini-DIN form factor.

The VESA connector is physically a three-pin IEC-1076-4-105 connector, also called a "mini-DIN."

The Stereo Sync signal uses TTL levels (= 0 if less than 0.8 volts, = 1 if more than 2.4 volts). It is set to 1 when the left image is displayed and 0 when the right image is displayed. The transition occurs during the first three scanning lines after the vertical sync pulse.

The VESA connector feeds +5 volt power to the connected equipment (goggles or infrared transceiver). The available current must be between 300 and 1000 mA. The connector is located on the source equipment, such as a PC or a graphics card, or more rarely a DVD player. It can also be integrated within a video monitor or TV set. There are still some OpenGL graphics cards for PCs equipped with the VESA connector (NVIDIA Quadro, ATI FireGL). A well-known model of VESA-compatible active glasses is CrystalEyes from VRLogic (*www.vrlogic.com*).

Note that the sequential 3D video display works with CRT (Cathode Ray Tube) monitors only. Even though current flat screens are still able to accept interlaced video signals, they store the received image in a digital memory that is copied one point at a time in the flat screen display panel. The very notion of progressive display no longer exists and any attempt to synchronize an LCD or plasma screen with a VESA connector is doomed to fail.

Pseudostereoscopic Displays

What to do when you want to give a feeling of depth without any special glasses or autostereoscopic screens? Computer monitors can display moving images. It is therefore tempting to use the "motion disparity" depth cue to replace stereoscopy and give some 3D look and feel to a still image.

Wiggle

This original 3D display method is called Wiggle, which could also mean "wriggling." The idea is to display the two views of a stereoscopic pair alternately in rapid succession, something that is easy to do with the .GIF or .PNG animated image format or with a Flash, Java, or JavaScript applet. An animated GIF is directly visible in any Internet browser. For Flash applets, the browser must support Flash and Java. Note that iPhones and other Apple products do not support Flash and will therefore be excluded from the experiment. Most people – even one-eyed people – are able to appreciate 3D with such images if the flash rate is carefully chosen between 2 and 4 changes per second. The effect is similar to observation of a nearby object while the head moves rapidly back and forth or left to right. As nearby objects have larger parallaxes, they will move more in the image plane, while distant objects remain stationary. Obviously this method is not a true stereoscopic one and is not suitable for printed media. In addition, details of objects in perpetual motion are difficult to assess.

Remark
Many animals that have little or no stereoscopic vision swing their heads periodically to obtain distance information that they lack.

Wiggle in an Animated .gif

Generating a pseudostereoscopic.GIF animated image is very simple with the StereoPhoto Maker program. You will find the latest version of the program on the Web at: ***http://stereo.jpn.org/eng/stphmkr/index.html.***

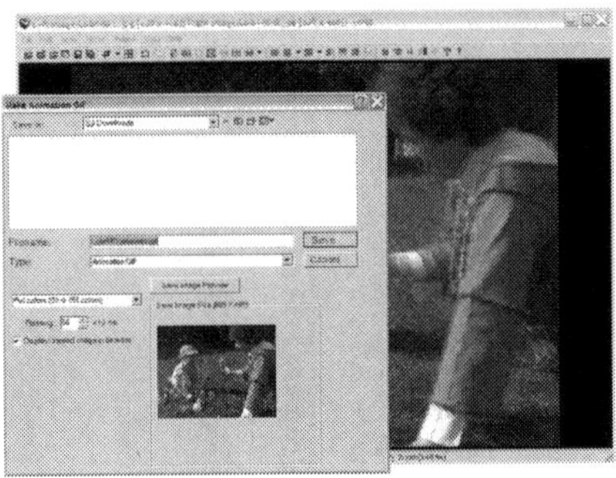

The StereoPhoto Maker application and its "Make Animation Gif" dialog box

Masuji Suto's website is available in Japanese, English, German, and French and offers, among other 3D-related applications, StereoPhoto Maker, a free program running under Windows. Once copied to your disk without any installation procedure, the application is ready for use. Open your two images with File>Open>Open Left-Right Images. Correct them if necessary (press Alt+A for the automatic mode), and then save the animated .GIF file with File> Make Animation Gif.

In the Make Animation Gif dialog box, you can modify the flashing rate with a precision of 1/100 of a second. It is also possible to reduce the number of colors to reduce the file size, although this has limited value.

Wiggle in a Web Page

There is free software allowing you to generate a JavaScript + Flash applet to upload a 2D-like wiggle picture. It is called "Wiggle Stereoscopic (3D) Viewer" and is available at: *sourceforge.net/projects/wiggle/files/*.

Once the .ZIP file containing the application is downloaded, unzip it into a directory on your computer and click on the HTML source file to launch the application. You should give the program the names of both the left and right image files, then define in each the farthest and nearest points of interest. You can also enter a title that will be placed below the image. The applet then generates a piece of HTML code that you can insert as is into a web page.

Piku-Piku

The Piku-Piku (Japanese term that translates as "wriggling") technique is similar to the Wiggle described above, but with a further refinement: In addition to displaying both left and right images, it also displays a series of intermediate images generated by interpolation. The number of images displayed in sequence then goes from two to five, or even more, and the effect is more pleasing to the eye. The Piku-Piku converter is available online on the Start 3D company's website (*www.start3d.com*), which provides the service free while secretly hoping that you will order lenticular paper to print your images. Indeed, the intermediate images generated by the Piku-Piku method are essential for the production of quality lenticular prints. Start 3D offers the ability to save your image on its website directly or to export it in a web page as JavaScript. During the creation operation, the Piku-Piku converter automatically applies a series of alignment disparity and brightness corrections, making the image look better.

Lenticular Printing

Although the general public is much more attracted to moving images, photos and their dissemination on paper still have a future ahead of them. The main advantage of distribution on paper is its low cost. All one needs to view conventionally printed 3D images is a pair of colored glasses. In the case of lenticular prints, 3D can be appreciated with the naked eye. Finally, printing stereo image pairs designed to be viewed with a stereoscope combines low printing cost and quality.

3D imagery on paper will never supplant traditional photography, as difficult compromises are required: the need to wear special glasses, the lower quality of anaglyphs, or choosing a very specific angle and adequate lighting for lenticular prints. Despite these drawbacks, 3D prints can be used in specific cases where depth perception is a real plus for the message being sent out. Thus, 3D printing has a place in fairs and exhibitions, or when traditional media want to report on the latest audiovisual 3D news. 3D printing on paper is divided into three categories: lenticular, anaglyph, and for stereoscopes.

Principle

Lenticular printing is the paper version of autostereoscopy. It is difficult and expensive to implement, yet extremely effective because very natural and does not require the use of any accessories, glasses, stereoscopes, or other devices. The lenticular images are printed on the back of a plastic sheet with tiny engraved parallel cylindrical lenses. The width of the lenses is between 0.25 and 1.6 mm, depending on use. Generally, the smaller lenses are used for small sizes seen up close and large ones for large visual displays, posters, etc.

A series of parallel vertical lenses called lenticules diffracts the image behind it so that each eye sees only a fraction of the image behind each lenticule. The part of the image seen by one eye or the other depends on the angle of incidence of the image and the type of lens. The lens is made of recyclable PET (polyethylene terephthalate, as is used to form soda bottles) plastic. Lenticular films are characterized by their size and thickness, and by the lenticules' pitch. If the lenses are of good quality and sufficiently large, not only two but many images can be displayed behind them. If they correspond to a series of views of the same scene taken with slightly different angles, you can get the impression of turning around the object as you move from left to right in front of the image. The best 3D effects are obtained with some ten shots. The minimum number is obviously two; the maximum number of images with the widest lenses is sixty-four.

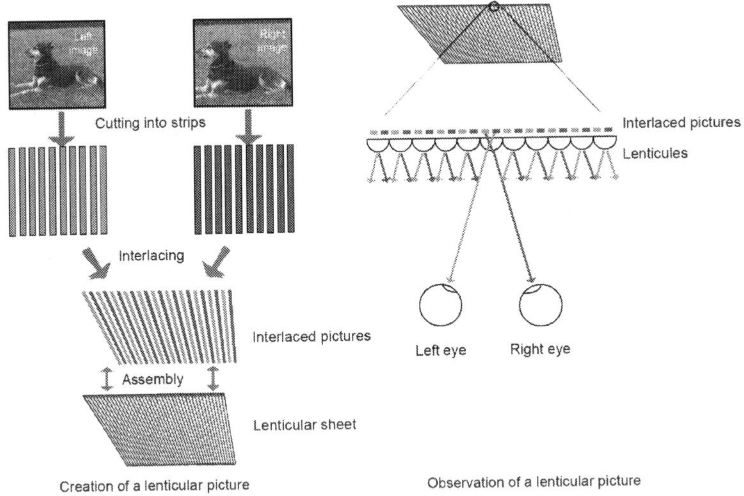

Creation and observation of a stereoscopic lenticular panel

Orientation

It is important to note that the lenticules must be oriented vertically so that the eyes can perceive different images simultaneously. Some lenticular images are mounted on horizontally oriented backings. Those images are not stereoscopic, but can be animated by a bottom to top movement which then causes a short sequence of images to pass behind the lenticules, making a 2- to 3-second movie.

Cutting into Strips

To achieve a good 3D effect, the printed image must consist of vertical strips alternately cut from the left and right images of a stereoscopic pair. The strips should be cut obviously at the same pitch as the lenticules of the sheet. Generally, printing is done by a service provider, who uses the appropriate software to cut the source images into strips corresponding to the pitch of the selected lenticular sheet. Given the extreme precision and finesse required when cuting the very narrow strips (well below a millimeter in width), it is advisable to provide very high quality computer files with a resolution of 600 dpi (dots per inch) or more.

Format

A wide range of sizes is possible. The maximum depends on both the sheet sizes available and the printing process. Sizes of the order of a square meter are common as the plates have a maximum size of 120 × 240 cm (3 x 6 feet), but it is possible to juxtapose several plates to make huge posters. The largest lenticular poster of the world, cataloged in the Guinness Book of Records, is located at the Mandalay Bay casino in Las Vegas, NV, USA. It was created by Big3D (*www.big3d.com*) in 2007 and is 33 meters long!

Printing Method

With lithographic printing it is possible to print directly on the back of a transparent lenticular film. Almost all lenticular displays in the world are printed by means of this process. As a rule, a six-color offset press is used, making it possible to follow the normal four-color printing with one or two layers of very opaque white covering so as to give the image a background with excellent contrast. The back of the sheet is then protected by laminating it with a polyester or paper film or by a screen-printed varnish.

The drying step of the lithography process should not be taken lightly. It is not uncommon to have to wait three full days before being able to pack a panel for transport.

Pitch: The lenticular pitch is measured in lpi (number of lenticules per inch). Choosing the pitch is a crucial step that will determine the conditions under which the 3D image will be observable. The following table will serve as a guide.

Variants of lenticular printing

lpi	Viewing Distance	Use	# of copies
15-30	from 2 to 8 m	Wall posters, exhibitions	A few copies
40	from 1 to 5 m	Posters, standees	Small series
60	< 3 m	Leaflets, A3 or A4 posters	Large series
70-80	< 1 m (handheld)	Postcards, ads	Mass dissemination
100	< 1 m (handheld)	Quality postcards	Mass dissemination

Software

Although the precision required to make lenticular images is not within everybody's grasp, many software applications preprocess 3D images and cut them into strips, for example, those developed by Triaxes (*www.3dmasterkit.com*) and Imagiam (*www.imagiam.com*). There is also an interesting open source program on SourceForge (*lentikit.sourceforge.net*).

Anaglyph Printing

Mass printing of stereoscopic images entails compressing costs as far as possible, and therefore using proven techniques. The only way to print a magazine, or even a daily newspaper, in 3D is to use the anaglyph process. Although historically anaglyphs have also been used in movies, their only application still in use besides paper prints is for occasional screen checks on a computer and on web pages. The paragraphs below cover the creation of anaglyphs for printing, as well as for on-screen display. The principle applies equally well to still images and to videos.

Principle of the Anaglyph

The idea is that two different images can be transmitted to our eyes even though they are superimposed on the same piece of paper (or on a screen, moreover), provided that the images are of different colors. A color filter placed in front of each eye reveals only the image of the corresponding color. If both images are part of a stereoscopic pair, the scene appears in 3D. Obviously, for the filtering to be effective it must minimize ghosting (also called crosstalk). To suceed, you must take care to use two colors that are perfectly complementary to each other. The human brain is extremely resoureful: It is capable not only of reconstructing the scene in 3D, but also of giving it all its colors, even if each eye has only been given partial color information.

The only thing left to do is to find a pair of filters letting only part of the spectrum go through for one eye, and the rest of the spectrum – neither more nor less – for the other eye, while keeping comparable brightness levels. Fortunately, many solutions are possible.

Keeping the brightness comparable in the two images is necessary to avoid excessive eye strain. Differences in brightness between two eyes will cause retinal rivalry that our brains are not used to in observing the real world.

Various Solutions

The ideal solution would be to let all colors go through to each eye, while still being able to differentiate the two views. Complicated multispectral methods can achieve such an outcome, but they require light sources containing very specific spectral lines. This is the case of the Dolby (formerly Infitec) system used in some movie theaters. Dolby filters let through half of the red part of spectrum in one glass, and the rest in the other, and they act the same with the blue and green parts of the spectrum. Thus, each eye gets red, green and blue, but not quite the same red, green and blue! Unfortunately, this method works only when the light source itself is controled, as in a movie projector. In the case of computer screens or printing on paper, that luxury is not possible and we must make do with filtering as best as we can a colored image with three basic components, *i.e.*, red, green, and blue.

Several color combinations have been tried over the years, but the red-cyan couple has become almost universal. The table on the next page shows the different variants of commonly-used filters with their advantages and disadvantages.

Advantages of the anaglyph method:

- easier to use than naked-eye methods such as voluntary strabismus;
- usable by anyone with normal color vision;
- the entire surface of the substrate is used for a single image, unlike the stereoscope displaying two images side by side;
- glasses are cheap, especially cardboard models;
- red-cyan glasses are readily available, often included in magazines or promotional flyers;
- anaglyphs can be printed on any paper or transparent sheet and are usable with projectors, TV sets, and ordinary computer screens.

Note
As this area is not standardized, some anaglyphs are created with the colors reversed from left to right relative to the next table, which presents the most common configurations.

Anaglyph printing methods and their characteristics

Anaglyph Type	Left Filter	Rigth Filter	Badly perceived image elements	Correctly perceived image elements
Red-Cyan	Red: Wratten 26	Cyan: Wratten 44	Sky, bright red objects	Skin tones, earth, vegetation
Amber-Blue	Amber: Lee 179 Chrome Orange, plus Lee 211 neutral grey	Tokyo Blue: Lee 071	Dark images	Skin tones, sky, well-lit buildings
Red-Blue (rarely used)	Red: Wratten 26	Blue: Wratten 38A	Green colors are invisible, the image gives a monochrome feeling	Architecture
Red-Green (rarely used)	Red: Wratten 26	Green: Wratten 58	Blue colors are invisible, the image gives a monochrome feeling	Sports on grass
Magenta-Green (rarely used)	Magenta: Wratten 32	Green: Wratten 58	Avoid for on-screen display	printed paper only
ColorCode	Amber (proprietary)	Blue (proprietary)	Predominantly blue scenes	Skin tones, sky, well-lit buildings
Dolby	Multispectral proprietary	Multispectral proprietary	Does not work for printing, TV, and computer screens	All colors, used in movie theaters only

Disadvantages of the anaglyph method:

- **Ghosts**: Quite often part of the left image is seen through the right glass and vice versa. This is due to the mismatch between the glasses and the observed image support. It is possible to fine-tune the image colorimetry if both the print (or screen) and glasses come from the same source. But in most cases, the producer of the image does not control the whole transfer chain, including the choice of media, printing process, and selection of glasses. Ghosts are less visible if parallaxes are kept low. What is lost in 3D effect is regained in comfort for a majority of users;

- **Retinal rivalry**: Each eye receives a different image, the brightness of which can be very different. If the differences are too great, the brain finds it difficult to merge the two images and headaches quickly ensue. This happens with red-cyan anaglyph images when observing red cars or vast light blue skies. Some unpleasantness may also depend on the type of anaglyph: Some methods, such as magenta-green or ColorCode, have a very dark glass. In such cases, the subjects should be very bright in the part of the spectrum that corresponds to the dark glass if you want to avoid eyestrain;
- **Unnatural colors**: Whatever the various anaglyph methods' merits may be, the colors are never correct. Some progress has been made since the anaglyphs of a hundred years ago, but the preceived colorimetry can never be perfect with such colored filters systems. Dolby glasses come closest to perfection, but their use is limited to movie theaters.

Remark
In the case of moving images on a screen, if retinal disparity is too great, a Pulfrich effect (see Chapter 2, p. 39) occurs, and its addition to the stereoscopic effect may be disturbing.

Why Use Anaglyphs?

Given the above-mentioned drawbacks, one might think that the usefulness of anaglyphs is very limited. However, in many cases, such as the mass distribution on print, they remain the only affordable solution. Their drawbacks may be minimized by choosing the method very carefully, checking printed proofs to refine the color balance, and avoiding excessive parallax. Small parallaxes, obtained by taking pictures with a small interaxial distance, generally give good results: The depth effect is shallower, but the resulting image is less tiring to look at with the glasses and more enjoyable for the reader who goes through a document without the glasses. A good anaglyph is usually the result of a difficult compromise that depends on its intended use.

Red-cyan Anaglyph

The most popular method is the red-cyan anaglyph. Despite its flaws, such as frequent excessive retinal rivalry, it is well suited to general subjects, excluding large red objects.

Here is the formula for producing a red-cyan anaglyph from a stereoscopic pair of color pictures:

RedCyan.R = Left.R
RedCyan.G = Right.G
RedCyan.B = Right.B

In the above formulas, Left and Right denote the pixels of the original left and right images; RedCyan denotes the pixels of the resulting image; and the R, G, and B suffixes of each pixel indicate its red, green, and blue components.

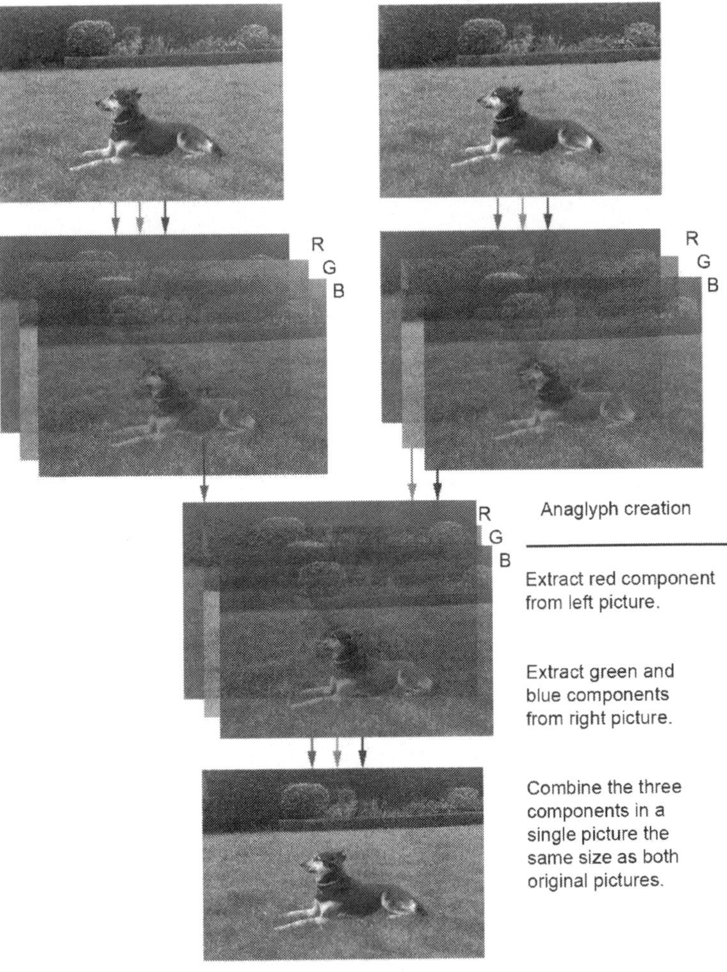

Creation of an anaglyph by overlaying left red, right green, and right blue components

Half-color Anaglyph

If the red-cyan anaglyph is rendered unusable by the presence of too much red in the scene, the best solution is often to reduce the image's color range, without bring it down to monochrome. This solution is called "half color." Retinal rivalry is reduced at the expense of contrast.

Here is the formula for producing a half-color anaglyph from a stereoscopic pair of color pictures:

Halfcolor.R = (0.299 × Left.R) + (0.587 × Left.G) + (0.114 × Left.B)

Halfcolor.G = Right.G

Halfcolor.B = Right.B

In the above formulas, Left and Right denote the pixels of the original left and right images; Halfcolor denotes the pixels of the resulting image; and the R, G, and B suffixes of each pixel indicate its red, green, and blue components.

Dubois Optimized Anaglyph

There are many rules of thumb to optimize colors' rendering and reduce retinal rivalry in anaglyph. The most commonly used method is called Dubois. Developed in 2001 by Eric Dubois, it decreases the ghosting effect (observed when part of the right image is seen by the left eye and vice versa) by subtracting from each image a portion of the opposite one. The parameters used were adjusted for CRT displays and red-cyan glasses manufactured by American Paper Optics, one of the largest makers of this type of article. The result on LCD flat screens and other types of glasses often remains very satisfactory.

With the Dubois method, colors are a little less saturated and flesh tones are too yellow-green; on the plus side, retinal rivalry is very much reduced. The advantage of the Dubois method is particularly noticeable for red or dark blue objects.

Dubois.R = (0.456 × Left.R) + (0.500 × Left.G) + (0,176 × Left.B) - (0.043 × Right.R) - (0.088 × Right.G) - (0.02 × Right.B)

Dubois.V = (-0.040 × Left.R) - (0.038 × Left.G) - (0.016 × Left.B) + (0.378 × Right.R + (0.734 × Right.G) - (0.18 × Right.B)

Dubois.B = (-0.015 × Left.R) - (0.021 × Left.G) - (0.005 × Left.B) - (0.072 × Right.R) - (0.113 × Right.G) + (1.226 × Right.B)

In the above formulas, Left and Right denote the pixels of the original left and right images; Dubois denotes the pixels of the resulting image, and the R, G, and B suffixes of each pixel indicate its red, green and blue components.

Given the negative or greater-than-unity coefficients, it is not impossible that some values fall outside the [0,1] range. It is then necessary to crop the values to bring them back between 0 and 1.

Black-and-white Anaglyph

Some anaglyph images with a strong cyan or red cast remain difficult to watch without eyestrain, despite all optimizations. In this case, a simple solution is to convert both pictures to black and white before the anaglyph conversion. Careful! A black-and-white anaglyph image contains color components that are absolutely required for depth perception. It is therefore always mandatory to print them in color and not in black and white!

Here is the formula for generating a black-and-white anaglyph from a stereoscopic pair of color pictures:

BlackAndWhite.R =
(0.299 × Left.R) + (0.587 × Left.G) + (0.114 × Left.B)

BlackAndWhite.V =
(0.299 × Right.R) + (0.587 × Right.G) + (0.114 × Right.B)

BlackAndWhite.B =
(0.299 × Right.R) + (0.587 × Right.G) + (0.114 × Right.B)

In the above formulas, Left and Right denote the pixels of the original left and right images; BlackAndWhite denotes the pixels of the resulting image; and the R, G, and B suffixes of each pixel indicate its red, green, and blue components.

ColorCode

ColorCode is a proprietary variant of the amber-blue anaglyph. The ColorCode method was patented by Sven Sorensen. In the UK, the TV station Channel 4 broadcast a series of ColorCode programs in November 2009. The system has also been used historically in the U.S. for commercials (SoBe, *Monsters vs. Aliens*) at the Super Bowl 2009 halftime that propelled 3D to the forefront of the American mainstream audience. The Danish company ColorCode promotes its system in broadcasting and encourages its use for web applications. The patented method covers not only the ColorCode 3-D filter, but also the accompanying encoding method, as well as the playout program on the PC, called respectively "ColorCode CX Pro Encoding" and "ColorCode Viewer." Printed versions of ColorCode images are also usable and have the advantage of being visible (without 3D) with relatively few annoying artifacts.

Patent
The corresponding U.S. patent is filed under #6,687,003. More information is available on the ColorCode website (*www.colorcode3d.com*).

The ColorCode system was originally designed to improve rendering on U.S. color television screens because the NTSC coding system is very unfaithful for red shades, thus reducing the effectiveness of red-cyan glasses greatly. Differences in brightness between the red and green parts on one side and blue on the other are the biggest problems faced by amber-blue anaglyphs, as the blue image is often very dark.

The ColorCode encoding process greatly improves the blue image brightness and minimizes this problem. In addition, the ColorCode encoder requires small parallaxes, leading to a shallow depth feeling. However, these small disparities provide a combined image with colors that are quite natural. Without glasses, there are blue and orange fringes on the lateral edges of objects, but the overall impression is much more pleasant than with traditional red-cyan anaglyphs.

In practice, and although the ColorCode method is not known in detail, it is estimated that the right (blue) image of a ColorCode pair contains approximately 66% of the blue component and 17% of the red and green from the original right image, thus avoiding the excessive retinal rivalry on objects containing very little blue. The left image, in turn, does not present any blue information from the original left image. The system banks on the fact that very few blue objects have very pronounced depth differences: they are often the sky or a water surface. This also explains why the system is less accurate for scenes with a strong blue cast, which, fortunately, are very rare in practice.

ColorCode.R = Left.R

ColorCode.G = Left.G

ColorCode.B =
(Right.R × 0.17) + (Right.V × 0.17) + (Right.B × 0.66)
(estimates)

To approximate the results of the ColorCode method, one should modify the image with a nonlinear increase in the blue component's overall brightness (Gamma> 1).

In the above formulas (which are rough approximations proposed by the author and not by ColorCode), Left and Right denote the pixels of the original left and right images; ColorCode denotes the pixels of the resulting image. The R, G, and B suffixes of each pixel indicate its red, green and blue components.

Many programs offer opportunities to create anaglyph pictures. The well-known StereoPhoto Maker is freeware. It not only transforms your stereoscopic image pairs to anaglyphs, but is also able to correct many imperfections, such as geometric and color disparities, automatically.

Anaglyph Glasses

Anaglyph viewing glasses are usually cheap because they are made of cardboard with filters cut from ordinary photographic film. Even in small quantities, the price of a pair is often under a dollar.

Tip
The *www.stereoscopyshop.com* website offers several brands of anaglyph glasses, along with other 3D-related products.

You can also make your own pair of anaglyph glasses yourself with cardboard and scissors using the template provided hereunder. The hardest part is to get good filters, which is possible at a professional photographer's shop. Ask for Wratten 46 (red) and 44 (cyan) filters, from which you will cut 60- by 36-mm rectangles. Reproduce the pattern below to scale on a lightweight cardboard, cut it out, and glue the filters on the inside, with the red one on the left and the cyan one on the right. Mark both folds and then bend the branches backward to produce excellent red-cyan glasses. The advantage of making your glasses yourself is of course that you can customize them at will, by altering their shape and dimensions, or by printing a variety of patterns on them.

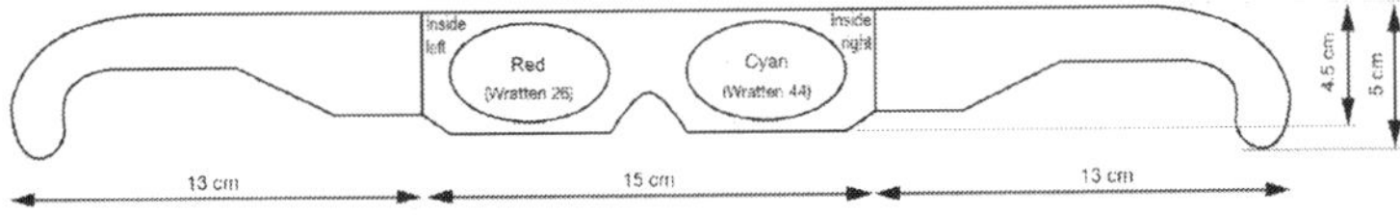

Cutout pattern for a pair of anaglyph glasses made of light cardboard (300 g/m^2)

Quality red-cyan glasses with hard plastic lenses and folding frames are also available. On the plus side you will find a better viewing angle thanks to large curved glasses. However, this kind of glasses is often criticized for poorer image separation than with true photo filters, such as those used in cardboard glasses. Before purchasing this type of glasses, always try them on some of your anaglyph prints.

Red-cyan glasses with a plastic frame

© www.yt3d.com

Printing for Stereoscopes

The stereoscope is more than 150 years old, and yet it sells and is still being built. In addition to the View-Master, still on the market after more than 50 years, there are easy-to-print-and-distribute folding paper stereoscopes. A typical example comes from the Dutch printer Amazing Card (*www.amazingcard.nl*). The product consists of a hexagonal tube made of light cardboard, inside which two stereoscopic images are bonded. On the opposite side two small converging lenses are inserted through which one can look at the pair of images while keeping the eyes focused at infinity. A variant of the same system offers the ability to insert various stereoscopic cards in the holder. The format of the images on the card is about 32 × 52 mm (1.25 x 2 inches) and the interaxial distance is 65 mm.

The folding stereoscope is used primarily as an advertising medium. As its exterior is made of lightweight cardboard, it is easy to customize at will with logos or other advertising slogans. This kind of product sells for 3 to 8 dollars per unit in small quantities, a price matching lenticular solutions. Note that the StereoMerger software comes with a script that generates directly pairs of JPG images that are sized and separated by the optimal distance for Amazing Card stereoscopes.

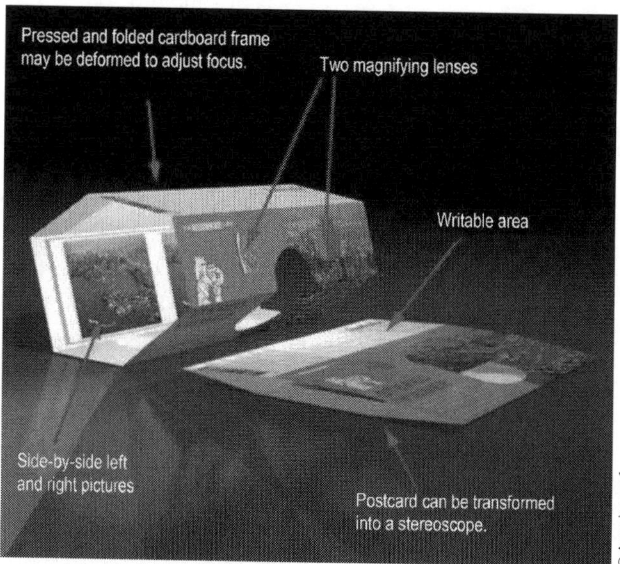

The folding stereoscope from Amazing Card

Broadcasting via the Internet

9

How should I encode a 3D movie to put it on line?

How can a 3D film be viewed easily on an online site?

Can I send out one of my 3D movies on YouTube?

How can I download a 3D movie I saw on YouTube onto my PC?

Is there a lot of 3D content on the Web?

Webcasting videos has become a universal means of communication. So, YouTube has already exceeded one billion daily views, which proves that it is impossible to discount the Internet as a broadcasting medium for moving pictures. Of course, the distribution of 3D images has not been forgotten: A growing number of sites display 3D content. Examples include *www.3dmovies.com*, *www.3df33d.tv*, and many others.

YouTube and the yt3d Mode

Whether you want to share 3D souvenir videos with friends or create viral marketing clips that people will spread for you, YouTube (*www.youtube.com*) is an easy-to-use transmission vector. YouTube accepts virtually all image resolutions up to dual full HD format. One of the main limitations, *i.e.*, duration limited to 10 minutes, was lifted recently. The maximum uploaded file size is

Key word
It is easy to find 3D videos on YouTube. Add the key word "yt3d:enable=true" in the search box. More than 10,000 3D videos are available on YouTube.

limited to 2 GB for standard users, but if you are using the latest browsers – especially Chrome – you will be able to upload files of up to 20 gigabytes.

Watching a 3D Video on YouTube

There are many 3D viewing modes on YouTube. The viewing options can be changed in the 3D menu at the bottom of the image. You must first choose the type of 3D display system that you are using. By default, YouTube proposes anaglyph mode with red-cyan glasses. You can easily change it by clicking on "Other options..." in the menu. A second screen then comes up with the following main choices: colored glasses, interlaced, side-by-side, no glasses, and HTML 5 stereo view (see the two figures below).

With anaglyph glasses you can also choose to dim the colors in order to reduce the retinal rivalry that occurs when objects on screen are red or cyan, and therefore misunderstood by one of the two eyes. You have three options: "Color," which keeps the full color range; "Optimized (Dubois)," which halves the colors' intensity and is often the best compromise; and finally, for the most serious cases, the "Grayscale" option, which removes the colors completely and gives a monochrome but more comfortable image.

The YouTube player and its 3D menu

Image © B. Michel

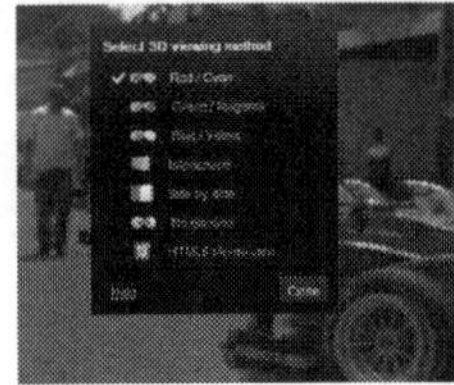

The YouTube 3D options menu
The NVIDIA 3D Vision mode is available only on the most recent HTML 5-compatible – browsers.

YouTube 3D Local Copy and Active Glasses

With a suitable browser, such as Mozilla Firefox 6, the HTML 5 YouTube mode is compatible with NVIDIA 3D Vision active shutter glasses. As 3D is always more enjoyable in high definition, the most suitable method is to proceed in two steps: First, download the file in the highest possible resolution onto a local hard drive, and then watch it with Stereoscopic Player. All existing display modes will then be available, even those that are not managed directly by YouTube.

Downloading a file from YouTube can be done by cutting and pasting the video URL to a specific download page (*e.g., www.keepvid.com*). A slightly more comfortable method is to add a download script offering the "Download to a YouTube page" option to your browser. Unfortunately, YouTube is engaged in a constant fight against such scripts, which makes them difficult to locate. The safest way to find them is still to search for the "YouTube download script" keywords on the web. The script is easy to install; a simple click is usually enough. Once the script is installed in the browser, a new Download menu appears under all videos, whether in 3D or not. It lets you choose which version you want to download. With few exceptions, one should always choose the highest available quality.

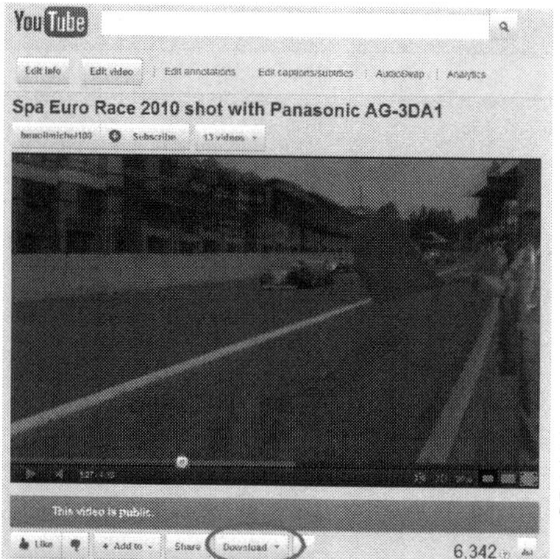

Image © B. Michel

The new Download menu appears in the YouTube page as soon as the download script has been installed in the browser.

Uploading a 3D Video onto YouTube

Creating a 3D video and broadcasting it on YouTube are simple to do. Here are the steps:

1. Use your favorite editing software to create a video where the left and right images are side by side (possibly compressed in width, but that is not mandatory). YouTube recommends positioning the left image on the right and the right image on the left so that the 3D image can be viewed by crossing the eyes, but the reverse is also possible. Use the maximum available resolution; there is no need to reduce the image quality artificially. If the running time exceeds 15 minutes, you will need to split the video into several episodes.
2. Compress the movie in one of the following formats: AVI (.avi), QuickTime (.mov), MPEG-4 (.mp4), Windows Media (.wmv), or HTML5 (.webm). Take care to limit the file size to no more than 2 GB, and then upload it to YouTube.
3. Log in to YouTube and go to your dashboard/upload page at *www.youtube.com/my_videos_upload*. Drag and drop your 3D video file anywhere on the page to start the upload process.
4. The upload interface lets you select a category and add a title, a description, and several key words called tags. You may also choose a category, a distribution license, and privacy settings. Adding "YT3D" in the tags list is recommended but not mandatory. Specific tags may help you fine tune the aspect of your 3D movie; the details are given below.
5. Click on the "Advanced Settings" tab, where you may give details such as the recording's location and date and the very important "3D Video" settings. Select "This video is already in 3D" from the four proposed settings. A second dropbox appears with four 3D presentation alternatives. Select the appropriate one (usually: "Side by side: Left-hand video on the left-hand side").

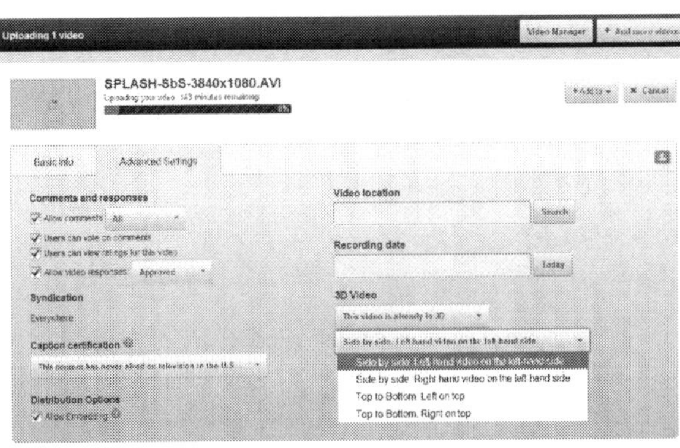

Videos longer than 15 minutes

Videos longer than 15 minutes are allowed for YouTube users whose accounts are in good standing (as per YouTube Community Guidelines) and if this has been verified over a mobile phone. In this case, videos can be up to 12 hours long. To enable this, click on the "Increase your limit" link at the bottom of the "*www.youtube.com/my_videos_upload*" page.

6. Now all you have to do is to give your audience the URL of the newly loaded video. Still, take the precaution beforehand to check that the online file is complete, the 3D menu is visible, the aspect ratio and sound are correct and, most important, the stereoscopy is not reversed!

Be careful

Using a copyrighted soundtrack in a YouTube movie causes advertisements to be overlaid on the image, with the side effect of preventing the 3D menu from appearing and thus creating a *de facto* ban on viewing the film in 3D.

In previous versions of the YouTube upload interface, specific tags had to be used to specify the exact 3D appearance of uploaded videos. Such tags are still allowed and are useful when something looks wrong, such as the aspect ratio of the image. If a 3D video doesn't appear in 3D even if the above settings were selected, you may try to add the "yt3d:enable=true" tag in the tags list.

YouTube 3D Tags

3D Tags are less important than before but may nevertheless save the day when something goes wrong with your uploaded 3D video. Tags are simply lists of key words allowing search engines to find videos on YouTube, but they also act as parameters for the YouTube broadcast engine. You can add multiple tags separated by spaces. Inserting the following tags will force YouTube to take specific actions:

yt3d:enable=true	Declares the video as already in 3D, in side-by-side format and with the right-hand video on the left-hand side by default.
yt3d:swap=true	Swaps the left and right images and declares that the left-hand video is on the left-hand side.
yt3d:aspect=16:9	Declares that the aspect ratio of the video is 16:9. Another typical value is "4:3".
yt3d:aspect=4:3	Declares that the aspect ratio of the video is 4:3.

Advanced Uploading with YouTube

Remark
Detailed explanations of the upload procedure are given through a "Help" button on the "youtube.com/my_videos_upload" web page. The button is usually in the bottom right-hand corner of the page.

For a safer upload, including the ability to resume when the transmission of a file is interrupted by mistake, YouTube offers the option of uploading larger files, as long as you are using the latest version of Google Chrome, Firefox 5 or higher, or a recent version of Internet Explorer with Silverlight enabled. This option is free, so use it! In this case, YouTube allows users to upload up to 20-gigabyte files. A Java program is integrated into the "*my_videos_upload*" web page on YouTube and transfers the file in pieces, ensuring that each piece of the file is received before sending the next one. In this way, if the network connection is lost during the transfer, it automatically resumes at the current part instead of restarting from scratch. Note that to use this advanced transfer, you need to install Java (version 1.5 or later) on your computer. You can download the latest version of Java at: *www.java.com*.

YouTube detects copyrighted sound content that has been used without permission and overlays a banner on your video. This banner is currently incompatible with the 3D menu. In conclusion, to upload 3D videos, you must make sure to use an original soundtrack that is not copyrighted. A typical example was recently detected in a wedding video: The surround sound in the background was coming from the reception room amplifiers and the CD played a well-known song. The YouTube robot detected recent commercial music with a level loud enough to trigger the ad banner and the 3D menu disappeared at once.

Searching 3D Videos on YouTube

If you go to *www.youtube.com* to search for videos with a specific keyword, such as «trailer», you will see in the address bar something like this:

www.youtube.com/results?search_query=trailer

To search for 3D videos, add the magic string "&threed=1" at the end of your query and the results will be filtered so that only 3D videos appear in the results list:

www.youtube.com/results?search_query=trailer&threed=1

Other Sources of Online 3D Videos

General Sites

Very few video sites host their movies locally. Most simply redirect to one of the most popular sites: YouTube, Dailymotion, and Vimeo. Their viewing options will therefore be the same as on these video repositories.

To search for 3D clips on YouTube, just use the "yt3d" key word, which will refer you to more than 100,000 references. On Vimeo, 3D clips are grouped in a 3D Stereoscopy channel that has more than 2,000 videos, including short films, demonstrations, tutorials, and music videos (see *vimeo.com/channels/stereoscopy*). On Dailymotion, several thousand videos are listed under the "3D" key word, but only a few are really in 3D and 3D viewing options are cruder than on YouTube, with Off, Side-by-side, Red/cyan, and Interlaced as the only options.

Stock Movies

The studios that provide general-purpose clips for cinema and television have also started to add 3D movies to their catalogs. Thus, Artbeats (*www.artbeats.com*), Mammoth (*www.mammothhd-3d.com*), and others provide hundreds of high-quality 3D clips, some even in 4K resolution: fly-overs of cities and rivers, aircraft take-offs and landings, landscapes, wildlife scenes, etc. Obviously, these 3D clips are for professional use and priced accordingly.

Pay Sites

Some sites are presented as specialized access points to, as a rule, paid content, or at least in part. 3DMovies (*3dmovies.com*) is one example.

Sites for Mobile 3D

3DeeCentral (*3deecentral.com*) offers paid content and some free trailers, but it is especially interesting because it offers 3D content for Apple's iPhone and iPad mobile devices. The 3DeeCentral platform is proprietary and runs on Windows 7 PCs. You thus have to install an application combining online shopping, downloading, and 3D visualization. The available modes are NVIDIA 3D Vision, screens or projectors with passive glasses, and anaglyph. On mobiles, you will need to install the free 3DeeCentral application and lay a soft lenticular layer provided by Spatial View,

3DeeCentral's parent company, on the screen. This soft lens serves as both a screen protector and an autostereoscopic lenticular separator. Depth perception is then possible without glasses.

Independent Sites

There are also independent initiatives offering side-by-side content viewable on 3D TV and various projection systems that accept this format natively. All you need is an Internet browser with full screen capability without borders.
One example is *www.stereoscopic-3d.co.uk.*

3D Web

There are currently two cutting-edge technologies for making 3D acces- on standard web pages, namely, MPP nology from Microsoft for browsers accept legacy HTML 4.0 and the new- HMTL 5.0 standard, which some browsers are starting to support. The pioneer the second category is Mozilla Firefox.

HTML version 5.0 logo

NVIDIA's 3D Vision Live

Designed specifically for users of NVIDIA shutter glasses, the 3D Vision Live site requires a Windows PC with an NVIDIA (GeForce or Quadro) graphics board and NVIDIA 3D Vision drivers, an HTML 4.0 (or later) browser, such as Internet Explorer, Firefox or Google Chrome, and the "Microsoft Media Platform Player Framework" (MMPPF) from Microsoft. A large range of screens is supported, including most of the 3D monitors and 3D TVs on the market, as well as projectors and laptops compatible with NVIDIA's 3D Vision kit (*www.3dvisionlive.com*). MPP has been free open source software since 2011, giving developers the ability to deploy scalable solutions on demand based on Silverlight technology and with full support for IIS Smooth Streaming and Microsoft PlayReady DRM. With the NVIDIA plug-in, they have all the tools to put 3D video streaming platforms on line.

Various video portals are being created on the Web using this method. They offer 3D videos in the most common formats, including, of course, 3D Vision. A pioneer is the *www.studio3d.tv* site.

The 3D Vision Live website from NVIDIA

WebM

The future standard for encoding video for the Web is called "WebM." The WebM open source development project is subsidized by Google, and the fact that it is free and royalty-free gives it a prominent place as a standard for Web video. WebM was designed from the start to mesh with HTML 5.0 and is about to be adopted as a single format for YouTube, which will gradually convert its huge content base to this format.

Playing WebM Format Videos

Websites offering WebM content provide links to download the plug-ins, codecs, and DirectShow filters necessary for their use. The installation procedure is sometimes tedious, but must be done only once. Once installed, the WebM DirectShow video filter integrates with applications such as Microsoft's Media Player, thereby automatically making your computer compatible with the new format.

Encoding WebM Format Videos

Encoding stereoscopic videos in WebM format can be done with a number of video encoders by selecting the appropriate option. You will find a list on the WebM project website (*www.webmproject.org*). The websites hosting videos in WebM all have the ability to host videos in many formats and do the transcoding automatically. One of the pioneers of HTML 5.0 and the WebM stereoscopic format is 3DF33D (*www.3df33d.tv*).

The WebM logo

© The WebM Project

Although many coding modes exist in the WebM 3D standard, the format of choice for WebM 3D videos is over-under, where both left and right images are superimposed in a double-height frame, with the left image on top. Thus, a 3D HD video with 1920 × 1080 pixels will be coded as a 1920 × 2160-pixel (W x H) frame.

Autostereoscopy

10

What is autostereoscopy?

Why aren't all TV sets autostereoscopic?

How does one produce images for glasses-free TV?

Is it easy to find autostereoscopic screens on the market?

3D without Glasses

Autostereoscopy is 3D viewing without glasses or any other artificial devices between the image and the viewer's eyes. Not having to put on a bothersome pair of glasses is an obvious advantage, especially in the case of screens located in places that people merely pass through from time to time. It is difficult to ask passers-by to use – and to return afterwards! – special glasses to glance at some advertising!

The Ideal Autostereoscopic Screen

The ideal autostereoscopic screen, also called a "holographic" screen (from the Greek *holos* "whole" and *graphein* "to write"), does not yet exist. The hologram technique, which produces images by means of laser interference, is one avenue of research but it still performs far short of ordinary television sets when it comes to color, brightness, and contrast.

The ideal autostereoscopic screen should show us a scene as it is in real life, with its depth, shadows, and shapes. To be totally realistic, the perspective must change, revealing objects' sides more or less as you move your head from left to right or back and forth. This is known as "motion parallax." Few autostereoscopic screens can offer convincing motion parallax.

An autostereoscopic screen must generate different views for the viewer's eyes and send them in the right directions without cross-interference. If several viewers are in front of the screen, two solutions are possible: generating only two views or generating a large number of views.

With only two views, getting the images to all the viewers' left and right eyes without error is far from simple. An active system has to detect the number and positions of all the eyes looking at the screen at every instant, even if they are hidden by normal glasses, hair, or a hat's brim, and use this information to direct the light beams. The complexity of such a system makes it practically unfeasible.

The second solution is to generate a large number of views and project them in many contiguous directions with the hope that the viewers' eyes will be positioned each time in the different zones. These multiple views must obviously depict the same scene, but each from a slightly different perspective.

Autostereoscopic pictures printed on paper overlaid by a lenticular sheet have been known for decades. With the high resolutions that are available on today's printing presses, a large number of views can be distributed behind lenses so small as to be almost imperceptible. Images presenting sixty-four views with microlenses that are only 0.4 mm wide already exist. Getting such high pixel density on an electronic screen is not yet possible, but various tricks use electronic resources to get around this limitation.

Two-View Autostereoscopy

Two-view autostereoscopic screens are usually intended for single viewers. Two images are sent out towards two fairly wide invariable sectors to the left and right of a point opposite the screen and over a distance of roughly from 30 to 50 cm. This simple process works very well for all the small screens that are used by users at close range, *e.g.*, telephones, movie camera viewer screens, game consoles, and even laptop computers. With such screens, the user orients the screen and places it at the right distance to optimize the depth effect. Their comfort zones are effectively rather narrow.

There are three autostereoscopic display methods on the market that give good results on small screens, namely, lenticular, parallax barrier, and time multiplexing.

Lenticular Screen

This is a simple sheet of plastic bearing vertical microlenses the size of two pixels on the screen's tile. Each microlens deviates the light from every other pixel on one side and from the other pixels on the other side, so that each eye receives a half-resolution image. Appropriate software is of course required to interlace the two images so that the even pixels all belong to the left view and the odd pixels to the right view. The advantage of this method is that it is easy to adapt to existing devices that were not initially designed for 3D: All you have to do is add the appropriate software and a plastic sheet in front of the screen. When it comes to its disadvantages, we can note of course the fact that the lenticular sheet blurs the image for non-3D applications and halves the resolution.

Parallax Barrier Screen

This system is more complex, for it requires an optical layer behind the display panel's pixels. This layer includes alternating transparent and opaque zones located halfway between the active liquid crystal panel and the light source. The solution is cheap but makes the display panel a little thicker. The advantages and disadvantages are the same as for the lenticular screen, but the image's appearance in non-3D mode is less disturbing: You notice the lack of resolution and brightness, but you don't have the moiré effect that is produced by the microlenses.

The three types of two-view autostereoscopic screen

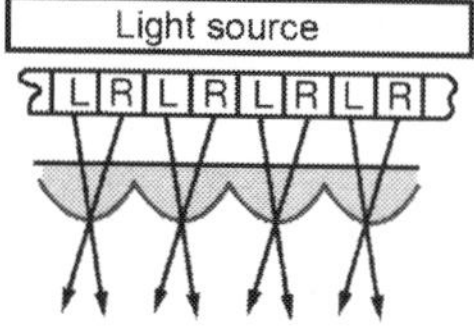

Lenticular

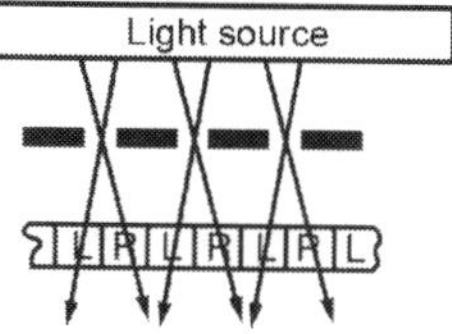

Parallax barrier

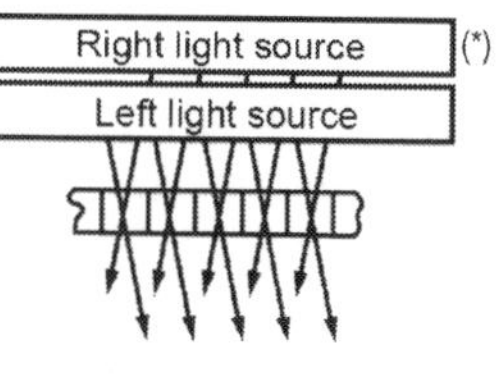

Temporal multiplexing

Multiplexed Screen

This is the most intelligent of the three systems. It works a little like the parallax barrier screen, but the light source is doubled. One light source lights the screen towards the left while the left image is displayed, then another light source directed more to the right lights up a fraction of a second later when the right image is displayed. If the refresh rate is fast enough (more than 100 Hz), the flicker will be imperceptible. This method uses all the pixels of the display's tile for each image. Its resolution is thus twice that of the two other methods at no greater cost. It is used for the Fuji W1 and W3 3D cameras' screens, among other devices.

Tracking

Mounting a webcam above the screen is a way to detect the user's head position. If the computer then generates a pair of stereoscopic images that shift slightly to the left or right, the impression of realism is accentuated by this motion parallax effect. Unfortunately, the illusion is shattered as soon as one eye crosses the boundary between the zones where the image is projected towards just one eye at a time. Too large a head movement towards the left will be enough for the right eye to find itself in the same display area as the left eye. In this case, the scene will lose all its depth. There is a solution for this problem. It consists in moving the parallax barrier or the lenticular network linked to the screen to follow the head's movements and thus keep the boundary between the projection zones in the middle of the viewer's face at all times. Such systems exist in laboratories, but since they involve mechanical translations, they are expensive and subject to the usual problems of wear and calibration that plague all mechanical systems. In practice, viewer tracking systems are thus not widespread on autostereoscopic screens. What is more, as the images must be generated at the moment that the viewer moves her/his head, the technique applies only to CGI, not to filming real scenes.

Multiview Autostereoscopy

The various limitations of the two-view systems can be circumvented by increasing the number of views. Let's imagine a screen that projects eight different images in eight contiguous angular sectors. The angle at which one can look at the screen will then be much wider than with two views only. Indeed, in this case each eye will perceive a different image

3DTV Solutions's eight-lens Cambox

The Alioscopy screen displays eight views.

from each of seven positions opposite the screen. In practice, these are not seven specific positions, but rather seven comfort zones where the viewers will perceive the depth without glasses. Of course, this means that eight rather than two images must be projected onto the screen, and thus shooting must be done with eight different movie cameras or various DP processes used to generate eight slightly shifted views of the same scene.

We can give the example of the 47-inch eight-view Alioscopy screen, which is 104 cm (40.9) inches wide. The number and size of the comfort zones depend on the optical system's geometry and the number of views. In the case of the 47-inch Alioscopy, there are sixteen 50-cm-wide comfort zones ranging from 3 to 9 meters from the screen's surface, with an optimal position at 4.4 meters, making it possible to accommodate up to forty viewers.

Oblique lenticular network on the surface of an Alioscopy screen

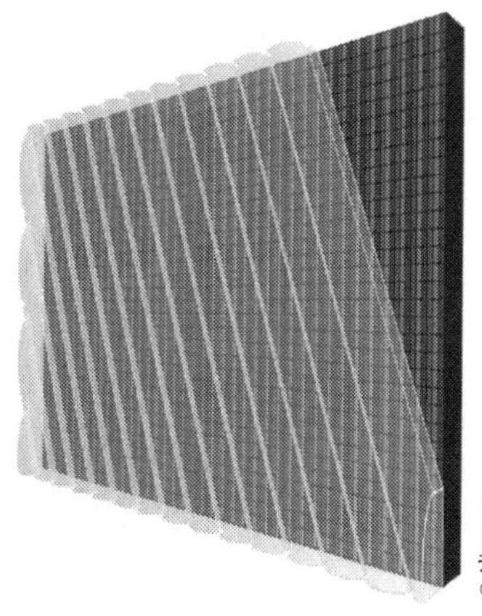

The eight images are generated by placing in front of the tile a lenticular network with a pitch of exactly eight subpixels, each subpixel being intended for one of the eight images. When we speak of "subpixels" in this connection we are referring to each of

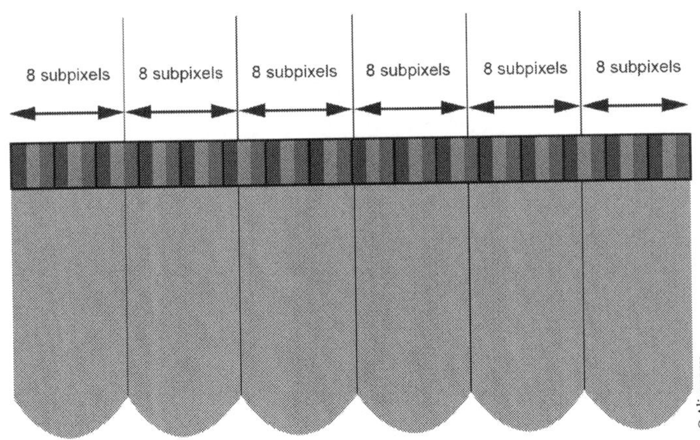

Horizontal section of an autostereoscopic screen. Each microlens covers 8 subpixels.

the light points on the screen, which are red, blue, or green, and three of them (R, G, and B) are needed to create all possible colors of a "true" colored pixel. That is why we speak of "subpixels" to refer to these light points on the screen. With a standard HD tile of 1920 x 1080 pixels, there are thus 5760 (1920 x 3) subpixels per light and 720 microlenses. Each microlens is 0.14 millimeter wide.

Each microlens disperses the light of eight subpixels in eight consecutive angular sectors. To make sure that each image does indeed contain equal amounts of R, G, and B information, the cylindrical microlenses are slightly slanted instead of being strictly vertical. In this way, each line of the image corresponds to another sequence of subpixels (for example, "R-G-B-R-G-B-R-V," and then "G-B-R-G-B-R-G-B," etc.) Of course, the display's driver must take each subpixel's link to each image into account if the views are to be consistent.

Loss of Resolution

The images' horizontal resolutions are obviously inferior to those of the panels used, since the displayer's pixels are shared out among all the views. One would be mistaken, however, in believing that an eight-view screen divides the perceived resolution by eight. While this is effectively what happens in purely mathematical terms, let's not forget that the brain – a major link in the visual perception chain – uses the information obtained from the two eyes to reconstruct an internal representation of the scene. The slightly different elements of objects seen by the two eyes thus give rise to a better internal representation than would be obtained by just one of the views. The resulting perceived resolution is thus not as bad as the number of views might lead one to believe.

In practice, the loss of horizontal resolution also results in a loss of depth resolution, since depth is given by the horizontal shift between images. University studies have shown that the depth resolutions of the best screens currently available are not greater than half the human visual system's power of resolution. Luckily, autostereoscopic screens are used mainly in the area of advertising and target passers-by rather than people who settle in for the long haul. In this context, the subjects that are presented are often characterized by great contrast and depth, with a lot of in-your-face effects, and the lack of depth resolution is not upsetting.

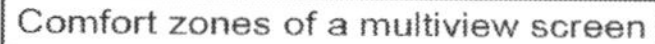
Comfort zones of a multiview screen

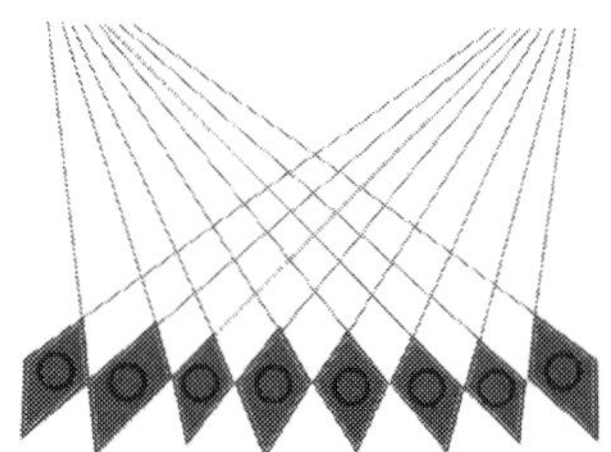
With a multiview screen, the comfort zones in which 3D is visible depend on the number of views and microlenses' curvature. If the zones are wide, they will have less depth.

Number of Views

Depending on the multiview system's geometry, the transition from one comfort zone to the next goes through a monoscopic zone where the two eyes see the same image. In other cases, the image may seem blurry. The multiview screens currently on the market offer five, eight, or nine views, depending on the make and size.

The Future

Increasing the number of views will make it possible to enhance the picture's resolution and the size and number of comfort zones. Two obvious paths are being taken: one is increasing the display's resolution, the other is time multiplexing. 4K displays have been announced and even built by various companies with a resolution of 4096 × 2048 pixels. However, the total lack of 4K content on the market is likely to slow such solutions' deployment considerably. The Canadian company Zecotek (*www.zecotek.com*) is one of a small number of companies around the world that are investigating time multiplexing using HD resolution displays. The idea is to display the various images in quick succession. The two problems to solve are the display's speed and changing the light beams' orientations. As we have seen, several makers have released 120 Hz two-view display panels. These panels, which are lit by LEDs that can keep pace with the 120 Hz refresh rate without problems, are elegant solutions to the problem. The brightness and contrast requirements of multiview display panels, however, are too high for today's LED sources.

Zecotek corporate logo

The Zecotek system offers ninety full HD resolution views in a viewing angle on the order of 50° with screens ranging from 10 to 50 inches (25-130 cm). To achieve this quality, a very high image refresh rate - more than 2,000 images per second! - is necessary. Micromirror DLP projectors such as those used in digital cinema projectors are the only display technology capable of achieving such speeds. Zecotek thus uses such a projector located behind the screen. Since a large number of images come from the same projector, each image is visible only a tiny fraction of the time: scarcely more than 1% of the time for a 90-view screen and 2% of the time for the 40-view version. The result is very low screen brightness, despite the use of very powerful lamps. The large number of views and high resolution offset this handicap.

The heart of the Zecotek system is obviously the optical device that deviates the pixel light at a different angle for each view. This device is called a "dynamic lens" and, since Zecotek's screens are as yet unmarketed prototypes, it is difficult to describe them in greater detail.

Multiview 3D Production and Editing

Shooting

The multiview images are usually generated by computers with animation software such as After Effects and a multiple viewpoint shooting plug-in. Real objects can be photographed by successive shots taken at different angles. Real scenes can also be filmed with a rig onto which the desired number of cameras is mounted. When you come out of this shooting phase you have multiple photographs and videos corresponding to the N points of view required for the screen that will be used to display the final content. All these files must be assembled without loss of synchronization and with carefully calibrated horizontal translations to achieve the desired depth effect. This thus requires software dedicated to this task.

Editing

Few software applications can generate multiview video files that are compatible with the various types of autostereoscopic screen on the market. These screens, moreover, are incompatible with each other. The makers all offer links on their websites to the right editing software for their products.

3D-Tricks

One of the rare software packages that take all of these constraints into account is 3D-Tricks, compositing software put out by 3DTV Solutions (Clichy, France, *www.3dtvsolutions.com*). 3D-Tricks lets you create animations from a wide variety of sources, such as 2D images, multi-PoV 3D images, 3D objects, text, and sequences of images, whether stereoscopic or not. 3D-Tricks lets you view the compositing underway on an autostereoscopic screen in real time. A powerful hardware configuration – fast hard disks and a top-of-the-line Nvidia graphics card – is recommended, however.

3D-Tricks is a fairly conventional editing package in terms of presentation, with a trim bin, multitrack time line, and viewing window. However, this window shows a three-dimensional space in which the various elements – stills, videos, and text – are assembled in the space opposite the viewer. You position the viewer, the 3D screen's frame, and the various elements of the scene in this space. The program then computes the final rendering and displays it on the connected autostereoscopic screen.

An embedded text generator lets you generate large amounts of text without going through external animation software. The associated 3D-Play software lets you run the generated sequences again on the target screen with a sound track, to play several clips over again in a loop, and so on.

File Format

The files are all reconverted into a special file format for image sequences, MultiView Sequence [*.mvs]. This format takes the shape of an XML document that includes the files' positions and generates the necessary information for optimal depth computation (distance to the convergence point, interaxial distance, cameras' focal lengths, etc.). The files are placed in N directories, one per point of view, with identical names to which the suffixes _cam01, _cam02, _cam03, etc., are added.

Autostereoscopic System Suppliers

Here is a non-exhaustive list of suppliers of autostereoscopic monitors, software, and associated services. This list does not include the countless research projects in the field that have not yet created marketable products.

Comparison of the Best-known Autostereoscopic Solutions

Company	Autostereoscopic products	Views	Country	Website
3D Fusion	42" screens 8" photo frames Video wall of 3 × 3 42" screens Plug-ins for 3ds Max, Maya Multiview shooting service, content creation, and compositing	9	United States	*www.3dfusion.com*
3D International	21", 40", 46", and 65" screens Multiview compositing and broadcasting software Plug-ins for 3ds Max, Maya Active X control Development tools Multiview shooting service	5-8	Malaysia	*www.visumotion.com*
Alioscopy	24", 42", and 47" screens Multiview mixing software with and without compression, plug-ins for 3ds Max, Maya, Softimage, and Cinema4D Multiview shooting service	8	France	*www.alioscopy.fr*
Dimenco	52" and 56" screens Plug-ins for 3ds Max and Maya Multiview conversion service	15	The Netherlands	*www.dimenco.eu*

Company	Autostereoscopic products	Views	Country	Website
Holografika	26", 32", 72", and 140" screens 2D-to-multiview conversion software Interactive viewing software Plug-in for 3ds Max OpenGL and OpenInventor drivers	n.a.	Hungary	*www.holografika.com*
Magnetic 3D	22", 26", 32",42", 46", and 57" screens Multiview compositing, remote transmission, and broadcasting software Windows DirectX 3D driver and Active X control	9	United States	*www.magnetic3d.com*
Tridelity	24", 26", 27", 42", 55", and 65" screens 21" portrait-format screens Multiview compositing and broadcasting software Plug-ins for 3ds Max, Maya, and dozens of other programs VRML and biomedical viewing tools	5	Germany	*www.tridelity.com*
Zecotek	10" to 50" rear-projection monitors, in the premarket phase	40-100	Canada	*www.zecotek.com*
Zero Creative	24", 46", 55", and 65" screens Multiview shooting service 2D-to-multiview conversion service	8	The Netherlands	*www.zerocreative.com*

Useful References

3D Publications

3D A-to-Z, an Encyclopedic Dictionary, Richard W. Kroon, McFarland.

3D Displays, Ernst Lueder, Wiley.

3D TV and 3D Cinema, Bernard Mendiburu (2011), Focal Press.

3D Movie Making, Bernard Mendiburu (2009), Focal Press.

Foundations of Stereoscopic Cinema, Lenny Lipton (1982), a period reference book available for free download.

EDCF guide to 3D Cinema, European Digital Cinema Forum.

3D Events

Dimension3, International 3D Forum; annual event in Paris, France (*www.dimension3-expo.com*).

3DStereoMedia, the European 3D Stereo Summit for Science, Technology, and Digital Art; annual event in Liège, Belgium (*www.3dstereomedia.eu*).

SMPTE Annual Conference on Stereoscopic 3D, scientific conference on stereoscopy, annual event in the USA (*www.smpte.org*).

Toronto International Stereoscopic 3D Conference (*3dflic.ca*).

ISU Annual Congress *(International Stereoscopic Union)* dedicated to 3D photography (*www.world-of-3d.com*).

3D Web

Clubs and Federations

Deutsche Gesellschaft für Stereoskopie (Germany): *www.stereoskopie.org*

International Stereoscopic Union: *www.isu3d.org*

National Stereoscopic Association (USA) : *www.stereoworld.org*

Nederlandse Vereniging voor Stereofotografie (the Nertherlands): *www.stereofotografie.nl*

Societa Stereoscopica Italiana (Italy): *www.ssi-3d.it*

Société suisse de stéréoscopie (Switzerland): *www.stereoskopie.ch*

Stereo Branch of the Photographic Society of Vienna (Austria): *www.stereoskopie.at*

Stereo Club of Southern California (SCSC): *www.la3dclub.com*

Sydney Stereo Camera Club (Australia): *www.oz3d.info*

The Stereoscopic Society (UK): *www.stereoscopicsociety.org.uk*

Union des professionnels de la 3D relief (France): *www.up-3d.org*

Information

Weekly 3D newsletter about stereoscopy by Benoît Michel: *www.stereoscopynews.com*

Software

Free software by Masuji Suto (Muttyan): *stereo.jpn.org*

Scripts for Canon still picture cameras: *chdk.wikia.com*

StereoDataMaker synchronizer: *apn3d.stereoscopie.eu*

MPO Toolbox, an .MPO file converter: *www.stereoscopynews.com/mpo*

Shooting Equipment

Novoflex tracks: *www.novoflex.com/en*

LANC Shepherd synchronizer: *www.ledametrix.com*

Stereoscopic Processors (SIP)

3Ality SIP2100: *www.3alitydigital.com*

Binocle Disparity Killer: *www.binocle.com*

JVC 2D3D1: *www.pro.jvc.com*

Miranda 3DX-3901: *www.miranda.com/3DX-3901*

Sony MPE-200: *www.sony.fr*

Stereographer's Wizard: *www.advanced3dsystems.com*

Color Pictures from this book on the web

Chapter 2 - page 65:
www.stereoscopynews.com/digitalstereoscopy/fig-02-22.jpg

Chapter 3 - Page 99:
www.stereoscopynews.com/digitalstereoscopy/fig-03-32.jpg

Chapter 3 - Page 110:
www.stereoscopynews.com/digitalstereoscopy/fig-03-43.jpg

Chapter 3 - Page 115:
www.stereoscopynews.com/digitalstereoscopy/fig-03-48.jpg

Chapter 3 - Page 117:
www.stereoscopynews.com/digitalstereoscopy/fig-03-52.jpg

Chapter 8 - Page 267:
www.stereoscopynews.com/digitalstereoscopy/fig-08-11.jpg

Chapter 8 - Page 268:
www.stereoscopynews.com/digitalstereoscopy/fig-08-12.jpg

Glossary

1080p-24 Hz: HDMI-compatible transmission format, 1920 × 1080 pixels, full frames, 24 or 23.98 images/s.

2D +Delta: Single image with metadata allowing reconstruction of the second view from the first. Delta may be a depth map, a compressed version of difference pixels, predictive data, and so on.

3D: In three dimensions. Also called "3D Cinema." Some use "S3D" or "S-3D" for stereoscopic 3D imagery.

3D DVD: DVD using an obsolete format to present stereoscopic content. Images are interlaced, thus losing half the vertical resolution for each eye. "*Ghosts of the Abyss,*" a movie by James Cameron, exists in that format.

720p-50 Hz: HDMI-compatible transmission format, 1280 × 1080 pixels, full frames, 50 images/s.

720p-60 Hz: HDMI-compatible transmission format, 1280 ×1080 pixels, full frames, 60 or 59.94 images/s.

Accommodation: Physiological process by which the eye changes its lens's shape and thus optical power to get a clear image of an object at a given distance.

Amblyopia (lazy eye): Impaired vision in one eye resulting in double, poor or blurry vision. In the case of suppression amblyopia, the brain "turns off" the visual processing of images from this eye, preventing double-vision and hampering stereopsis. Around 5% of the human population is amblyopic.

Anaglyph: Color filter separating light in two non-overlapping parts so that each eye perceives a different image. Many variants exist, the most common being red-cyan and amber-blue.

Alpha: The "alpha layer" is the image component defining transparency. PCX, JPG, TIF, and TGA files may contain an optional alpha layer. These images are often designated as "RGBA," as opposed to "RGB" to designate images without a transparency layer.

Artifact: Unwanted and disturbing element of an image. The most disturbing stereoscopic artifacts are present in only one of the two images, causing retinal rivalry.

Autostereoscopy: 3D visualization without glasses. The screen sends a series of images (usually five to nine) to the audience in precise directions so that each viewer perceives two slightly different images.

AVC *(Advanced Video Coding)*: A video image compression method that is part of the MPEG-4 standard; also called "H.264" instead of "MPEG-4 AVC." The AVC standard is published as ISO/IEC 14496-10. AVC is used to compress images in Blu-ray discs and many HDTV channels.

Binocular: With both eyes. Binocular vision enables us to use stereoscopy to give depth to an image. It is most useful for objects located closer than 10 meters.

Black bars: Black bands placed on a display showing content with a different aspect ratio. See letterbox, pillarbox, and windowbox.

Blur: Image blur is a lack of small-scale details (mathematically speaking, the image lacks high frequency components). Blur can be caused by an optical defect or strong compression, or it may be added on purpose in postproduction to conceal details.

Cardboarding: A stereoscopic effect where different objects appear to be well separated in depth but the objects themselves seem flat. It is often caused by using lenses with excessive focal lengths during the shoot.

Comfort zone: Range of parallax values, both positive and negative, in which stereoscopic fusion occurs effortlessly.

Convergence: Angle between the two eyes' lines of sight when they both focus on a nearby object. For an object observed at infinity, convergence is also at infinity.

Crosstalk: Imperfect physical dissociation of images aimed at the left and right eyes during stereoscopic viewing. Also called "ghost" or "ghosting."

CRT *(Cathode Ray Tube)*: Name of the old TV tubes using electron beams to display images line by line. Still interesting as a timing measurement tool.

Deghosting: Ghosting removal. When you know in advance that the projection system will show the left eye a fraction of the right image, the deghosting process can subtract the negative of the future ghost from the image. In projection the two effects then cancel each other out and the perceived image is almost perfect.

Depth budget: Distance between the nearest and farthest objects in a scene. The depth budget is often expressed in terms of parallax as a percentage of the image width. Typical value: 3%.

Depth map: An additional image component, such as the alpha layer for transparency, but the value of which represents the distance between the observed pixel and the camera.

Depth script: Planning of depth budget development over time. There are intense moments and quiet moments, corresponding to the various emotional atmospheres during the whole movie.

Diplopia *(stereoblindness)*: A vision defect in which fusion of the images perceived by the two eyes does not occur, causing double vision.

Disparity: Difference between the two images of a stereoscopic pair. Many kinds of disparity exist: color differences; object positions; horizontal, vertical, or rotational shift; etc.

Divergence: An unnatural, even impossible, situation for humans in which each eye has to look at one point of a pair separated by more than the intraocular distance. Always to be avoided.

DLNA *(Digital Living Network Alliance)*: A standard decreed by a group of companies to ease multimedia content sharing at home. A DLNA server is able to send 2D and 3D pictures and videos to a compatible TV set.

DLP *(Digital Light Processing)*: A light beam modulating method using micro-mirrors inside a projector. There are single-DLP projectors with a color wheel and three-DLP projectors where each DLP sends one of the R, G, and B color components to the screen through a series of prisms.

Dubois: An anaglyph generation method invented by Eric Dubois (from Ottawa University, Canada) that lessens the amounts of pure red and cyan in the image to reduce retinal rivalry. See *www.site.uottawa.ca/~edubois/anaglyph/*.

DVI: A connector for digital monitors carrying the same signals as HDMI, usually offering HDCP security.

Dwarfism: See Hyperstereoscopy.

Floating Window: A virtual frame located in front of a stereoscopic object that is itself presented out-of-screen. A floating window reduces or even suppresses the discomfort caused by window violations.

Fourcc code: A four-letter code located at the very beginning of .avi and .wmf video files and defining the codec used to encode the video. The code indicates to the decoder which method to use to decode data properly.

Genlock (*GENerator LOCK*): A signal transmitted by a time base. Genlock is used to synchronize various video sources, usually cameras and recorders/players. It is extremely useful to synchronize the two cameras of a 3D shooting rig.

Gigantism: See Hypostereoscopy.

Ghosting: See Crosstalk.

HDCP *(High-bandwidth Digital Content Protocol)*: The HDCP protocol is a data encryption method used by HDMI and DVI connections. HDCP codes only protected contents, not free and unprotected content.

HDMI: A standardized communication link between a video source and a TV set. Version 1.4 supports common stereoscopic formats. See www.hdmi.org for more details.

HIT *(Horizontal Image Translation)*: The horizontal displacement of the images in a stereoscopic pair, sometimes called "reconvergence."

Horopter: The locus of points in space whose images stimulate corresponding points on the two retinas and thus yield a single image, *i.e.*, all those points have zero parallax. When shooting, the horopter is not a plane if the cameras converge.

Hyperstereoscopy: Exaggerated stereoscopic effect caused by an interaxial distance larger than the normal interocular space. It causes dwarfism and gives the impression of looking at a scale model.

Hypostereoscopy: Minimized stereoscopic effect caused by an interaxial distance smaller than the normal interocular space; causes gigantism. For macro shots, the interaxial distance may be reduced to a few millimeters only.

Immersion: Feeling of being part of the action on screen; often described by the expression "suspension of disbelief" or "willing suspension of disbelief."

Interocular: Distance between the centers of both eyes, usually between 63 and 65 mm in men. The range of current values goes from 48 mm (for children) up to 72 mm.

Interaxial: Distance between the lenses of a stereoscopic camera pair. By varying the interaxial distance, one can amplify or reduce the 3D effect.

Stereoscopy inversion: Exchange of left and right images; all depth feeling in the scene is reversed. Also called pseudostereoscopy.

IOD *(Interocular Distance)*: Horizontal distance between the eyes, very often (but wrongly) used to describe the interaxial distance between the two cameras of a stereoscopic pair.

JPEG and JPEG2000: Compression methods for still images. The JPEG standard, the oldest, is available in two stereoscopic variants called "JPS" and "MPO." JPEG2000, the last incarnation of JPEG, is a scalable multi-layer codec, so it can encode multiple images (including stereo) and variable resolutions. Thus, one can decode the same JPEG2000 file in 4K or 2K, in color or black and white.

JPS: An image coded in JPS format is a JPEG file containing a conventional stereo pair made of two side-by-side images, the left one being on the right, and vice versa. Just rename a .JPS image file to .JPG to display it as is.

Keystone: See Trapeze.

Lens flare: Optical aberration or halo generated by reflections of a point light source inside the lens, often causing disparities between the two images.

Letterbox, letterboxing: Letterboxing is a way to display movies with a broader aspect ratio than 16:9, which must be displayed with black bars above and below the image.

LUT *(Look-Up Table)*: Correspondence table between an input value and an output value. Generally applied to pixel brightness, a LUT allows for various lighting condition corrections, for example following a gamma curve.

3D-LUT: A three-dimensional LUT is a correspondence table with three inputs and three outputs. Applied to the three color values R, G, and B of a pixel, a 3D-LUT can switch from one colorimetry definition to another, for example to display images prepared for DLP-Cinema on a HD TV.

Monoscopic: Using only one image.

Motion Parallax: Horizontal disparity between two images due to a time delay between the two images; causes an erroneous feeling of depth change depending on the motion's direction.

MPEG-2: Video compression standard used since 1994 for many digital TV transmissions, as well as for DVD encoding. MPEG-2 is also used to compress HDTV channels, but is less effective than the most recent AVC standard. MPEG-2 TS (Transport Stream) is an encapsulation method enabling MPEG-2 to contain an MPEG-4 AVC stream.

MPEG-4: Recent video compression standard, including AVC (for high definition, also known as "H.264") and MVC (for stereoscopy and multiview) modes.

MPO: The MPO file format is created by concatenating two JPEG files containing the two views of a stereoscopic pair. Just rename an .MPO image to .JPG to display the left view (the right view will be ignored). Special software such as MPO Toolbox is required to extract the secondary image from an MPO file.

Multiview: A multiview image includes several variants of the same scene, usually shot with slightly different perspectives. There may be two views for a stereoscopic image, or eight or more views for images that will be displayed on autostereoscopic screens. The MVS (Multi View Sequence) and MVI (Multi View Image) formats are specifically multiview.

Occlusion: Depth cue given by an object that conceals another one behind it.

Orthostereoscopy: 3D shot taken with an interaxial distance equal to the average interocular distance (65 mm). See also Hypo- and Hyperstereoscopy.

Out-of-screen: Depth effect observed when objects are perceived in front of the screen; corresponds to negative parallax values.

Parallax: Horizontal disparity between homologous points from a stereoscopic pair. The parallax determines the distance of an object: A positive parallax corresponds to objects located beyond the screen plane and a negative parallax corresponds to objects located in front of the screen plane.

Pillarbox, pillarboxing: Effect occurring on widescreen video displays when black bars are placed on either side of an image because its aspect ratio is squarer that the 16:9 screen.

Polarization: Filtering of the light according to its polarization. There are light-polarizing filters with circular (left or right) or linear (90° between the two directions, both generally inclined at 45° to the horizontal) polarization. Polarization allows for left and right images' separation using inexpensive passive glasses; requires a silver screen for projection.

Proprioception: Self-perception; awareness of the position of various parts of one's own body relative to gravity. The fusion of stereopsis and proprioception builds the mental representation of the world in our brains.

Pseudostereoscopy: Left and right views switch of a stereoscopic pair. Also called inverted stereoscopy.

Pulfrich: Stereoscopic effect obtained when one view is darkened, slowing its transmission to the brain; anecdotal, but sometimes a source of discomfort in a stereoscopic image.

RAID *(Redundant Array of Independent (originally Inexpensive) Disks)*: Group of matched disks operating in parallel and seen by the system as one large, high-capacity disk. There are several types of RAID: RAID 0, the fastest, stripes data across all disks in parallel. RAID 1 duplicates data on two drives that mirror one another. RAID 5 uses a portion of storage (one out of four or five disks) to introduce redundancy, thereby securing data at the cost of reducing available size and speed compared with RAID 0.

1/30 Rule: Rule of thumb that is not always valid but serves as the basis of the interaxial computation (IOD = 1/30 of the distance to the nearest object).

RGB: Color coding of one pixel with its three components: red(R), green (G), and blue (B).

Rig: A 3D rig is a mechanical assembly coupling two cameras for 3D shooting. Rigs come in two flavors: parallel and mirror. The latter allow closer positioning of the two cameras' optical axes. The key qualities of a 3D rig are rigidity, ease of adjustment, optical quality of the mirror, and overall weight.

Retinal rivalry: Inconsistency between the two views of a stereo pair where corresponding points in the two images do not represent exactly the same object. Examples: a visible reflection in one of the images only; vertical offset between the two images.

S3D: Stereoscopic 3D cinema, often also abbreviated as "3D" or "S-3D."

SAN *(Storage Area Network)*: A SAN is a high-speed network for pooling a large number of hard drives. Each workstation sees the SAN disks as internal drives and accesses them as fast as or even faster than if they were physically present locally. The connection technology generally consists of optical fibers (fiber channel).

SbS *(Side by Side)*: Frame-compatible transmission mode where left and right images are side by side inside a single "container" image.

SIP *(Stereoscopic Image Processor)*: Electronic device correcting raw stereoscopic 3D images in real time at the output of a 3D rig.

SIRD *(Single Image Random Dot Stereogram)*: Commonly called "stereograms." These 2D images contain 3D patterns, with depth effects visible to the naked eye, and require a few visual focusing exercises.

SSD *(Solid State Disk)*: Storage disks without any moving parts, generally offered in 2.5-inch format. They are in fact extremely fast non-volatile semiconductor memories.

Stereopsis: Perception of objects' distances and volumes due to the two visual stimuli provided by the eyes.

Stereoscopic: Using two images to give a depth feeling through binocular vision.

Stereoscopic fusion: Process by which the brain transforms a left and a right view into a three-dimensional representation of the scene.

Stereoscopic window: Frame in which the image is projected on a physical surface. This frame is always in the screen plane.

TaB *(Top and Bottom)*: Frame-compatible transmission mode where the left and right images are superimposed above each other; also called "Over-Under."

Trapeze: Image distortion in a shot taken with two convergent cameras; the viewpoints' rotation causes eyestrain due to vertical disparities in the corners.

Video frame: One image from a sequence of moving images. The frame includes the image itself (in 2D or 3D), but also various metadata characterizing the image and the way it should be displayed.

View-Master™: The world's best-selling stereoscope, displaying seven 3D views stored on cardboard disks.

VfW *(Video for Windows)*: Microsoft Windows component for viewing videos. VfW includes the Media Player application and various utilities. It uses the .AVI file format.

VPF *(Virtual Print Fee)*: VPFs are a way for studios to cofinance the transition to digital for movie theaters. For each digital movie screened, the studio gives back a share of the box office revenue to the operator or to a third-party company to help finance the equipment.

Window violation: Stereoscopic image in which objects in front of the screen plane appear cut by the edge of the screen. This situation is unnatural and causes visual discomfort. It is corrected by using a floating window.

Windowbox, windowboxing: Black bars surrounding content that is displayed on a video display that does not have the same aspect ratio.

YCbCr: Another name for YUV color coding. Y is the brightness symbol, while Cb and Cr designate the two color components.

YUV: Coding a pixel's color by its brightness (Y) and chrominance (U and V). The U signal carries Y-R information and the V signal carries Y-R information, thus making it possible, by recombination with the Y signal, to reconstitute the R, G, and B values of the corresponding RGB coding.

Zero: An object is located "at zero" when its parallax is nil; it is in the screen plane; also called "ZPS" for "Zero Parallax Setting."

StereoscopyNews, editor

www.stereoscopynews.com

Made in the USA
San Bernardino, CA
06 September 2016